A Sustainable Approach to Land Use

Scrivener Publishing
100 Cummings Center, Suite 541J
Beverly, MA 01915-6106

Publishers at Scrivener
Martin Scrivener (martin@scrivenerpublishing.com)
Phillip Carmical (pcarmical@scrivenerpublishing.com)

A Sustainable Approach to Land Use

Kundan Meshram

and

Chatrabhuj

Scrivener Publishing

WILEY

This edition first published 2026 by John Wiley & Sons, Inc., 111 River Street, Hoboken, NJ 07030, USA and Scrivener Publishing LLC, 100 Cummings Center, Suite 541J, Beverly, MA 01915, USA

For more information about Scrivener publications please visit www.scrivenerpublishing.com.

Wiley Global Headquarters
111 River Street, Hoboken, NJ 07030, USA

For details of our global editorial offices, customer services, and more information about Wiley products visit us at www.wiley.com.

The manufacturer's authorized representative according to the EU General Product Safety Regulation is Wiley-VCH GmbH, Boschstr. 12, 69469 Weinheim, Germany, e-mail: Product_Safety@wiley.com.

AI disclosure statement: The author confirms that no artificial intelligence (AI)-generated content or AI-assisted writing tools were used in the preparation of this chapter (including summaries and the abstract). All text, analysis, and conclusions are the result of the author's original work.

Library of Congress Cataloging-in-Publication Data

ISBN 9781394303250

Cover image: Generated with AI using Adobe Firefly
Cover design by Russell Richardson

Set in size of 11pt and Minion Pro by Manila Typesetting Company, Makati, Philippines

Printed in the USA

10 9 8 7 6 5 4 3 2 1

Contents

List of Figures

List of Tables

Preface

Sustainable land use planning is necessary for reasons like social, ethical values, environment, human development and economic aspects. Because global challenges such as fast urban growth, shifts in climate, unavailability of resources, and inequality are increasing rapidly, the way we plan and manage land has never been more consequential. The purpose of this book is to offer a structured, comprehensive, interdisciplinary approach to understand best land management practices with scope, challenges and solutions.

The motivation for writing this book arises from the increasing need for a multidisciplinary aspect of sustainable land use. It incorporates socioeconomics, environmental science, technology, ethics, and community participation. As global challenges rise, like changes in climate, rapid urbanization, unplanned development, deforestation, food insecurity, loss of biodiversity and degradation of land, there is a dire need to formulate policies and strategies to guide society towards long-term sustainability.

This book will serve as a vital tool for students, researchers, scientists, policymakers, and government in different fields like environmental science, economics, sociology, technology, ethics, and public policy to understand and implement effective and sustainable land use policies. It is developed to combine theory and practice by incorporating tradition and modern techniques.

Every chapter features theoretical points, handy tools and advice for further reading, making the book a great aid for academics, policy development and fieldwork. The chapters in the book each concentrate on a unique aspect of making our land use more sustainable.

- Chapter 1 introduces the key principles and concepts of sustainability, follows how they have changed over time, explains their benefits and problems and explains the role they play in planning for land use.

- Chapter 2 reviews ways land is being used, the main reasons behind those uses and the part that technology plays in noticing changes.
- Chapter 3 covers guidelines for sustainable land use, pointing out how it affects all aspects of society—economy, society and environment.
- Chapter 4 discusses how economics plays a role, especially in the impacts of land use change, creating policies and coordinating success with sustainability.
- Chapter 5 highlights matters related to society by highlighting cultural heritage efforts, examining how planning improves the well-being of communities and discussing how close-knit communities affect planning.
- Chapter 6 examines how land use patterns affect the environment, looking at ecosystem services, environmental policies and the role of planning.
- Chapter 7 examines the role of technology, including how GIS, remote sensing, AI and UAVs help with decision-making, city planning and working with residents.
- Chapter 8 outlines the policies that can ensure sustainable land use at home and abroad, focusing on the role of zoning and leadership.
- Chapter 9 lists the difficulties in adopting sustainable land use such as opposition to change and the unavoidable friction between quick profits and lasting sustainability.
- Chapter 10 presents a range of global case studies from Costa Rica, China, Kenya, the Sahel, the United States and India that highlight what has been done well and what has been learned.
- Chapter 11 discusses how involving the community helps with decisions, participatory planning and collecting feedback digitally.
- Chapter 12 introduces remote sensing as a way to extract land use patterns. It is then demonstrated in environmental monitoring, disaster response and urban planning.
- Chapter 13 provides a detailed analysis of satellite images, covering the steps of image processing, enhancing data, using classes and learning from machine algorithms.
- Chapter 14 focuses on using AI and machine learning for features on land, model building and handling huge datasets.

- Chapter 15 looks at actual urban land usage in Bhopal and Raipur, explaining the methods used, where the data came from and the results.
- Chapter 16 explores analysis and change detection of Indore city with an accuracy assessment and alignment of land use pattern with UN SDG goals.

1

Sustainability Principles and Dimensions

Abstract
This chapter presents the broad idea of sustainability, investigating its concepts and dimensions across various fields. It explains the interdependence of sustainability principles and their dimensions. Tracing the historical trajectory, it emphasizes the emergence of sustainability discourse and the importance of its adoption. The chapter explains why sustainability is important, outlines its numerous benefits, and forecasts its future significance. It explores sustainability's past, present, and future trajectories, focusing on the dynamic interplay of issues and solutions. Furthermore, it investigates the concept of sustainable land use, defining its essence and significance within the broader sustainability framework. Engaging with these issues, the chapter presents a complete grasp of sustainability concepts and aspects, delivering insights critical for encouraging sustainable practices in terms of land use patterns.

Keywords: Dimensions, environment, economic, land use pattern, social, sustainability

1.1 Introduction

The ability to meet the demands of the present generation without compromising the needs of future generations is sustainability. It calls for using resources carefully and in a balanced manner, considering social, economic, and environmental factors, as depicted in Figure 1.1. It ventures to fabricate a harmonious system that encourages long-term well-being for nature and humans [1].

Sustainability can be attained in various sectors like commerce, agriculture, power generation, land use patterns, policy development, monitoring and transportation. Sustainable practices are included in energy, minimising waste, resource conservation, ethical operations and socially responsible activities. Its concept has affected global issues like climate change, resource management and social inequality [2]. Many individuals,

Kundan Meshram and Chatrabhuj. A Sustainable Approach to Land Use, (1–24) © 2026 Scrivener Publishing LLC

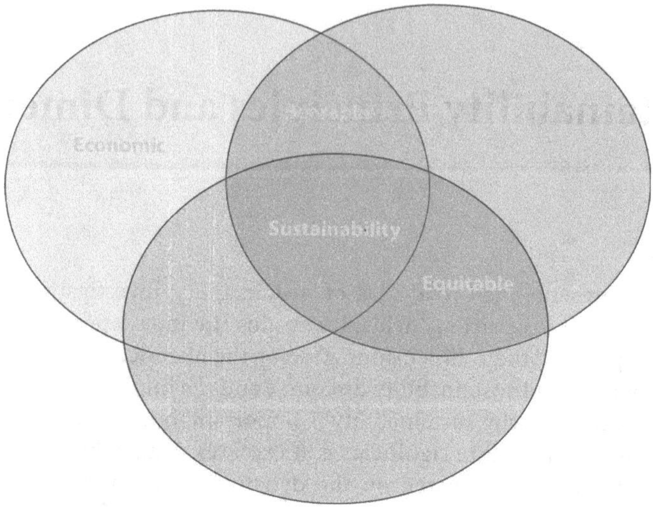

societies and governments focus on adopting sustainability concepts into their daily operations.

1.2 Sustainability and Dimensions

Sustainability is a complex concept focusing on three dimensions—economic, social, and environmental—together known as "Triple Bottom Line." The objective of sustainability is to balance and integrate these characteristics. Below is an overview of each dimension.

i. The environmental dimension focuses on the protection of ecosystems, biodiversity and resources. Further goals include monitoring of climate and greenhouse gas, promotion of green energy, safeguarding natural habitats, promoting a resilient environment, conservation of forests, prevention of pollution and waste mitigation strategies.
ii. The social dimension focuses on a better life, social injustice, equality and inclusivity. Social equality is the fair allocation of resources and opportunities among individuals. It supports human rights and also ensures access to basic services like education, clean water, healthcare, jobs and sanitation.

iii. The economic dimension focuses on financial development, stability and prosperity for all. It enhances resource efficiency and reduces waste through sustainable production techniques. Inclusive economic growth ensures that economic advantages are distributed evenly across all parts of society, creating opportunities in sectors that promote environmental sustainability, such as renewable energy and conservation. Financial responsibility involves long-term economic planning for future generations.

iv. The cultural dimension involves recognizing and maintaining variety, heritage, and traditional knowledge. Cultural respect includes appreciating diverse perspectives, customs, and traditions. Recognizing and respecting the rights of indigenous communities is crucial for sustainable resource management. Cultural heritage conservation involves preserving historical and cultural locations, objects, and practices.

v. The governance and institutional dimension develops effective governance structures, policies, and institutions to promote sustainability. It deals with policy framework and integrating sustainability concepts into decision-making processes, ensuring legal frameworks that support sustainability and comply with environmental and social regulations. It also promotes open and accountable government through transparent reporting and accountability procedures.

The interconnectivity of these factors emphasizes the significance of an effective approach to sustainability. To achieve sustainability, a comprehensive understanding of the linkages between economic activity, social well-being, and environmental health is needed. Efforts to address difficulties and promote sustainability should take into account the interdependence and trade-offs between these elements, which are shown in Figure 1.2.

The principles of sustainability guide individuals, organizations, and societies toward practices that promote long-term well-being, balancing economic, social, and environmental considerations [3]. While specific principles can vary, the following are commonly recognized as foundational to sustainability:

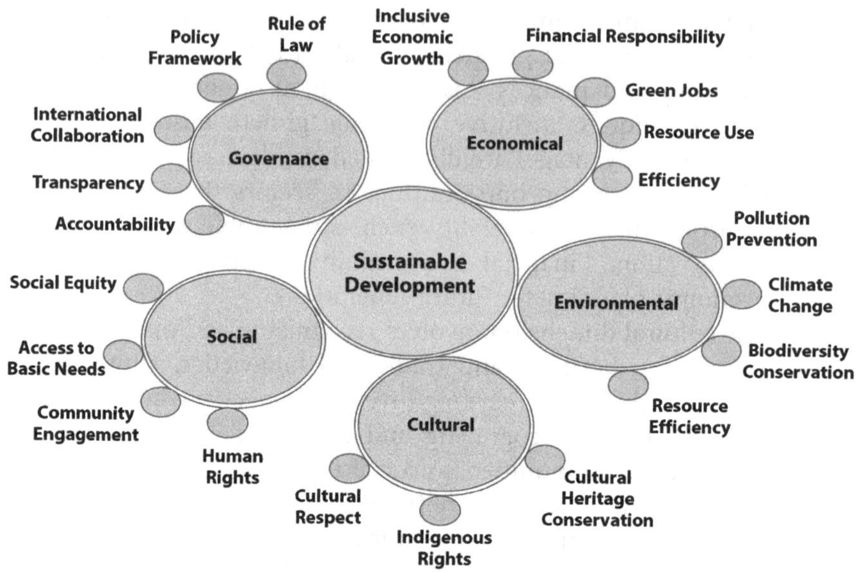

Figure 1.2 Sustainable development.

a. Interconnectedness: Interconnectedness refers to the dense web of relationships between social, economic, and environmental systems. Mapping all objectives of sustainability, it is evident that one objective is connected to another in a web-like form. If one sector is influenced, that will cause a ripple effect that will affect multiple areas.
b. Adaptability and Resilience: This refers to the ability of government, society, and the system to adapt to change that favours sustainability, like climate change, policy, laws and regulations. It showcases the requirement of a flexible, malleable, and resilient system that can absorb shock and disturbance. Sustainable development that is more flexible and resilient can benefit both humans and nature in the long term.
c. Cultural Respect and Diversity: It points to the essence of greeting, conserving, celebrating and enjoying the diversity of human culture and customs which have developed over the ages and passed on from our ancestors. This sustainability idea recognizes the inheritance of cultural and traditional wisdom. It focuses on respect and culture to be more inclusive, equitable, and acceptable.

d. Transparency and Accountability: This is one of the main aspects of better government and efficient decision-making. It highlights the need for transparent communication, engagement of stakeholders and accountability of officials to make sure decisions made are in favour of people, society, and government. It boosts trust, legitimacy, faith and effectiveness by the local communities.
e. Long-Term Perspective: This emphasizes the requirement of sustainability. Sustainability with short vision and ideas can be fatal for future outcomes. So, every action must be taken with planning and consideration of all factors to make sure the sustainability cycle repeats itself. It should focus not only on the demands of the current generation but also on those of upcoming generations. Resources are limited, so a forward-thinking mindset should be used to preserve them.
f. Economics: This is the most influential factor; even government policies and decisions are made on this. It affects families, society and government on a larger scale. So, optimal policies should be made in such a way that they do not hamper economic feasibility. Policymakers should establish a balance between economic growth and sustainability in a way that ensures resilience, stability, and equal prosperity. By making economically feasible policies, one can achieve long-term benefits.
g. Social Equity: This advocates for fair treatment, justice, social respect, dignity, liveability, equal opportunities for all, and distribution of resources. It promotes the equal distribution of ideas and resources, reducing gaps, and strengthening social cohesiveness. It prioritizes social justice to solve problems and build a better society.
h. Ecological Integration: This comprises the combined health of an ecosystem, which is essential for the majority of life forms of flora and fauna on Earth. It focuses on conservation, preservation, mapping, monitoring, guarding ecosystem services, and preventing habitat degradation. It makes sure many life forms other than humans also thrive naturally on Earth.
i. Polluter Pays Principle: This principle is a key component; it emphasises that individuals must take responsibility for proper disposal without polluting the environment. It encourages people to adopt cleaner and sustainable methods

in their daily lives and tries to promote eco-friendly products, policies and laws that do not harm the ecosystem and human health by imposing financial penalties on polluters.

j. Stewardship: This is defined as the appropriate conservation of natural resources focusing on long-term sustainability. It includes waste reduction, biodiversity protection and better decision-making for environmental and social factors.
k. Participation and Collaboration: This shows the need for government, local bodies, communities, stakeholders, and individuals to create an efficient decision-making process, aligning with sustainability. This principle is both inclusive and collaborative, and it has resulted in stronger, equitable and more effective solutions. It encourages each community at each level to participate in totality to deal with problems related to sustainability.

These principles collectively achieve the goals of sustainability, assisting a balanced and harmonious relationship between humans and nature.

1.3 History of Sustainability

The history of sustainability has evolved based on social, economic and environmental factors and their development over time. They have influenced the cultural development based on human activities with nature. A quick summary of the history of sustainability is presented below.

a. Sustainability has its roots in nineteenth-century conservation initiatives. Concerns about deforestation, soil erosion, and wildlife protection led to the formation of early environmental ethics.
b. Industrial Revolution and Environmental Awareness (Late 19th – Early 20th Century): The Industrial Revolution led to considerable changes in production and consumption habits. As environmental degradation became more apparent, individuals such as John Muir and Theodore Roosevelt in the United States fought for the protection of natural resources and the construction of national parks.
c. Modern ecology emerged as a scientific study in the mid-twentieth century. Pioneering ecologists, like Rachel Carson, raised awareness about the effects of human activity

on ecosystems. Carson's book *Silent Spring* (1962) focused on the environmental effects of pesticides and is widely regarded as launching the contemporary environmental movement.

d. United Nations and Sustainable Development (1970s–1980s): The 1972 UN Conference on the Human Environment in Stockholm was an important step towards addressing global environmental challenges. The concept of sustainable development was formulated in the 1980s, particularly with the publication of the Brundtland Report in 1987.
e. The 1992 UN Conference on Environment and Development (Earth Summit) in Rio de Janeiro led to the creation of Agenda 21, which is a plan for sustainable development.
f. Kyoto Protocol and Climate Change: In 1997, the international community pledged to reduce greenhouse gas emissions to address climate change. It underlined the importance of a global approach to environmental sustainability.
g. Growing Business Influence (2000s-Present): Sustainability is becoming a top priority for businesses, governments, and organizations. Concepts such as Corporate Social Responsibility (CSR) and triple-bottom-line accounting (economic, social, and environmental impacts) gained traction.
h. In 2015, the UN developed 17 Sustainable Development Goals (SDGs) as part of the 2030 Agenda. It establishes a framework for addressing poverty, inequality, water, soil, cities, communities, environmental degradation, and other global issues.

The history of sustainability represents a continual change in our understanding of the intricate interactions between human activities, the environment, and social well-being. It emphasizes the significance of implementing policies that promote long-term ecological balance, social fairness, and economic development.

1.4 Need for Sustainability

The demand for sustainability arises from the awareness that our present way of living, with unrestricted resource use, environmental degradation, and social inequity, is unsuitable in the long run [4, 5]. Several factors highlight the need for sustainability, which are depicted in Figure 1.3.

The diverse nature of sustainability has shown the multifaceted nature of its related factors. Climate change is influenced by the greed of humans, who harness resources for their profit, like consumption of fossil fuel, deforestation, mining, greenhouse gas emissions, and encroachment of natural green land. Likewise, biodiversity loss is a vital threat to the local ecosystem and the essential services it provides. Natural resources are available in limited numbers, so their conservation and optimized utilization must be done. Social inequality has emerged as an important aspect of unsustainable development, leading to financial disparities and injustices. It should focus on better resource distribution and injustices like labour rights and gender inequality [3]. Ethics must be considered in designing every step of sustainability. It should also focus on public health, long-term planning, and its interconnected links. By embracing these interconnected principles, humanity can chart a course toward a more equitable, resilient, and ethical future, ensuring the well-being of both present and future generations and the planet as a whole. Addressing the need for sustainability requires a shift towards responsible and mindful practices in areas such as

Figure 1.3 Need for sustainability.

energy, agriculture, transportation, and urban planning. Sustainable development seeks to balance economic prosperity, environmental stewardship, and social well-being for the benefit of present and future generations.

1.5 Benefits of Sustainability

Adopting sustainable development has numerous environmental, social, and economic benefits [6–8]. Table 1.1 explains some important benefits connected with implementing sustainable practices.

The benefits of sustainability are interconnected and help to construct a more resilient, fair, and environmentally friendly world. As societies, businesses, and individuals increasingly appreciate these benefits, there is a growing impetus toward incorporating sustainable practices into various sectors of life.

Table 1.1 Benefits of sustainability.

Broad area	Sub area	Description
Environmental Benefits	Conservation of Resources	Sustainable practices promote the responsible use of natural resources, reducing depletion and environmental degradation.
	Biodiversity Conservation	Sustainable approaches help protect ecosystems, preserve biodiversity and maintain the balance of natural habitats.
	Climate Change Mitigation	Sustainable techniques, like utilization of renewable energy and energy-efficient technologies, will reduce the emission of greenhouse gases.

(*Continued*)

Table 1.1 Benefits of sustainability. (*Continued*)

Broad area	Sub area	Description
Social Benefits	Improved Quality of Life	Sustainable development also focuses on the well-being of humans by considering factors like clean water, better air quality, liveability in the city, proper sanitation, education and a better healthcare system.
	Social Equity	Sustainable practices promote fair distribution of resources and opportunities, reducing social disparities and fostering inclusivity.
	Community Engagement	Involving communities in sustainable initiatives encourages local participation, empowers individuals, and builds social cohesion.
Economic Benefits	Long-Term Economic Stability	Sustainable practices aid in long-term benefits by developing efficient resource utilization, waste reduction, innovation, awareness, and policy development.
	Cost Savings	Energy and resource efficiency measures often lead to cost savings for businesses and individuals in the long run.

(*Continued*)

Table 1.1 Benefits of sustainability. (*Continued*)

Broad area	**Sub area**	**Description**
	Market Competitiveness	Sustainable businesses may gain a competitive edge as consumers increasingly value environmentally and socially responsible products and services.
Health Benefits	Reduced Pollution	Sustainable practices, such as transitioning to clean energy and improving waste management, contribute to reduced air and water pollution, positively impacting public health.
	Safe Food and Water	Sustainable agricultural practices prioritize the use of safe and environmentally friendly methods, contributing to healthier food and water supplies.
Resilience to Climate Change	Adaptation Strategies	Sustainable development includes policies to reduce the impacts of climate change, enhancing resilience for communities and ecosystems.
	Disaster Risk Reduction	Efficient urban planning and development can reduce vulnerability to natural disasters, protecting lives and property.

(*Continued*)

Table 1.1 Benefits of sustainability. (*Continued*)

Broad area	Sub area	Description
Innovation and Technological Advancement	Research and Development	The pursuit of sustainability often drives innovation and research in green technologies, leading to technological advancements with broader applications.
	Green Jobs	The transition to sustainable practices creates opportunities for employment in fields like renewable energy.
Global Cooperation and Diplomacy	International Collaboration	Sustainability is a global concern, and addressing it requires international cooperation. Shared environmental challenges encourage countries to work together for collective solutions.
Long-Term Viability	Intergenerational Equity	Sustainability ensures that resources are managed responsibly, considering the needs of current and future generations.

1.6 Past, Present and Future

1.6.1 Past Views

Views on sustainability were not always as prevalent or explicit as they are now. The concept of sustainability has changed over time, and awareness of environmental and social challenges has only increased.

a. Prior to the Industrial Revolution, civilizations relied on traditional and local customs, which were often naturally sustainable. Agriculture, for example, followed seasonal cycles, and communities had a strong connection to their natural surroundings.

b. The Industrial Revolution represented a dramatic transformation in societal practices. While it resulted in economic progress and scientific improvements, it also exacerbated resource exploitation, pollution, and social inequality. In this time frame, environmental consequences due to industrial activities were not a matter of concern.
c. Conservation movements were created due to the ill effects of industrialization. Scientists like Theodore Roosevelt pressed for conservation efforts.
d. The World War II era was a period of rapid expansion and high demand in consumption. This led to the reconstruction and development of infrastructure, factories, and industries at a high rate with low regard for environmental and social sustainability factors.
e. To spread awareness regarding the harmful effects of pollution, such as a rise in temperature, natural cycle, loss of biodiversity and the need for environmental protection, a day dedicated to these objectives was established in 1970, known as Earth Day.
f. The concept of sustainability gained momentum in the 1980s, with the publication of the Brundtland Report in 1987. The report showed that sustainable development is a type of development that meets the demand of the current population without hampering the needs of future generations. This report was very much accepted by the general public with its idea of long-term sustainability.
g. Later on, many International conferences, such as the Kyoto Protocol, the Paris Agreement, and the 1992 Earth Summit in Rio de Janeiro, played a vital role in the formulation of laws and policies related to sustainable development. These were accepted by many countries that felt the global need for coordinated efforts to address environmental concerns.

1.6.2 Present and Future Trends

a. Environmental: In the present scenario, individuals are more ecologically sensitive towards issues like forest degradation, rise in pollution, and loss of biodiversity. Preserving natural environments and wildlife is of high priority, and future generations are likely to be more conscious of the severity of climate change. Increased awareness has affected people's perspectives on sustainable practices.

b. Climate Change: The present generation has taken an effective role in climate action, engaging in global initiatives and recommending urgent measures to reduce carbon footprint. This rise in awareness and commitment to sustainability sets the stage for future generations to look for alternative lifestyles that prioritize environmental responsibility. By adopting simpler living standards, society can move toward a resilient and sustainable future.
c. Social Justice: Sustainability also deals with social standards of society like fair labour, working standards, diversity, inclusivity and equality for all. Upcoming generations are likely to see society, equality and the justice system as crucial components. They may advocate for inclusive and equitable solutions that address the concerns of marginalized communities.
d. Digital Activism and Social Media Influence: Social media platforms have a tremendous impact on moulding and raising awareness about sustainability concerns. Activism and awareness initiatives, frequently fuelled by social media, can swiftly gain popularity and impact public opinion. Activism and grassroots movements are likely to continue affecting the perspectives of future generations. Social media and digital platforms may still be strong instruments for mobilizing and advocating for sustainability.
e. Demand for Corporate Responsibility: There is an expectation that businesses and corporations would act ethically and sustainably. Many consumers choose to support companies that use transparent and responsible business methods. Future generations, when the world is highly interconnected, will focus attention on cooperation that addresses environmental concerns.
f. Innovation and Technology: It is viewed as a vital tool for addressing many challenges. The main aim is to develop sustainable technologies, renewable solutions, and innovations that foster a sustainable future. With time technological advancements continue, and they are expected to influence future generations' perspectives.

g. Education: Education has played an important role in influencing the perspectives of current generations. Many educational institutions have included sustainability in their curriculum. The present generation actively participates in the reform of policies, practices and laws in different areas. Educational institutes have included concepts not only on a higher degree level but also at the school level.

1.7 Challenges vs. Solutions

1.7.1 Challenges in Achieving Sustainability

Sustainability is a complicated situation where each component is interlinked, and achieving it is difficult. There are major hurdles in attaining sustainability, and major solutions are required at the local, national, and global levels. Here are the major challenges in achieving sustainability:

a. Lack of Integrative Approaches: Many of the problems are interconnected, and influencing one factor also affects other factors. There is a need for combined decision-making, but often efforts are made to tackle only one dimension, ignoring other factors. Decisions must be taken based on comprehensive sustainability, an integrated approach that takes account of economic, social, and environmental variables.
b. Climate Change: In the present scenario human activities have resulted in temperature rise, extreme weather conditions, deforestation, pollution and a rise in sea level, all of which have obstructed sustainability initiatives. There is a dire need to mitigate climate change and formulate strategies to adapt and reduce climate change, which will be in favour of humans in the long run. Solutions may include monitoring greenhouse gas and imposing fines on industries, afforestation, adaptation of green energy, ecosystem protection, climate activism and awareness campaigns.
c. Biodiversity Loss: This is a global crisis that affects the planet's ecosystems, human well-being, and economic stability. To a larger extent biodiversity is lost due to habitat destruction by human activities, and pollution of water, air and soil. This has led to economic loss, and protection is needed

against floods, droughts, and soil erosion. Urgent action is required to conserve and restore local biodiversity.

d. Short-Term Focus: This is a fault in planning, either due to financial or political issues. To achieve a short-term gain, policymakers, developers and governments ignore long-term gains. In many conditions, short-term issues overshadow long-term sustainability objectives. It is very difficult to influence decision-makers to select a long-term sustainable approach.
e. Resource Depletion: Many of the natural resources are limited; once used, they perish. With current technologies, humans are very much dependent on non-reusable resources like fossil fuels, minerals, coal and fresh water. For long-term sustainability, strategies are required to manage and shift from non-renewable resources to renewable resources.
f. Pollution: Human activities have led to pollution in air, water, and soil that harms human health, the environment and local biodiversity. Pollution causes health issues, environmental damage, and economic and social impact; toxic chemicals in soil reduce crop productivity and lead to desertification and loss of arable land. It is crucial to manage pollutants and pave the way for a cleaner, more efficient and healthier ecosystem.
g. Social Inequality: This is one of the major hurdles. Uneven distribution of resources, opportunities, and basic services limits economic growth, social well-being and overall development. Disparities in income, education level and healthcare systems affect marginalized communities, escalating poverty and exclusion. Gender inequality, racial discrimination, and injustice further widen the gap.
h. Overconsumption and Consumerism: Driven by mass production and never-ending demand for goods, this has led to unsustainable resource use and environmental deterioration. Constant natural resource extraction contributes to pollution, climate change and ineffective waste management. It also fuels economic variations as corporations profit while low-wage workers fall behind. Shifting towards sustainability requires collective efforts like promoting minimalism, supporting ethical products, implementing a

circular economy, enforcing strict eco policy and endorsing consumer awareness.

i. Waste Management: Effective waste management is essential to tackle non-biodegradable garbage like plastic, electronic waste, glass and metals, which pose a significant threat to the environment. Rapid urbanization and industrialization have fuelled demand and supply, leading to an increase in waste generation. Poor waste management compounds climate change by methane emissions from landfills and degradable waste. Addressing this challenge requires a multi-faceted approach, like promoting recycling, waste segregation, sustainable production, and consumption habits.

j. Economic Inequity: Economic systems that perpetuate inequality and promote short-term gains over long-term well-being constitute a threat to sustainability. A more inclusive and equitable economic framework is required to accomplish sustainable development objectives.

To address these difficulties, governments, corporations, communities, and individuals must work together in a revolutionary and transdisciplinary manner. It entails implementing sustainable technology, legislative frameworks, and behavioural changes that prioritize the well-being of both present and future generations, as represented in Figure 1.4.

1.7.2 Solutions for Sustainability

To address challenges, sustainability requires an integrated approach that considers economic, social, and environmental factors to achieve long-term balance. It also requires collaboration of governments, corporations, communities, and individuals, where everybody plays a crucial role in implementing sustainability practices. By aligning policies, laws, economic, and technological innovations with sustainability goals, societies can move towards more resource efficiency.

a. The government should make policies to achieve sustainable goals and a framework for all to guide decision-making at every level. Education is the most necessary for awareness and sustainable principles, since it empowers individuals for sustainable behaviours.

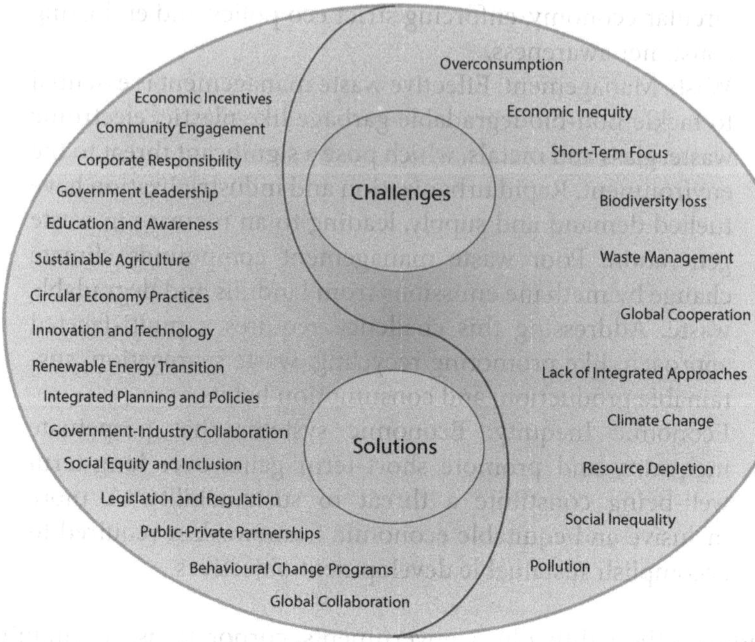

Figure 1.4 Challenges and solutions for sustainability.

b. Technological innovations are also required to reduce environmental impact, improve resource efficiency, and promote sustainability across all areas. To develop modern, efficient, economical, practical and innovative solutions government should invest in research and development.
c. A shift to a circular economy is very much possible. A circular economy helps in waste reduction and conservation efforts by using techniques of reduction, reuse, and recycling. It deals with the development and design of products that maximise efficiency and reduce environmental damage.
d. Renewable energy resources are a better alternative compared to perishable resources. When fossil fuels are burned, they emit energy and a lot of waste and greenhouse gases are generated, which harm the climate. On the other hand, renewable sources like wind, solar, geothermal, biomass and tidal are in nature. Harnessing them can be costly, but they are eco-friendly and beneficial for all in the long term.
e. Sustainable agriculture deals with plant growth, biodiversity, water conservation and soil health, which is crucial for

food security. Sustainable agriculture deals with harmful practices like excessive chemical use, deforestation, and soil depletion. Sustainable methods like crop rotation, organic farming, drip irrigation, and natural pest control help maintain soil fertility, reduce water utilization, reduce carbon footprint and promote biodiversity.

f. For long-term sustainability commitment, community engagement is a must. It ensures the local population is also engaged in decision-making. Local communities have valuable knowledge regarding the environment, resources, and challenges. This is done so that they feel a sense of ownership over projects and ensure that solutions are fair.
g. Aligning economic incentives with sustainability is important for addressing environmental challenges. Policies such as green subsidies, sustainable finances, green loans, tax incentives and carbon pricing can be added to taxes on pollutants. Combined with well-structured economic incentives and better global partnerships, they are important for long-term sustainability.
h. The public-private partnership will further facilitate innovation, resource management, finance, and policies. Ensuring social fairness in sustainability is vital to prevent marginalized communities from taking on disproportionate environmental and economic burdens. Laws and regulations must be strictly enforced in all sectors, making sure of fair and responsible operation across all fields.

Combining all these qualities will help in achieving sustainability goals. It demands a well-structured, collaborative effort from governments, local communities and individuals to attain sustainability in all respects together [9].

1.8 What is Sustainable Land Use?

Sustainable land use is a multifaceted approach towards responsible land management, conservation, and development so that land resources meet current and future demand. It is based on the principles of eco-friendly social and financial approaches in prioritizing sustainable land use for a better ecosystem, forest cover, wetlands, wildlife and humans so that both coexist without harming each other [10, 11]. From environmental concerns, it also holds a social aspect of making sure everyone has access to resources, protecting local livelihoods, and making land-related decisions. Efficient land use management requires the local community's involvement. It supports the local population by eco-friendly agriculture, ecotourism, and agroforestry while safeguarding local knowledge and cultural heritage. It also enhances food security, balances urban and rural development, integrates green/open spaces, climate resilience and essential infrastructure. Addressing climate change dynamics and its impact is also an integral part of sustainable land use. It takes essential steps that reduce greenhouse emissions and improve carbon-capturing techniques. It emphasises stewardship and governance for land management by policies, laws, regulations, and incentives for sustainable practices.

There are many benefits in terms of finance. Sustainable land use influences economic growth, resource efficiency, and benefits in long-term cost saving. It promotes eco-friendly agriculture, soil conservation, renewable energy integration, and better urban planning, leading to long-term profits and the creation of jobs. Better sustainable land use attracts green financing, carbon credits, and government incentives, which benefit both corporates and governments. Efficient planning improves property values,

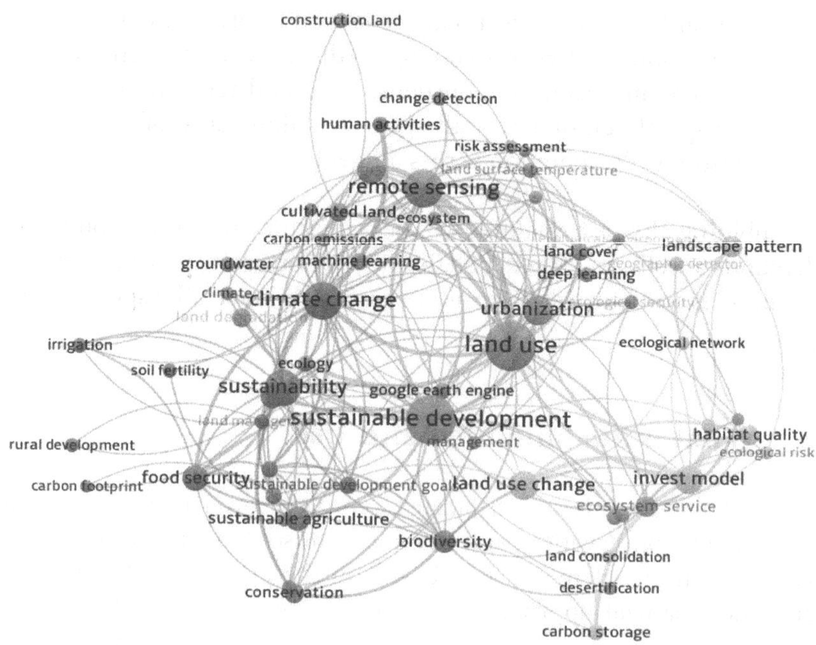

Figure 1.5 Links of sustainable land use and its applications.

reduces infrastructure costs, provides liveable conditions, and ensures long-term food and water security.

Sustainable land use acknowledges the consequences of decisions and actions in the long run, recommending better strategies that consider the combined environmental, social, and economic factors. In current scenarios, challenges posed by climate change can be tackled by sustainable land use methods by analysing environmental conditions and societal needs, which are depicted in Figure 1.5.

Engagement of local communities in decision-making regarding land use is very important. It ensures that diverse viewpoints are heard and local knowledge is incorporated in the process of developing effective and culturally sensitive land use strategies [12, 13].

1.9 Significance of Sustainable Land Use

The importance of sustainable land use lies in its impact on the environment, society, and the long-term well-being of current and future generations. Here are some key aspects of the significance of better land use:

a. Environmental Conservation: Sustainable land use is important for environmental conservation because it considers eco-friendly practices, preserves biodiversity, maps climate change, conserves soil health, and manages water resources. In current scenarios, strategies like habitat conservation, agroecology, afforestation and zoning will reduce habitat destruction, emission of greenhouse gases, soil erosion, climate variability and natural disasters. It can be useful in the repair of degraded ecosystems, conservation of endangered species, and species migration. It conserves soil, water, and agricultural products, and maintains the well-being of both ecosystems and humans. These techniques improve the long-term health of the ecosystem and ensure the availability of crucial resources for present and future generations.
b. Climate Change Mitigation: Sustainable land use patterns effectively capture carbon dioxide through techniques like afforestation, agroforestry, reforestation, and sustainable agriculture. Planting local trees in open land and agricultural land captures carbon, acting like a carbon sink. To avoid the release of stored carbon into the atmosphere, lands like peatlands, wetlands, and other carbon-rich ecosystems

are conserved. Adding climate-smart land use practices helps promote sustainable development.

c. Preservation of Natural Resources: Sustainable land use preserves natural resources for upcoming generations. It is essential to protect natural resources such as biodiversity, habitat, flora and fauna, soil, water, minerals, fragile ecosystems and forests. Furthermore, it encourages responsible use of natural resources and makes sure that they are extracted wisely without exploitation.
d. Resilient and Adaptable Ecosystems: Sustainable land use develops a resilient system that acts as a buffer zone that resists environmental changes. By using strategies like habitat restoration, ecosystem adaptation, and biodiversity conservation ecosystems can endure stress from climate change. Operations like afforestation, coral restoration, and wetland rehabilitation improve ecosystem health, maintaining their ability to thrive and benefit both humans and nature.
e. Economic Stability and Resource Efficiency: It optimizes resource utilization with low waste generation. Economic stability is achieved by strategies of the development of eco-friendly products that lower production costs, and making diverse income sources. Eco-industrial development promotes resource efficiency and waste reduction through industrial symbiosis, in which waste from one industry becomes a resource for another, decreasing production costs and environmental effects. Furthermore, sustainable land use methods improve resource efficiency by reducing land degradation, soil erosion, and water pollution, so maintaining natural capital and lowering the need for costly remedial efforts. By incorporating economic factors into land management decisions, sustainable land use secures the long-term viability of economic activity while improving resource efficiency and supporting economic stability.
f. Improves Ecosystem Services: Sustainable land use supports human well-being and maintains ecological equilibrium. Ecosystems can provide many benefits to humans and the environment by employing sustainable land management methods such as habitat preservation, soil conservation, and water resource management. Habitat preservation is the protection of key habitat hotspots and safeguarding vulnerable species. Integrating ecological principles in sustainable

land management decisions can enhance ecosystem services, resilience and sustainability.

g. Cultural Values: Sustainable land use also incorporates cultural landscape; it respects cultural identities and histories by preserving heritage sites, sacred sites, archaeological relics, and traditional land use practices. It also deals with techniques of eco-tourism and landscape restoration. It can help to revive local cultures and economies by using traditional crafts, cultural festivals, and sustainable tourism activities. It promotes the diversity of human culture with the intrinsic value of nature.
h. Long-Term Planning: Sustainable land use protects and enhances natural resources, ecosystems, and cultural heritage for upcoming generations by using sustainable techniques like conservation, restoration, and responsible partnership. With long-term planning of land management decisions, it assures that today's decisions will impact tomorrow's well-being.

Sustainable land use lies with the harmonizing approach of humans and nature, to protect the environment, social well-being, and economic stability on smaller and larger scales [14].

1.10 Conclusion

Sustainability is vital in all fields; it fosters environmental, social and economic solutions, especially in the field of land use planning. Land is a finite resource, and using it carefully is the demand of today. Land holds the majority of living species flora and fauna, biodiversity, wetland, forest cover, agricultural land and urban settlement, so a harmonious relationship is required for the co-existence of all. This chapter explores all the complexities, challenges, solutions, and opportunities for achieving better land use for all. It highlights the need for sustainable development for current and upcoming generations and the requirement of collaborative efforts taken by individuals, society, corporations, policymakers, and governments in attaining sustainable land use. It helps others understand the concept of sustainability and its impacts through discussions and action, promoting communal responsibility and fostering sustainability practices.

References

1. Adolphe, L., A simplified model of urban morphology: application to an analysis of the environmental performance of cities. *Environ. Plann. B Plann. Des.*, 28, 183–200, 2001, https://doi.org/10.1068/b2631.
2. Nechyba, T.J. and Walsh, R.P., Urban Sprawl. *J. Econ. Perspect.*, 18, 4, 177–200, 2004, https://doi.org/10.1257/0895330042632681.
3. Mouratidis, K., Urban planning and quality of life: A review of pathways linking the built environment to subjective well-being. *Cities*, 115, 1–12, 2021, https://doi.org/10.1016/j.cities.2021.103229.
4. Lemoy, R. and Caruso, G., Radial analysis and scaling of urban land use. *Sci. Rep.: Nat.*, 11, 22044, 2021, https://doi.org/10.1038/s41598-021-01477-y.
5. Clifford, J.P., *et al.*, The relationship between city size, decentralisation and economic growth. *J. Econ. Stud.*, 2022, https://doi.org/10.1108/JES-03-2022-0146.
6. Fichter, K., Lüdeke-Freund, F., Schaltegger, S., Schillebeeckx, S.J., Sustainability impact assessment of new ventures: An emerging field of research. *J. Cleaner Prod.*, 384, 135452, 2023.
7. Bressanelli, G., Perona, M., Saccani, N., Challenges in supply chain redesign for the Circular Economy: a literature review and a multiple case study. *Int. J. Prod. Res.*, 57, 7395–7422, 2019.
8. Song, Y., Merlin, L., Rodriguez, D., Comparing measures of urban land use mix. *Comput. Environ. Urban Syst.*, 42, 1–13, 2013, https://doi.org/10.1016/j.compenvurbsys.2013.08.001.
9. The United Nations, Sustainable Development Goals [SDGs]. United Nations: Sustainable Development, 2022, https://sdgs.un.org/goals.
10. Agrawal, A., *et al.*, From environmental governance to governance for sustainability. *One Earth*, 5, 615–621, 2022, https://doi.org/10.1016/j.oneear.2022.05.014.
11. Wang, Z., Fu, H., Liu, H., Liao, C., Urban development sustainability, industrial structure adjustment, and land use efficiency in China. *Sustainable Cities Soc.*, 89, 104338, 2023, https://doi.org/10.1016/j.scs.2022.104338.
12. Lazurko, A., *et al.*, Boundaries of the future: A framework for reflexive scenario practice in sustainability science. *One Earth*, 6, 12, 1703–1725, 2023, https://doi.org/10.1016/j.oneear.2023.10.027.
13. Tambe, S., *et al.*, Bridging science, policy and practice for sustainability: Towards a conceptual framework. *Environ. Sci. Policy*, 145, 208–216, 2023, https://doi.org/10.1016/j.envsci.2023.04.007.
14. Kanda, W., Klofsten, M., Bienkowska, D., Henry, M., Hjelm, O., Challenges of circular new ventures: An empirical analysis of 70 cases. *J. Cleaner Prod.*, 141103, 2024, https://doi.org/10.1016/j.jclepro.2024.141103.

2

Understanding Land Use Patterns

Abstract

It is essential to understand land use patterns to address various challenges of sustainability and its dimensions, to promote social fairness, health, and environmental protection. It is crucial to understand the factors driving land use patterns in order to build more resilient, inclusive, and sustainable communities. This chapter highlights the significance of technology in mapping, monitoring, evaluating, and classifying land use patterns. It differentiates between land use patterns and land use cover, showing the importance of understanding these terms. It also emphasizes the role of advanced technologies like Remote sensing, Geographic Information Systems, Artificial Intelligence, and Machine learning to classify and determine land use, which are beneficial to urban planners, policymakers, developers, environmental impact assessment, researchers, and stakeholders. It shows the importance of technology in informed decision-making, the implementation of sustainable techniques, achieving long-term goals, and paving the way for future land management policies.

***Keywords*:** Decision-making, land use pattern, management, mapping, sustainability, technology

2.1 Introduction

Land use patterns are an intricate arrangement of land based on human activities; these patterns are influenced by the interaction of natural and man-made causes. The study of land use patterns focuses on the arrangements of land for various purposes, like residential, industrial, commercial, agricultural, and conservation areas. It is essential to study these complex patterns to understand development and urban planning. Effective land use helps to attain balanced development, reduce deforestation, and protect biodiversity [1–3]. It also enables resource management, policymakers, and urban planners to make informed decisions for sustainable development and to attain a balance between economic progress and environmental conservation.

Kundan Meshram and Chatrabhuj. A Sustainable Approach to Land Use, (25–44) © 2026 Scrivener Publishing LLC

2.2 Factors Affecting Land Use Pattern

There are many factors that affect land use. Over time, the arrangement and distribution of human activity in the region forms the land use pattern. This development is driven by many elements like changes in population, economics, technology, culture, environment, and political scenarios. The main factors are further subdivided into natural and anthropogenic factors, which are depicted in Figure 2.1. The natural factor consists of local topography, climate, soil quality, and hydrology; with a favourable climate settlement becomes easy; availability of water and fertile soil in plain areas are most suitable for residential purposes. Key factors in human activities that affect land use patterns are economic opportunities, population dynamics, culture, and government policies. Urbanization is mainly due to the search for better financial opportunities for jobs and businesses [4–6]. Current population density and cultural values also affect the land use pattern. If a community of a particular culture is present, then individual trends align with that community. The type of government policy is also a deciding factor; individuals prefer a peaceful city with enforced laws and regulations rather than an unstable city.

a. Settlement and urbanization: Land use patterns often begin with the formation of settlements and urban centres. Human populations congregate in specific areas because of variables such as access to water sources, arable land, transit routes, facilities, comfort, and economic possibilities. As urbanization proceeds, these settlements increase and expand,

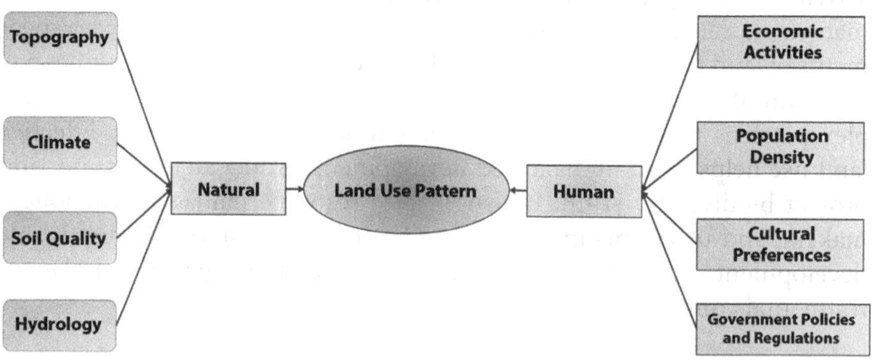

Figure 2.1 Factors influencing land use pattern.

resulting in a concentration of residential, commercial, industrial, and institutional activities within cities.

b. Infrastructure development: As cities grow, infrastructure is crucial for supporting human activity and connectivity. Infrastructure development encompasses the construction of transportation networks (roads, highways, railroads, airports, and ports), utilities (water supply, sewage systems, electrical grids), and communication systems (telecommunications, internet lines). The architecture and design of infrastructure networks influence land use patterns by determining accessibility, mobility, and the spatial organization of human settlements.
c. Economic growth and diversity influence land use patterns: Changes in land use are driven by the creation of new sectors, market demand changes, technical advancements, and adjustments in consumer preferences. For example, the expansion of manufacturing businesses may result in the creation of industrial zones, whereas the growth of service sectors may result in the proliferation of commercial and office buildings. Economic activities have an impact on residential development patterns, with housing markets responding to demand for various types of housing (single-family homes, apartments, condominiums) and amenities.
d. Land use planning and regulation: To promote organized development of land use patterns, effective planning, rules, laws, and enforcement are necessary. Techniques like zoning, planning, building codes, environmental codes, and development policies should be implemented at the local, regional, and national levels. Individuals prefer to stay in cities that are built systematically, with bylaws and regulations.
e. Cultural and social dynamics: The impact of culture on land use policies cannot be overlooked. It is a natural tendency of an individual to settle with their community. Demographic, lifestyle preferences, cultural values, and social dynamics all influence housing and community development patterns. It adopts a particular architecture, urban planning, and public places.
f. Environmental considerations: In recent years, where societies have prioritized sustainability, they are concerned about changes in climate, depletion of resources, loss of habitat, pollution, and degradation of local biodiversity. So, they

are looking for sustainable land use patterns and a harmonious relationship between nature and human settlement. Nowadays, they prefer open green areas, natural water bodies, local biodiversity, forest areas, and city gardens, which must be incorporated into the city rather than being just cramped concrete jungles.

2.3 Types of Land Use Pattern

Land use patterns are distinguished by dominating activities in particular areas. These patterns show the spatial arrangement of various land use activities [7, 8]. It is essential to determine the type of land use pattern to set policy for land regulation and development. Some of the primary types are described below.

a. Residential:
   Residential is a type of land use allocated for housing, resulting in residential zones. It has a wide range of options like single houses, residential buildings, apartments, and colonies. These are considered residential zones, which are further divided into urban, suburban, and rural areas based on population, urbanization, development, and planning. Residential zones are placed near commercial and industrial zones for better economic activity and the comfort of the population residing in them. Urban areas are packed with dense population cover with only expanding areas outwards. Suburban areas are less compact; they have free or open space between cities. A rural area is an integration of housing with natural cover. Residential land impacts the local environment and social dynamics; it fosters community engagement, formation of culture and heritage, and social norms.
b. Commercial:
   Commercial is a type of land that is used for commercial purposes; it encompasses a wide range, from retail to business. It includes shopping malls, offices, commercial corridors, corporate, financial institutions, and professional hubs. It is more concentrated in regions with significant economic activity and consumer demand. These strategies maximize

customer and business, take advantage of innovation, and promote economic progress.

c. Industrial:
This type of land is determined by allocating land use for industrial use, like manufacturing, logistics, warehouses, distribution centers, and industrial parks. Industrial land use reflects infrastructure for the production, storage, and distribution of goods. It ranges from heavy industries like steel, vehicles, and chemicals to light industries like clothes, electronics, and food processing, which require a lot of space. It is essential to understand and place industries effectively for economic development, job creation, planning, and environmental consideration and to prevent negative consequences like pollution, congestion, high surface temperature, and land degradation. Sustainable industrial land incorporates land use zoning and environmental remediation activity to attain a balance between economic growth and community well-being.

d. Agricultural:
Agricultural land is the land allocated for growing crops, livestock, and other cultivated activities. It also includes crop production, orchards, vineyards, and agroforests. Agriculture shapes the cultural landscape, and traditional farming links land with community. Sustainable agricultural techniques use conservation methods, reduce environmental degradation, and maintain the long-term viability of food production.

e. Recreational Area:
Recreational areas are the areas of land that are allocated for leisure and outdoor activities, such as city parks, national parks, wildlife areas, and protected natural areas. It is an essential component of urban planning and landscape. It provides opportunities for individuals and communities for leisure activity, relaxation, recreation, connection with nature, community interaction, and a sense of well-being. It ranges from green areas, botanical gardens, and sports fields to picnic areas.

f. Institutional:
Institutional land areas are those that are occupied by any public or private institution that serves educational, healthcare, governmental, religious, and community functions.

Some common examples included for educational purposes are schools, colleges, universities, and research centers, and for healthcare, clinics, hospitals, and medical care units. Government institutes include courts, police stations, fire stations, city hall, and other emergency services. Religious institutions are temples, mosques, churches, and synagogues. Community services include museums, culture centers, recreational facilities, and libraries.

g. Mixed Land Use:
In mixed land use, a land area integrates multiple land use types like residential, commercial, industrial, institutional, and recreational areas in one single area. This integrated technique promotes utilization of land in effective ways, reduces distance, and improves economic activities. Smart land use integrates smart growth and sustainability concepts, making it an alternative to traditional suburban growth. Mixed development, when planned carefully, promotes inclusive, resilient, and vibrant landscapes that enhance human health, take an eco-friendly approach, and result in economic prosperity. Some examples include urban downtown, smart city, and transit-oriented development.

These are some of the primary land use patterns observed in different regions worldwide, as given in Table 2.1 [9–12]. Lands are further classified into sub-types based on their uses and characteristics. Agricultural land is used for cultivation, livestock, and plantation like coffee, tea, and fruits. Farmland comprises cropland and grazing areas for feeding livestock. Orchards are focused on fruit-bearing trees like apples, oranges, berries, and many more, whereas vineyards are specialized for grape cultivation. Land near water and wetlands includes lakes, rivers, and ponds. A junction between land and water is called wetland, which is low-level standing water that serves as an essential habitat for many species. These lands are moss, swamps, and bogs. Urban areas are further classified into built-up areas, industries, parks, and roads. Natural land is land that is not influenced by human activities. It consists of forest, grassland, gardens, natural landscapes, bushland, shrubland, etc. There are some other lands that are common to other lands like sand, coastal areas, open land, historical locations, and religious places.

Table 2.1 Types of land use.

Land type	Sub-type	Description
Agricultural Land	Plantation	Cultivated areas for crops such as fruits, tea, and coffee.
	Farmland	Cropland and other agricultural areas for cultivation.
	Grazing Area	Land used for livestock grazing.
	Orchards	Areas cultivated for fruit-bearing trees such as apple orchards, citrus groves, etc.
	Vineyards	Land planted with grapevines for wine production.
	Plant Nurseries	Areas for cultivating and propagating plants for sale or transplantation.
Water-Related Land	Waterbodies	Includes lakes, rivers, ponds, and other bodies of water.
	Wetland	Areas with saturated soil or standing water.
	Swamp	Wetland characterized by trees and shrubs.
Urban Land	City	Areas of dense human habitation and infrastructure.
	Urban	Developed areas with buildings and urban infrastructure.
	Built-up	Land covered by buildings and structures.
	Industrial Zones	Areas designated for industrial activities and manufacturing facilities.
	Parks	Land designated for recreational activities such as sports, picnics, and leisure.

(*Continued*)

Table 2.1 Types of land use. (*Continued*)

Land type	Sub-type	Description
	Campus	Land belonging to a specific institution or organization, is often used for educational, research, or administrative purposes.
Natural Land	Grassland	Areas covered mainly by grasses.
	Green Land	Vegetated areas include parks, gardens, and natural landscapes.
	Fallow Land	Agricultural land left uncultivated for a period to regain fertility.
	Forest	Areas dominated by trees, including deciduous forest and woodland.
	Rural Area	Areas outside urban centres with low population density.
	Bushland	Areas with dense vegetation, including shrubs and small trees.
	Shrubland	Areas dominated by shrubs rather than trees.
Miscellaneous Land	Sand Area	Areas predominantly covered by sand.
	Open Land	Areas predominantly covered by sand.
	Road	Land designated for transportation infrastructure.
	Archaeological Sites/Heritage Sites	Areas containing historical artifacts, ruins, or remnants of past civilizations.

2.4 Land Use Pattern vs. Land Use Cover

It is essential to understand the difference between land use patterns and land use cover; they are classified based on human activity and natural ecosystem. Land use pattern is an arrangement and distribution of land within a given landscape, like residential, commercial, agricultural, and industrial regions. It gives insight into how land is used for different purposes.

Land use cover, on the other hand, refers to the physical qualities or surface cover types that exist on land, such as forests, grasslands, urban areas, bodies of water, and barren terrain. Land use cover focuses on the physical look or composition of the land surface, regardless of the specific human activities that take place on it.

While land use pattern focuses on the practical aspects of human land use, land use cover informs us about the biological and physical properties of the land surface. Together, these ideas provide a complete knowledge of how human activities and natural processes alter the landscape, influencing land management decisions, environmental legislation, and sustainable development initiatives [13–16]. Their characteristics are given in Table 2.2 below.

Table 2.2 Land use pattern and land use cover.

	Land use pattern	Land use cover
Definition	Land use pattern refers to the spatial arrangement and distribution of various human activities and land uses across a given area or region.	Land use cover, also known as land cover, refers to the physical and biological coverings of the Earth's surface, including natural and artificial features.
Focus	It emphasizes the organization and layout of human activities on the land, including residential, commercial, industrial, agricultural, recreational, and conservation activities.	It focuses on the physical characteristics and natural vegetation types, as well as human-made structures and infrastructure, that cover the land surface.

(*Continued*)

Table 2.2 Land use pattern and land use cover. (*Continued*)

	Land use pattern	**Land use cover**
Characteristics	Land use patterns explain the types, intensities, and geographical distributions of various land uses within a given area. It is explained how humans utilize land and for what purpose.	It is the physical features of land surface like grassland, wetland, marsh, lakes, mountains, forest open land, etc.
Analysis	Analyzing land use patterns consists of mapping of distribution of various land use, studying the dynamics of development, change in pattern, agricultural development, urbanization, investigation, and determining the forces that drive change in land use patterns.	It is the categorization of and mapping of various natural features on land; it can be done by remote sensing techniques. It accesses land cover over time, mapping habitat loss, and degradation of natural lands. It understands ecological and environmental impacts.
Examples	Residential neighbourhoods, commercial districts, industrial zones, agricultural fields, recreational areas, and protected reserves.	Forests, croplands, grasslands, water bodies, urban areas, barren lands, and transportation corridors.

Land use pattern is a spatial arrangement formed by human activities, whereas land use cover is the physical qualities and natural elements that cover land. It is essential to understand both concepts as they affect the environment, land management policies, and planning.

2.5 Why Land Use Patterns Must Be Studied

Studying land use patterns is important for many reasons, spanning socio-economic development to environmental conservation [17, 18]. The study of land use patterns is expanding all over the world. Figure 2.2 shows the countries active in research on land use patterns.

There are several key reasons why the study of land use patterns is crucial, as explained below.

a. Natural Resource Management: Land use pattern directly affects the availability and management of resources like water, soil, forest, biodiversity, and minerals. Disorganized planning will accelerate soil erosion, deforestation, water pollution, and destruction of habitat. Effective water resource management helps with groundwater recharge and monitoring of contamination, while conservation methods like

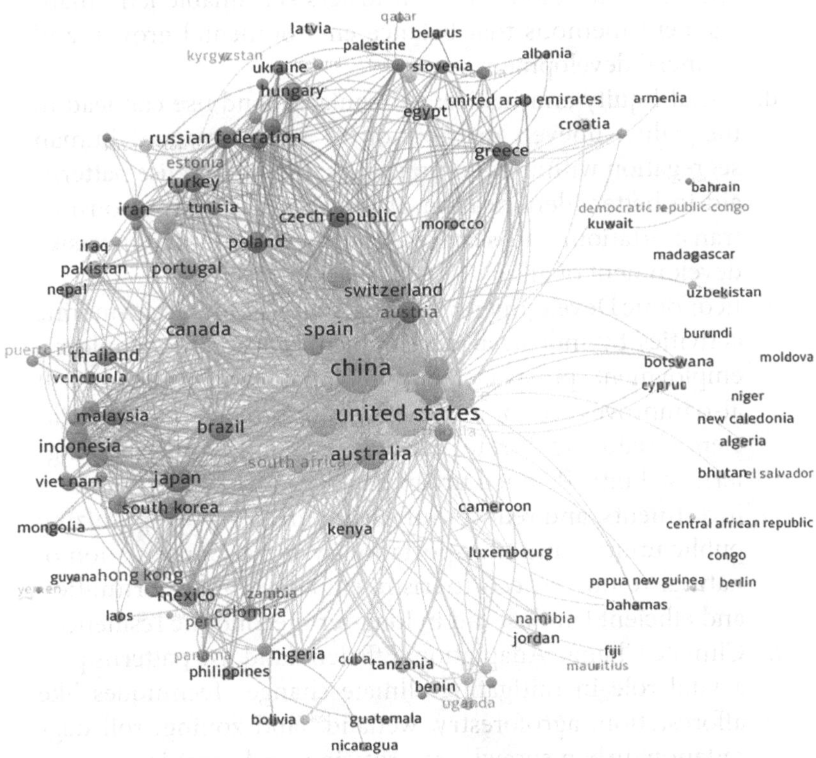

Figure 2.2 Countries engaged in land use patterns.

crop rotation, afforestation preserve soil fertility. Sustainable urban planning incorporates risk reduction strategies, wetland preservation, mitigating floods and landslides, and agroforestry. By integrating these strategies, policymakers ensure the long-term availability of natural resources.

b. Disaster Risk Reduction: Sustainable land use patterns aid in reducing environmental degradation, reducing vulnerability, and enhancing resilience. Better planning and zoning reduce flood-prone and landslide-prone areas. It also preserves wetlands, forests, and mangroves, which act as a buffer for floods and tsunamis. Eco-friendly urban planning reduces heat-related risks by integrating ecological and community participation.
c. Sustainable Development: Land use patterns affect the environment by resource utilization, conservation of habitat, and climatic resilience. It helps in the identification of sensitive areas, reduces deforestation, prevents soil degradation, and promotes biodiversity. It fosters sustainable land management methods that balance environmental growth and financial development.
d. Social Equity and Justice: Unregulated land use can lead to inequality, uneven distribution of resources, and human segregation within its communities. Better land use patterns aid in better identification of social vulnerability, housing, transportation, and services, and they encourage inclusive development of all.
e. Economic Development: Land use patterns impact economic activities by industrial development, corporate complexes, employment prospects, and regional competitiveness. It also improves economic benefits by improving resource efficiency, reducing disaster-related costs, and promoting long-term stability. Planned urban city has more property values, investments, and reduces infrastructure costs by optimizing public utilities and transportation systems. Conservation of natural resources, eco-tourism, employment opportunities, and efficient land use aid in long-term economic resilience.
f. Climate Change Adaptation: Efficient land use patterns play a vital role in mitigating climate change. Techniques like afforestation, agroforestry, wetland, land zoning, soil degradation, urban sprawl, crop rotation, and organic farming.

Integrating these techniques ensures a balanced and adaptive response to climate change.

g. Infrastructure Planning: Land use patterns optimize land allocation, reduce environmental impacts, and enhance long-term efficiency. By using techniques like smart zoning and mixed-use development, sustainable land use ensures that infrastructure and public transportation are efficiently distributed. It prioritizes climate-resilient infrastructure, like flood-resistant drainage systems, renewable energy integration, and eco-friendly building designs.

2.6 Relevant Departments and Stakeholders in Land Use Pattern Research

Analysis of land use patterns and their related driving factors is important for various fields for its wider benefits and application in society and humans, especially in the fields of engineering, science, formulation of policy, and sustainability, as described below.

a. Society: Sustainable land use patterns are essential for better living conditions for a society, as improved amenities like green space, open areas, water bodies, and basic services, which improve the quality of life, aid it. In addition, analyzing these patterns reveals various dynamics, resource allocation, and access to opportunities that promote fairness and justice in society. Land use patterns identify sensitive areas for natural hazards, aid in disaster management, and strengthen the community.
b. Human well-being: Land use patterns to a greater extent affect human well-being by influencing health, air, water, open space, physical activity, environment, and housing. Understanding these complex patterns helps in policy formulation, decision-making, planning, availability of housing, comfortable temperatures, and rich diversity, contributing to overall well-being.
c. Engineers: For infrastructure planning and resource management to develop better city planning, transportation networks, utilities, terrain mapping, and crucial infrastructure are developed by engineers. Research on land use patterns

contributes to sustainable development, assuring long-term goals, landscape planning, redeveloping cities, and conservation of soil, water, air, and energy.

d. Scientists: Scientists play a vital role in the protection of the environment and analyzing climate change by understanding land use patterns. They examine the land pattern and assess its impact on the local ecosystem, biodiversity, and habitat, leading to conservation policies and ecosystem management.
e. Policymakers: Land use patterns are most important in determining urban planning and policy formulation. These patterns are related to planning laws, zoning guidelines for cities and their expansion in a sustainable way. Land use policies are essential for environmental conservation by protecting natural resources and biodiversity. Policies are necessary to enforce the law against illegal activities, encroachment, and degradation of natural lands, protecting green lands and water bodies, which helps in balancing human growth and environmental conservation.
f. Urban planners: Land use trends provide critical insight for urban planning; these patterns enable sustainable growth of cities and regions through planning, policies, zoning regulations, and standards. It also protects natural resources, biodiversity, and environmental degradation, assuring long-term sustainability.
g. Resource management: Land use patterns improve resource efficiency and distribution by using methods like effective land use, waste reduction, environmental protection, and uniform distribution of land. It is important to understand these patterns that build resilient communities.
h. Real estate developers and investors: Effective land use patterns are indications of better development and attraction of investors, providing insight into market trends, demand for property, and investment opportunities. Developers identify suitable regions for residential, commercial, and industrial areas, aiding in informed decision-making. If a city is well developed and designed, investors tend to invest in those cities rather than unplanned cities.
i. Tourism: Effective land use patterns attract tourists to the city. Effective land use incorporates cultural sites, ancient architecture, open green areas, natural areas, and better

transportation systems. Tourists like to visit natural areas and coastal regions, where people are welcoming, have better transport systems, less congestion, suitable climate, lodging, and recreational facilities. By understanding and utilizing land use patterns, stakeholders can make a tourist-friendly city and generate revenue from it. This will aid in economic growth and cultural enrichment of tourist locations.

j. Educators and researchers: Analysing land use provides critical insight in fields like geography, environmental science, landscape, civil engineering, and economics. It investigates human interaction with the environment, urbanization, deforestation, water resource management, night light for development, natural landscape, development of road networks, rainfall patterns, impact of global warming, melting of icebergs, and many more. It enables the development of informed decision-making and building sustainable and resilient communities.
k. Healthcare professionals: Land use patterns are indirectly linked to health. Effective land use planning results in an abundance of sunlight, fresh air, a clean environment, sustainable farming techniques, free from pesticides and harmful fertilizers, better living conditions, and access to clean water, which regulates the chances of disease and promotes the health of the individual and community.
l. Transportation planners: Sustainable land use patterns consider the need for an efficient transportation system. It considers congestion, vehicular movement, pavement conditions, mobility, and network design. It considers local topography, traffic, and develops an efficient route for movement. By incorporating components of land use patterns engineers create transportation networks that are efficient and sustainable.
m. Heritage professionals: Understanding land use patterns is essential to conserve and preserve cultural heritage sites, historic artifacts, and traditional landscapes that were created by ancestors. It highlights local customs and legacy assets through cultural tourism programs. Specialists use their knowledge to safeguard and promote culturally significant places.
n. Disaster management: By understanding land use patterns, one can detect areas of risk like river lines, floodplains, and

risk zones. This helps in reducing risk, disaster preparedness, and better recovery. Proper zoning mitigates construction in disaster-risk zones like floodplains, seismic zones, and green areas, and planned development reduces the effect of hazards. By analyzing disaster risk zones, emergency services plan emergency shelters, evacuation routes, and a quick response.

All the above departments, sectors, stakeholders, and professions must study land use patterns, which are significant in solving multi-dimensional problems and developing interrelated solutions for sustainable development.

2.7 Use of Technology for Mapping Land Use Patterns

Land use patterns provide solutions to many problems in various fields, so determining them effectively becomes a major concern. The traditional method of manual survey is time-consuming, costly, labour-intensive, and prone to error. To overcome all those problems the latest technologies are used. Major techniques to determine land use patterns effectively are Remote Sensing and Geographic Information Systems (GIS) techniques. These technologies map, monitor, and analyse land use effectively in less time and low cost with the help of satellite images.

With satellite images, one can determine urbanization trends, water bodies, agricultural lands, deforestation rate, habitat loss, and other land use dynamics. Various software and tools like ArcGIS, QGIS, and Google Earth Engine (GEE) are used to process images. Emerging technologies like machine learning (ML) and artificial intelligence (AI) are incorporated with processing tools to improve accuracy in feature extraction and classification from satellite images.

In present times, many tools are used to extract land use from satellite/drone images. A sample of classified images is shown in Figures 2.3 and 2.4.

In the below images, land use is classified and represented in various colours. Red represents buildup areas, yellow is for agricultural land, green is for vegetation cover, and grey is for road surface.

In determining land use patterns, satellite images play a vital role and have a wide range of applications. Integration of GIS aids in the creation of detailed maps, providing a better visual representation. A satellite-based

Figure 2.3 Classified land use pattern.

Figure 2.4 Classified image.

system facilitates change detection and monitoring of urban sprawl, deforestation, agricultural areas, water bodies, and open areas.

Figure 2.5 depicts the role of satellite images and remote sensing in analyzing land use and provides insight into understanding spatial and temporal dynamics and human activities for better decision-making and sustainable development.

In the field of urban development, it aids in infrastructure mapping, demographic patterns, zoning areas, and informed land use decisions. It also facilitates environmental impact assessments by combining geospatial data, precision agriculture, crop monitoring, habitat mapping, conservation of biodiversity, transportation planning, and infrastructure development. Satellite images are effective in disaster management by identifying vulnerable regions, aiding evacuation plans, and assessing post-disaster damage. They also help in monitoring land degradation by evaluating soil quality and vegetation cover. Satellite imagery also plays a crucial role in

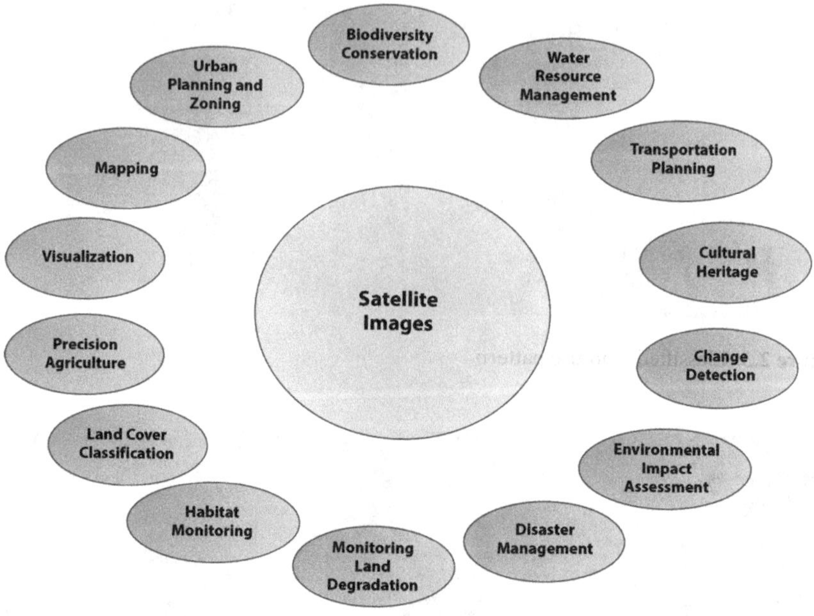

Figure 2.5 Role of satellite images in land use pattern.

water resource management, availability, distribution, preservation of cultural heritage, archaeology, historical sites, monitoring land use changes, and cultural landscapes.

2.8 Conclusion

It is vital to understand the dynamics of land use for better decisions and the implications of land management techniques. Land use patterns are influenced by many factors like geography, socioeconomic, legislative bodies, and government decisions. By analyzing land use patterns, one gets insight into the distribution, trends, and dynamics of land use, which are important in tracking environmental challenges, management of natural resources, and development of sustainable communities. To determine land use effectively, technologies like remote sensing, GIS, AI, and ML are used. These techniques aid in analyzing, mapping, monitoring, visualization, and evaluating land use patterns in areas of land management and urban planning. Integration of advanced techniques helps in more accurate and effective feature identification and informed decision-making, aiding in sustainable land use and informed policy formulation. Mapping land use patterns helps decision-making about urban design, transportation infrastructure, agricultural techniques, conservation of the environment, and management of disasters. Understanding land use is very critical to attain sustainable development goals and preserving long-term goals for both humans and nature.

References

1. López-Vargas, A., Ledezma, A., Bott, J., Sanchis, A., IoT for Global Development to Achieve the United Nations Sustainable Development Goals: The New Scenario After the COVID-19 Pandemic. *IEEE Access*, 9, 124711–124726, 2021, doi: 10.1109/ACCESS.2021.3109338.
2. Guzinski, R., Nieto, H., Sánchez, J.M., López-Urrea, R., Boujnah, D.M., Boulet, G., Utility of Copernicus-Based Inputs for Actual Evapotranspiration Modeling in Support of Sustainable Water Use in Agriculture. *IEEE J. Sel. Top. Appl. Earth Obs. Remote Sens.*, 14, 11466–11484, 2021, doi: 10.1109/JSTARS.2021.3122573.
3. Quayson, M., Bai, C., Sarkis, J., Technology for Social Good Foundations: A Perspective From the Smallholder Farmer in Sustainable Supply Chains. *IEEE Trans. Eng. Manage.*, 68, 3, 894–898, June 2021, doi: 10.1109/TEM.2020.2996003.
4. Huang, X., Liang, Q., Feng, Z., Chai, S., A TOD Planning Model Integrating Transport and Land Use in Urban Rail Transit Station Areas. *IEEE Access*, 9, 1103–1115, 2021, doi: 10.1109/ACCESS.2020.3047207.
5. Riekki, J. and Mämmelä, A., Research and Education Towards Smart and Sustainable World. *IEEE Access*, 9, 53156–53177, 2021, doi: 10.1109/ACCESS.2021.3069902.
6. Bai, and Sarkis, J., The Water, Energy, Food, and Sustainability Nexus Decision Environment: A Multistakeholder Transdisciplinary Approach. *IEEE Trans. Eng. Manage.*, 69, 3, 656–670, June 2022, doi: 10.1109/TEM.2019.2946756.
7. Cao, M., *et al.*, Multi-Scenario Simulation of Land Use for Sustainable Development Goals. *IEEE J. Sel. Top. Appl. Earth Obs. Remote Sens.*, 15, 2119–2127, 2022, doi: 10.1109/JSTARS.2022.3152904.
8. Parvin, K., *et al.*, Intelligent Controllers and Optimization Algorithms for Building Energy Management Towards Achieving Sustainable Development: Challenges and Prospects. *IEEE Access*, 9, 41577–41602, 2021, doi: 10.1109/ACCESS.2021.3065087.
9. Vali, Comai, S., Matteucci, M., Deep Learning for Land Use and Land Cover Classification Based on Hyperspectral and Multispectral Earth Observation Data: A Review. *Remote Sens.*, 12, 2495, 2020, doi: 10.3390/rs12152495.

10. Douass, and Ait Kbir, M., Deep learning approach for land use images classification. *ICIES'22, E3S Web Conf.*, 351, 2022, DOI: 10.1051/e3sconf/202235101043.
11. Goel, L., Gupta, D., Panchal, V.K., Hybrid bio-inspired techniques for land cover feature extraction: A remote sensing perspective. *Appl. Soft Comput.*, 12, 2012, DOI: 10.1016/j.asoc.2011.10.006.
12. Al Arjani, Modibbo, U.M., Ali, I., Sarkar, B., A new framework for the sustainable development goals of Saudi Arabia. *J. King Saud Univ. Sci.*, 33, 6, 2021, DOI: 10.1016/j.jksus.2021.101477.
13. Yigitcanlar, T., Agdas, D., Degirmenci, K., Artificial intelligence in local governments: perceptions of city managers on prospects, constraints and choices. *AI Soc.*, 38, 3, 2023, DOI: 10.1007/s00146-022-01450-x.
14. Han, B., Jin, X., Sun, R., Li, H., Liang, X., Zhou, Y., Understanding land-use sustainability with a systematical framework: An evaluation case of China. *Land Use Policy*, 132, 106767, 2023, https://doi.org/10.1016/j.landusepol.2023.106767.
15. Roy, S.K., Alam, M.T., Mojumder, P., Mondal, I., Kafy, A.-A., Dutta, M., Ferdous, M.N., Al Mamun, M.A., Mahtab, S.B., Dynamic assessment and prediction of land use alterations influence on ecosystem service value: A pathway to environmental sustainability. *Environ. Sustainability Indic.*, 21, 100319, 2024, DOI: 10.1016/j.indic.2023.100319.
16. Anthony, T., Shohan, A.A.A., Oludare, A., Alsulamy, S., Kafy, A.-A., Khedher, K.M., Spatial analysis of land cover changes for detecting environmental degradation and promoting sustainability. *Kuwait J. Sci.*, 51, 2, 100197, 2024, DOI: 10.1016/j.kjs.2024.100197.
17. Liu, J., Jin, X., Xu, W., Yang, F., Wang, S., Zhou, Y., Assessing trade-offs and synergies among multiple land use functional efficiencies: Integrating ideal reference and key indicators for sustainable landscape management. *Appl. Geogr.*, 158, 103037, 2023, DOI: 10.1016/j.apgeog.2023.103037.
18. Guo, Y., Tang, J., Liu, H., Yang, X., Deng, M., Identifying up-to-date urban land-use patterns with visual and semantic features based on multisource geospatial data. *Sustain. Cities Soc.*, 101, 105184, 2024, DOI: 10.1016/j.scs.2024.105184.

3
Principles of Sustainable Land Use

Abstract
This chapter introduces the fundamental concepts of sustainability in land management to justify research on economic development, social justice, and the environment. It operates under the triple-bottom-line model, which means that organizations need to consider options that would be of benefit to society and nature as well as the financial aspect. It demonstrates how society reached a balance between natural environments, social organizations, and the economy to benefit people. It indicates how the natural world can be protected and how the services that sustain life can be maintained. Sustainable practices are studied in terms of their economic viability through analysis of investments and benefits, as well as assessment of created employment opportunities and economic growth potential. Through educational initiatives, the focus is on building sustainability awareness alongside developing capacities and training leaders who will dedicate themselves to sustainable land practices. The last section discusses policy integration across various governance levels by highlighting the necessity of consistent guidelines for supporting sustainable practices.

***Keywords*:** Decision-making, community engagement, cost-benefit analysis, conservation, sustainable land use practices

3.1 Introduction

In the era of rapid urbanization, rise in population, and environmental challenges, there is a need for sustainable land use techniques. Human actions have seriously impacted the ecosystem and nature, which calls for a necessary and balanced approach to benefit humans and ecology [1–3].

The concept of Triple Bottom Line is the paradigmatic approach, which includes taking decisions in such a way that it not only considers economic success but also social equity and environmental stewardship. Taking a stance on this multidimensional approach gives stakeholders a common aim to identify a harmonious balance for long-term resilience and well-being in many sectors of society [4–6]. Importantly, the current

Kundan Meshram and Chatrabhuj. A Sustainable Approach to Land Use, (45–66) © 2026 Scrivener Publishing LLC

processes indicated the importance of economic success, social equality, and environmental stewardship, then set the framework for comprehension of sustainability in land management. Knowing the intrinsic worth of natural assets and the essentials of all kinds of biodiversity in maintaining the process of ecological balance is a significant piece of the conservation and preservation measures. Resource management experiments are investigated, and they point out the crucial role of preserving critical ecosystem services like soil fertility, water purification, and climate regulation dedicated to humans as well as the environment.

The focus of this chapter is on exploring the dynamic interaction between indigenous methods of knowledge of the earth (traditional ecological knowledge) and modern technical breakthroughs [7]. The focus is on adaptive management strategy and monitoring systems to respond to the environment problems and requirements of the stakeholders. The goal of this groundwork is to help in conducting a comprehensive investigation into the many aspects of sustainable land use and the relatedness of economic, social and environmental issues [8–11]. Through holistic, systems-based thinking and the promotion of partnership in collaboration, stakeholders can work together to establish a course for a future that is sustainable and inclusive for all.

3.2 Balancing Economic, Social, and Environmental Objectives

It is essential to balance economic, social, and environmental objectives in land use patterns for sustainable development. Economic factors that drive land use decisions prioritize urban development and infrastructure growth. It is also noted that unregulated city growth can lead to uneven income, community displacement and degradation of the environment. One of the most important parameters to ensure that all communities have access to housing, public space and essential services is social equity. This can be attained by the development of policies, effective urban planning and zoning regulations for all to prevent social exclusion of one community from other communities. Environmental sustainability is also one of the most important parameters to conserve ecosystem biodiversity, combat climate change and reduce over-exploitation of resources, deforestation, and pollution. There is a need for the integration of green infrastructure, sustainable agriculture, and eco-friendly approaches in the development of a land use pattern of a city. It is very important to achieve this balance

Table 3.1 Benefits of balance between economic, social, and environmental factors.

Area	**Sub-area**	**Description**
Community Well-being	Social Cohesion	Sustainable development strengthens community relationships by encouraging inclusion, equity, and participation in decision-making processes. The feeling of belonging contributes to community resilience and cohesion.
	Quality of Life	Balanced development increases community members' quality of life by providing access to critical services like education, healthcare, and clean water, as well as expanding recreational and cultural opportunities.
Ecosystem Health	Biodiversity Conservation	Sustainable practices help to conserve and restore ecosystems, preserving biodiversity and the ecological services they offer, like pollination, purification of water, and climate regulation.
	Ecosystem Resilience	Communities that preserve and restore natural habitats improve ecosystem resilience, allowing ecosystems to tolerate shocks like climate change, habitat fragmentation, and invasive species.

(Continued)

Table 3.1 Benefits of balance between economic, social, and environmental factors. (*Continued*)

Area	**Sub-area**	**Description**
Economic Prosperity	Green Jobs	Sustainable development work opportunities arise in renewable energy systems while helping the tourism industry and constructing ecological facilities to promote job growth and economic progress.
	Cost Savings	Implementing sustainable practices, such as energy efficiency measures and reduction of waste techniques, allows businesses and households to save money and resource usage.
	Innovation and Competitiveness	Supporting sustainability leads companies to create new ideas and builds market advantage through eco-friendly product development.
Public Health	Improved Air and Water Quality	By adopting sustainable development methods, such as clean energy adoption and pollution reduction, communities improve environmental health and help prevent health issues for the public.
	Enhanced Access to Green Spaces	Investing in effective urban planning with green space promotes physical activities, wellness and community engagement, benefiting overall public health.

(*Continued*)

Table 3.1 Benefits of balance between economic, social, and environmental factors. (*Continued*)

Area	Sub-area	Description
Social Equity	Reduced Inequalities	Sustainable development efforts seek to eliminate social disparities by addressing discrepancies in access to resources, services, and opportunities, fostering social justice and inclusion.
	Empowerment and Participation	Through decision-making participation, marginalized groups improve their influence, which leads them to establish policies and programs that affect their lives, resulting in more equitable outcomes.
Climate Mitigation and Adaptation	Reduced Greenhouse Gas Emissions	Sustainable activities adopt green techniques that aid in the reduction of greenhouse gas emissions.
	Resilience to Climate Impacts	By effective land use planning, communities avoid chances of extreme weather, water scarcity, disasters and prolonged vulnerabilities.

among all these three objectives through collaborative effort by participation in better decision-making and by formulating adaptive policies.

By incorporating all these three objectives, a smart sustainable land use pattern is developed that ensures the long-term sustainability of a city. However, one comes at the cost of the other. As economic growth progresses, it damages the environmental aspect [12]. The developer must find innovative solutions like green economic initiatives, mixed-use development of land, better transportation networks, and agroforestry to harmonize a balanced approach. There are many benefits of integrating economic, social, and environmental factors, for the community and for the ecosystem, which are represented in Table 3.1 above.

By considering this balanced approach and the numerous benefits and interconnectedness, one can work to achieve better outcomes from land use patterns and promote resilience, equality, and prosperity for all communities and the environment.

3.3 Integrating Traditional and Modern Practices

The transmission of traditional knowledge among indigenous and local groups develops a crucial system for sustainable land use due to various necessary elements as described below.

a. Traditional knowledge systems maintain holistic views of ecosystems by understanding the linked nature of human beings with water, land, plants and animals. By adopting this perspective, one can gain knowledge for implementing sustainable land management techniques that support ecosystem health.
b. A community knowledge system integrates itself within the specific landscape features where people live. The knowledge helps communities apply land use methods that match their unique environmental elements, such as various soil types and climate patterns alongside biodiversity elements to enhance their overall resistance against environmental change.
c. Local communities possess an extensive understanding of the identification and biological characteristics of their native plants, along with animal species as well as their habitats, behavioural patterns and ecological functions. The information serves biodiversity conservation activities because it guides both habitat protection measures, species management practices and restoration plans.
d. Community resilience together with social cohesion comes directly from traditional knowledge, which deeply connects with cultural traditions, social norms and community governance. Traditional methods of land management enhance both community resistance, social group unity through communal decision systems and communal assistance along with intergenerational knowledge exchange.
e. Imparted by indigenous and local knowledge systems is the ethical perspective, which entails nature respect along with

reciprocal and interdependent relations. The land use practices receive direction from these ideals, which prioritize sustainability.

f. Traditional knowledge serves for sustainable land use because it provides proven methods which foster both natural ecosystem protection and human-environmental accordance. The combination of traditional knowledge, scientific comprehension and contemporary technology enhances sustainable land management objectives by making them more effective and valid.
g. Organizations should unify Traditional Ecological Knowledge (TEK) with contemporary procedures because this constitutes an essential approach to maintain sustainable land use management. Acquired through observation combined with experimentation and environmental adaptation, indigenous and local people store their evolved knowledge systems known as traditional ecological knowledge. TEK integration with contemporary approaches to management delivers multiple advantages, which establish improved resilience and sustainability and cultural value [13] as Figure 3.1 illustrates.

Indigenous communities have an extensive understanding of their native ecosystems because they carefully know all plant and animal species, together with ecological procedures and seasonal patterns. Such information

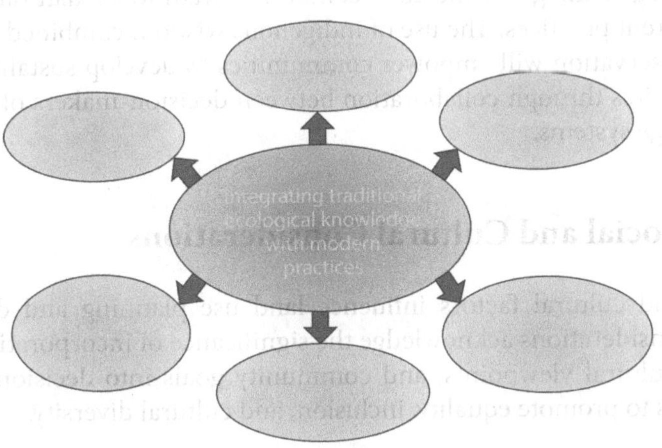

often possesses specialized characteristics that focus on particular local situations. Decisions about land use planning and management become more effective when decision-makers add indigenous wisdom to their process because this wisdom has traditionally sustained both ecosystems and human occupations for centuries [14]. Several strategies to address environmental changes, droughts, floods and catastrophic events are commonly found in traditional ecological knowledge systems. The effective local strategies lead ecosystems, together with communities toward better resistance when facing climate change and additional stressors. Traditional indigenous agricultural systems, including terracing together with crop diversification and water collection, have performed for many generations to combat irregular rainfall along with soil erosion.

Traditionally managed lands rest on sustainability principles because they aim to preserve ecological stability together with natural resource conservation across generations. Traditional ecological knowledge that receives integration within land use governance systems protects cultural customs through ensuring the indigenous and local population's cultural survival [15, 16]. Land management sustainability goals receive support from decision-makers who protect indigenous traditions because these efforts simultaneously develop cultural diversity and social unity. Land use planning becomes more successful because indigenous and local communities get involved in decision processes, and their cultural knowledge creates shared partnerships to manage conservation efforts. Researchers and professionals can develop environmentally sustainable remedies by creating stronger connections between traditional knowledge systems and scientific networks [17]. A sustainable method toward achieving land use management goals includes combining traditional natural wisdom with current practices. The use of indigenous wisdom combined with cultural preservation will empower communities to develop sustainable and resilient sites through collaboration between decision-makers of different knowledge systems.

3.4 Social and Cultural Considerations

Social and cultural factors influence land use planning and decisions. These considerations acknowledge the significance of incorporating social values, cultural viewpoints, and community goals into decision-making processes to promote equality, inclusion, and cultural diversity.

a. Community engagement and participation:
Involving local populations, customers, and indigenous groups in land use planning and decision-making processes ensures that their opinions are heard, their knowledge is recognized, and their interests are protected. Engaging in meaningful communication, consultation, and engagement with communities to establish trust and cooperation leads to the co-creation of solutions that meet their needs and ambitions.

b. Cultural preservation and heritage conservation:
This involves identifying cultural values, traditions and heritage of local communities, such as their connections towards land, traditional land use practices and policies. Incorporating cultural factors into land use planning and management safeguards sacred sites, cultural landscapes, and traditional territories, as well preventing the erosion of cultural identity.

c. Social equity and environmental justice:
Overcoming inequities in access to assets, possibilities, and decision-making processes ensures that land use decisions are equitable and beneficial to all members of society, particularly marginalized and vulnerable communities. It involves recognizing and addressing potential social and environmental inequalities, such as disproportionate pollutant exposure, limited access to green places, and community displacement as a result of development initiatives.

d. Indigenous wisdom:
It is essential to identify the importance of local and indigenous communities and their traditional knowledge. Local communities have local ecosystems, land management techniques and environmental change. Integrating traditional knowledge in land use planning and monitoring techniques aids in the resilience, protection of biodiversity, and fosters harmonious culture.

e. Social impacts:
Social impacts include identifying approaches to reduce negative social impacts while increasing positive results, such as establishing chances for local employment, promoting sustainable livelihoods, and improving the welfare of the community.

f. Cultural diversity and placemaking:
   Celebrating cultural variety and advocating for inclusive placemaking practices respect varied communities' distinct identities, traditions, and histories [18]. Developing inclusive public places, cultural facilities, and recreational opportunities promote social cohesiveness, cultural interchange, and a sense of belonging among all citizens.

Social and cultural factors influence land use planning and management decisions. By incorporating social values, cultural views, and community interests into decision-making processes, stakeholders can promote fairness, inclusion, and cultural diversity while also creating more sustainable and resilient landscapes that will satisfy the requirements of future generations, as shown in Figure 3.2.

Local communities have essential knowledge and expertise about their surroundings, including traditional practices, natural processes, and sociocultural values. Engaging communities tap into this vast pool of local knowledge, increasing the relevance, efficacy, and sustainability of decision-making processes and outcomes. When local people make decisions about natural resource management they develop greater protection,

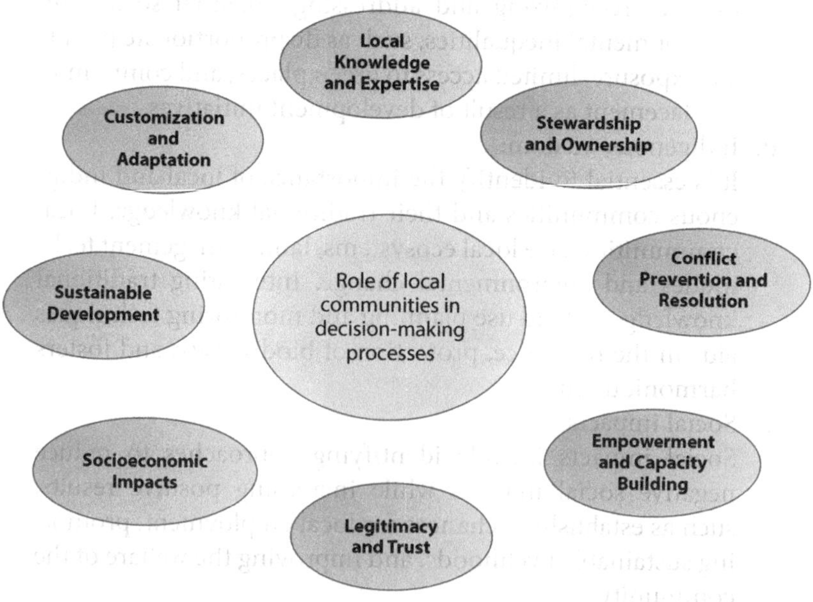

Figure 3.2 Benefits of engaging local communities in decision-making processes.

wilful care and full responsibility regarding local ecosystems. Communities that contribute to policy development, along with program development and management practice design demonstrate higher levels of program implementation fidelity and enforcement, which leads to better conservation outcomes together with sustainable management [5]. Critical management choices regarding natural resources bring extensive social and economic impacts to a neighbourhood's populations through their effects on standard of living and food supply, along with traditional cultural heritage. By allowing communities to participate in making decisions, their requirements together with their goals and worries will be recognized, thus decreasing adverse effects. Decision participation from local communities creates a mechanism to avoid conflicts and settle disputes regarding natural resource management. Steps toward participatory decision-making help different groups of stakeholders build trust while working together, thereby reducing conflicts and strengthening social unity. People in local communities develop unique needs together with distinct goals and perspectives through their cultural environment, as well as natural resources and historical stories. Local circumstances determine how communities take part in decision-making processes that create appropriate answers that suit specific cultural patterns while maintaining social acceptance. Local communities gain strength through decision-making participation because they receive both authority and power and the ability to determine matters affecting their daily lives. Engagement procedures develop community abilities as well as leadership skills and civic empowerment, which enables residents to advocate for their needs, negotiate with authorities and maintain accountability of decision-makers.

Decisions made with broad involvement of local communities gain higher status as well as higher transparency and accountability. Decision-makers establish processes that involve open communication, information sharing and local stakeholder participation to build trust and receive credibility in order to enhance decision-making legitimacy. Sustainable development goals that integrate environmental, social and economic factors need local community participation in decision-making processes for their establishment. When decision-making integrates local objectives with community values and specific knowledge it produces better outcomes that combine sustainability with resilience for social and environmental elements [8, 15]. Local communities need to participate in decision-making for effective and fair long-term resource management as well as development because this ensures democratic administration and equality.

3.5 Economic Viability of Sustainable Practices

Sustainable practices show their financial strength when they produce earnings while promoting fairness between all people and defending natural systems. The combined use of land and acceptance of sustainable strategies needs financial success across all land planning and management areas to reach long-term sustainability. Sustainable practices show these economic features in their operations.

a. Cost savings and efficiency gains:
   At the same time, sustainable practices bring down operating costs while creating a more efficient operation system for the future. Firms and homes that buy sustainable technologies such as LED lights and solar panels lower their energy usage expenses. Sustainable farm management practices, including conservation tillage and integrated pest management lower production expenses (mineral and fuel) reducing operational expenses of farmers and boosting their profitability levels.
b. Enhanced market access and premium pricing:
   Sustainable implementation by companies enables them to enter distinct markets and achieve higher prices for their environmentally conscious products. The rising customer base is willing to spend more money on items that were produced using sustainable methods and ethical procedures. The market gives premium prices to differentiated products that earn certification from sustainable farming and fair-trade programs, thus giving economic motivation to producers to adopt sustainable practices.
c. Reduced regulatory compliance costs:
   Companies, especially those operating in industries such as manufacturing and mining, find it expensive to maintain compliance with environmental standards. Businesses that adopt sustainable practices before regulatory requirements take effect reduce their chances of non-compliance and minimize associated penalties together with fines and legal expenses. Companies that fund pollution prevention and waste reduction solutions can decrease their regulatory restrictions and demonstrate their environmental

commitment to customers, which leads to market success and an improved reputation.

d. Long-term asset value and resilience:
Sustainable practices enhance future durability along with the increased value of assets from buildings to basic infrastructure and natural resources. Green building design and construction methods bring together energy efficiency and cost-saving benefits, which generate comfort together with productivity throughout buildings until they reach their life expiration. Communities together with enterprises can benefit economically from investing their assets in natural elements such as forests and wetlands to obtain ecological advantages including flood protection as well as water filtration and carbon sequestration.

e. Access to green financing and investment opportunities:
Today institutions and investors actively pursue sponsorship deals for businesses and sustainable projects that bring positive environmental and social outcomes. Businesses can acquire funding at favourable levels through green financing instruments which include ecological bonds and sustainability-linked advances. Investors find renewable energy projects alongside environmentally safe infrastructure development opportunities to be attractive investments.

f. Economic diversification and job creation:
Investing in sustainable practices can help broaden regional economies and create new jobs in developing industries like renewable energy, eco-tourism, and green technology. Renewable energy projects, such as wind farms or solar systems, generate jobs in production, installation, and operation. Sustainable agriculture approaches, such as organic farming or agroforestry, can help rural communities and generate job possibilities in cultivating, processing, and selling sustainable products.

g. Innovation and competitiveness:
Supporting sustainability promotes innovation and competitiveness by pushing organizations to create new products, services, and models that respond to changing consumer tastes and market demands. Companies that innovate and adapt to sustainability trends get a competitive edge by

setting themselves apart from competitors, attracting talent, and securing market share in emerging sustainability-focused industries.

The economic viability of sustainable practices includes a variety of benefits such as cost savings, market potential, regulatory compliance, asset value, financing access, job creation, and innovation. Businesses, communities, and legislators may speed the transition to a more sustainable and prosperous future by taking these economic considerations into account and demonstrating their financial benefits.

3.6 Potential of Job Creation in Sustainable Land Use

Sustainable land use has significant potential for job creation and economic development across various sectors, as explained below.

a. Agriculture and forestry: Sustainable agricultural approaches, such as organic farming, agroforestry, and permaculture, generate jobs for farmers, farmworkers, and rural residents. Sustainable agriculture creates jobs along the agricultural value chain, from production to distribution and retail, by fostering diversified cropping methods, value-added processing, and direct marketing channels.
b. Ecotourism and recreation: Sustainable land management strategies, such as conservation area management, restoration of habitat, and nature-based tourism, promote economic growth in rural and peri-urban regions by encouraging ecotourism, outdoor recreation, and cultural tourism. Sustainable land use preserves natural landscapes, biodiversity, and cultural places, which creates chances for eco-lodges, guided tours, adventure activities, and handmade products, drawing tourists, creating profits, and supporting local businesses.
c. Renewable energy: Sustainable land use encourages the growth of energy sources that are renewable such as solar, wind, hydro, and biomass, which generates employment in the development, setup, operation, and maintenance of energy facilities. Sustainable land management promotes the transition to low-carbon economies, lowers greenhouse gas emissions, and improves energy security, all the while

supporting creativity, entrepreneurship, and job creation in the renewable energy industry.

d. Green infrastructure and resilience: Sustainable land use encourages funding for green infrastructure projects like green roofs, urban parks, and floodplain restoration, which generates employment in construction, landscaping, and ecosystem restoration. Green infrastructure projects contribute to economic development, property values, public health, and quality of life by strengthening urban and rural communities' resilience to climate change impacts, natural disasters, and environmental hazards, while also providing numerous co-benefits such as improved air and water quality, minimized heat island effects, and higher levels of biodiversity.

e. Ecosystem services and payments: Effective land management creates economic value by providing ecosystem services such as carbon sequestration, water purification, and pollination, which can be monetized through Payments for Ecosystem Services (PES) programs, carbon markets, and biodiversity offset programs. By encouraging landowners and communities to protect and restore natural habitats, sustainable land use establishes revenue streams, job opportunities, and income diversification options, particularly for rural landowners, indigenous peoples, and local communities that rely on natural resources for their livelihood.

f. Sustainable infrastructure and urban development: Sustainable land use encourages investment in sustainable infrastructure projects like green buildings, public transit, and smart cities, which generate jobs in the construction, engineering, and technology industries. Sustainable land management promotes economic growth, social equity, and environmental sustainability in cities by incorporating sustainability, resilience, and inclusivity principles into urban planning and development processes, as well as reducing consumption of resources, waste generation, and degradation of the environment.

Sustainable land use has the potential to significantly increase job creation and economic development by encouraging entrepreneurship, innovation, and investment in the agricultural, forestry, tourism, energy, infrastructure, and ecosystem services sectors. Sustainable land management can help to promote inclusive, equitable, and sustainable development that benefits both people and the environment by aligning economic incentives with environmental and social goals.

3.7 Educational Initiatives

Educational programs are required on sustainable land use to teach all affected groups valid information about protecting our planet and achieving both social success and economic growth. Our educational programs teach different stakeholder groups including scientists, teachers, decision-makers, property owners and the public, to develop sustainable practices that result in long-term positive outcomes. The educational programs for sustainable land use patterns use these main program features.

Students learn better sustainability methods at each education level from basic to university when environmental education combines with curriculum to help them solve environmental issues and improve land usage. Students gain knowledge for environmental problem-solving when they learn scientific techniques through practical experiences and scientific public participation activities. The program teaches people about different methods of land preservation and effective farming practices while helping communities restore their natural environments during training events. Sustainable land use programs require training plans to empower experts and leaders alongside practitioners who learn how to effectively work with sustainable resources.

Pursing educational steps and public outreach teaches people how to see sustainability by sharing eco-friendly ideas. Different types of communication platforms, like social media, mass media, and social events allow us to reach diverse audience segments when sharing sustainability information. Human beings develop deeper environmental links when they study ecology by experiencing nature during outdoor education and experiential learning activities. Community monitoring programs give volunteers the chance to study nature as scientists and help people collect data about environmental health. Residents can use these programs to collect data that helps the environment by planning better decisions and strengthening natural systems. The lifelong education system helps students maintain their knowledge of sustainable land use fields and discover new environmental information. Sustainability-based industries can attend webinars and classes and take part in professional workshop training that leads to certificate programs through these educational programs.

Sustainable land use training programs help stakeholders understand sustainable practices better while sharing useful knowledge to embrace better sustainable choices. Local governments and business sectors should fund training for community members to learn environmental protection practices which leads them to support sustainability actions, resulting in better health outcomes for both people and nature.

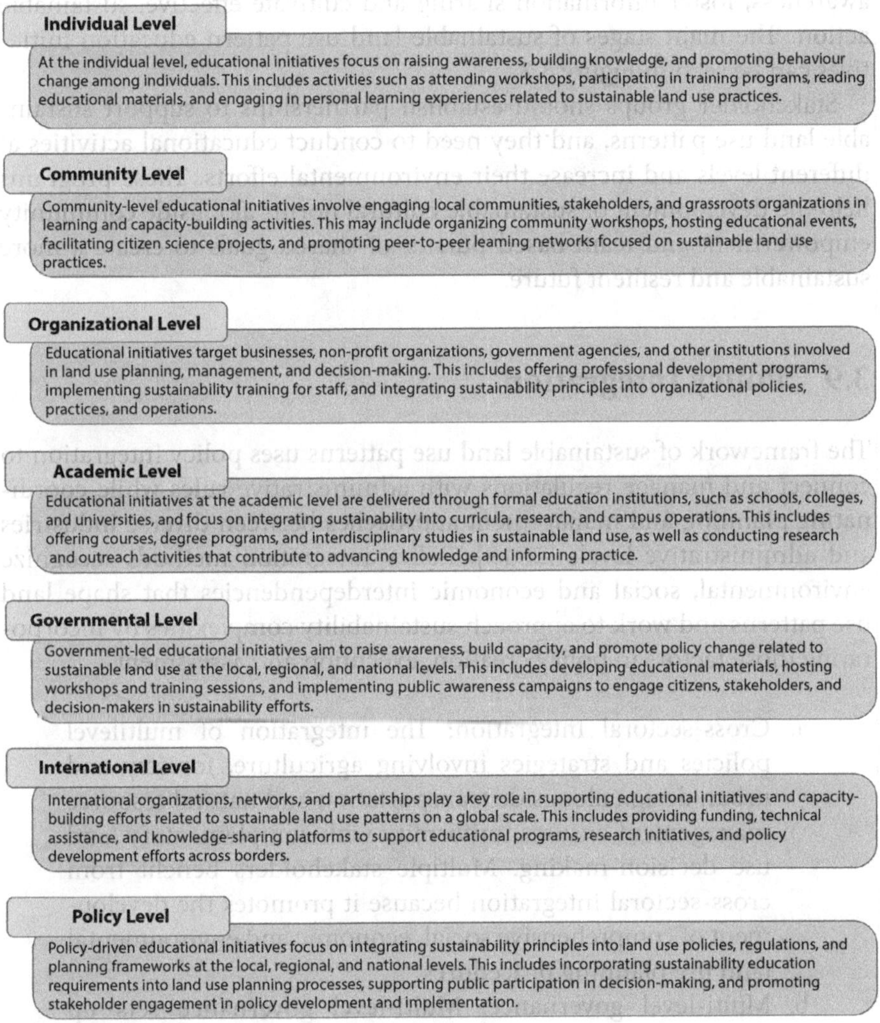

Figure 3.3 Different levels of education.

3.8 What are the Different Levels of Educational Initiatives?

Educational programs that promote sustainable land use patterns operate across multiple levels starting from individual initiatives up to community-led projects as well as institutional and policy-based programs. The different educational programs operate together across multiple levels to spread awareness, foster information sharing and cultivate effective, sustainable action. The main stages of sustainable land use pattern education initiatives can be seen in Figure 3.3.

Stakeholder groups should establish partnerships to support sustainable land use patterns, and they need to conduct educational activities at different levels and increase their environmental efforts. These programs help the development of sustainable cultural norms alongside community empowerment and team-based pursuit of shared goals to create a more sustainable and resilient future.

3.9 Policy Integration

The framework of sustainable land use patterns uses policy integration to connect and manage regulations with administrative rules while coordinating planning and management approaches between diverse industries and administrative levels [2, 19]. Policy integration methods recognize environmental, social and economic interdependencies that shape land use patterns and work to approach sustainability complexities by incorporating those factors in policy creation, execution and assessment.

a. Cross-sectoral integration: The integration of multilevel policies and strategies involving agriculture, forestry and urban design alongside transportation and natural resource management ensures uniformity with complementary land use decision-making. Multiple stakeholders benefit from cross-sectoral integration because it promotes the development of comprehensive social, economic and environmental land use planning procedures.
b. Multi-level governance: Multi-level governance sets up partnerships between government departments at different administrative levels together with civil and private sector

groups who design and make decisions regarding land use policies. Multi-level governance promotes comprehensive coordination between upper and lower government levels and laterally for data exchange and joint decision processes aimed at managing sophisticated land-use challenges including urban growth, forest destruction and water contamination.

c. Policy coherence: Policy coherence requires the creation of uniform and harmonious relations between land use planning and management policies throughout multiple policy sectors. The coordination of land use policies through policy coherence makes them more sustainable and effective because it reveals beneficial connections and negative relationships between different goals while promoting the use of unified policy structures.
d. Mainstreaming sustainability: The main promise of sustainability mainstreaming is its ability to organize both sustainability elements and priority issues within standard policy systems during decision processes. National and regional policy frameworks should adopt the sustainable development goals through mainstream processes while environmental impact assessments should become part of land use planning procedures and sustainability criteria.
e. Adaptive governance: The foundation of adaptive governance comprises adaptable policies that modify their procedures because of how stakeholder preferences shift and societal requirements and environmental factors are transformed. Policy adoption of adaptive governance provides policymakers with tools to promote inventive measures as well as experimental learning and knowledge creation. Multiple policy adjustments together with steering corrections would become possible through the ability to adapt in response to new problems or unexpected results in land use planning and management.
f. Participatory decision-making: Stakeholder involvement in policy development starts from formulation through execution to oversight and ensures participation by communities, indigenous populations and marginalized groups, local communities, and civil society entities. Participatory approaches enhance legal compliance of land use rules through solid

stakeholder relationships, which stimulate multiple groups to adopt sustainability initiatives due to transparent ownership structures.

Sustainable land use patterns can be achieved through the integration of methods that are credible, adaptive and integrated for land use planning and management. Policies focus on intricate sustainability problems while creating strengthened and more balanced sustainable land use results through encouraging unified government action between different sectors at various administrative levels and active stakeholder participation.

3.10 Conclusion

Sustainable land use demands careful attention to all three factors of economic development, social equality, and environmental aspects. As cities expand quickly during population growth and face environmental dangers, the community requires sustainable land practices to protect nature for lifetime health. The chapter showed that combining ecological wisdom from the past with new technology creates strong and tough plans to manage land. Organizations working together can save resources by planning urban spaces well and using ecologically friendly systems that benefit nature and society. Official education programs and policy working together support everyone involved in becoming knowledgeable participants in land use sustainability work. Putting resources into responsible environmental care creates new employment opportunities and leads to fresh ways to develop. Through joint work between decision-makers and environmental experts, the smart urban development model will succeed in meeting the needs of today and tomorrow. Effective land use sustainability requires permanent supervision and adjustment plus actions from all stakeholders. Through complete systems thinking, societies will build a future that achieves sustainable development of economy, society and nature together.

References

1. Bizoza, A.R. and Opio-Omoding, J., Assessing the impacts of land tenure regularization: Evidence from Rwanda and Ethiopia. *Land Use Policy*, 100, 104904, 2021, https://doi.org/10.1016/j.landusepol.2020.104904.

2. Bourgoin, J., *et al.*, Toward a land zoning negotiation support platform: 'Tips and tricks' for participatory land use planning in Laos. *Landsc. Urban Plan.*, 104, 2, 270–278, 2012, https://doi.org/10.1016/j.landurbplan.2011.11.008.
3. Carrilho, J. and Trindade, J., Sustainability in Peri-Urban Informal Settlements: A Review. *Sustainability*, 14, 13, 7591, 2022, https://doi.org/10.3390/su14137591.
4. Chigbu, U.E., *et al.*, Combining land-use planning and tenure security: a tenure responsive land-use planning approach for developing countries. *J. Environ. Plan. Manag.*, 60, 9, 1622–1639, 2017, https://doi.org/10.1080/09640568.2016.1245655.
5. Hendriks, B., *et al.*, Pro-poor land administration: Towards practical, coordinated, and scalable recording systems for all. *Land Use Policy*, 81, 21–38, 2019, https://doi.org/10.1016/j.landusepol.2018.09.033.
6. OECD & Sahel and West Africa Club, Africa's Urbanisation Dynamics 2020: Africapolis, Mapping a New Urban Geography, OECD, 2020, https://doi.org/10.1787/b6bccb81-en.
7. Othengrafen, F. and Reimer, M., The embeddedness of planning in cultural contexts: Theoretical foundations for the analysis of dynamic planning cultures. *Environ. Plan. A*, 45, 6, 1269–1284, 2013, https://doi.org/10.1068/a45131.
8. Palmer, C., *et al.*, Environmental Ethics. *Annu. Rev. Environ. Resour.*, 39, 1, 419–442, 2014, https://doi.org/10.1146/annurev-environ-121112094434.
9. Sahana, M., *et al.*, Where Is the Peri-Urban? A Systematic Review of Peri-Urban Research and Approaches for Its Identification and Demarcation Worldwide. *Remote Sens.*, 15, 5, Artic. 5, 2023, https://doi.org/10.3390/rs15051316.
10. Sante Riveira, I. and Crecente Maseda, R., A Review of Rural Land-Use Planning Models. *Environ. Plann. B: Plann. Des.*, 33, 165–183, 2006, https://doi.org/10.1068/b31073.
11. UN-Habitat, U.N.-Women, & others, SDG GOAL 11 Monitoring Framework: A guide to assist national and local governments to monitor and report on SDG Goal 11 indicators, 2016.
12. Verburg, P.H., *et al.*, Land System Science: Between global challenges and local realities. *Curr. Opin. Environ. Sustain.*, 5, 5, 433–437, 2013, https://doi.org/10.1016/j.cosust.2013.08.001.
13. Yakubu, I., From a cluster of villages to a city: Housing politics and the dilemmas of spatial planning in Tamale, Ghana. *Land Use Policy*, 109, 105668, 2021, https://doi.org/10.1016/j.landusepol.2021.105668.
14. Ali, E.B., Agyekum, E.B., Adadi, P., Agriculture for sustainable development: a SWOT-AHP assessment of Ghana's planting for food and jobs initiative. *Sustainability*, 13, 628, 2021, https://doi.org/10.3390/su13020628.
15. Asante-Yeboah, E., *et al.*, Analyzing variations in size and intensities in land use dynamics for sustainable land use management: a case of the

coastal landscapes of South-western Ghana. *Land*, 11, 815, 2022, https://doi.org/10.3390/land11060815.
16. Bretagnolle, V., *et al.*, Towards sustainable and multifunctional agriculture in farmland landscapes: lessons from the integrative approach of a French LTSER platform. *Sci. Total Environ.*, 627, 822–834, 2018, https://doi.org/10.1016/j.scitotenv.2018.01.142.
17. de Jong, L., *et al.*, Understanding land-use change conflict: a systematic review of case studies. *J. Land Use Sci.*, 16, 223–239, 2021, https://doi.org/10.1080/1747423X.2021.1933226.
18. Wiggering, H., *et al.*, Indicators for multifunctional land use—linking socio-economic requirements with landscape potentials. *Ecol. Indic.*, 6, 238–249, 2006, https://doi.org/10.1016/j.ecolind.2005.08.014.
19. Tran, D.X., *et al.*, Developing a landscape design approach for the sustainable land management of hill country farms in New Zealand. *Land*, 9, 185, 2020, https://doi.org/10.3390/land9060185.

4
Economic Aspects of Sustainable Land Use

Abstract
The practice of sustainable land use serves as a vital method to resolve multiple economic problems that emerge from uncontrolled land management activities. This chapter looks at sustainable land use economics to build sustainable environmental protection parallel to economic development alongside social prosperity. The study examines basic principles of sustainable land use economics and explains how various land use patterns are initiated while evaluating their extensive economic consequences for different industrial areas. The analysis includes investigations of structures that combine sustainable land practices with financial stability and social equity. It examines how economic goals face challenges while balancing environmental sustainability before presenting new solutions to resolve these conflicting priorities. Sustainable urban development plays a crucial role in sustainable land use practice because it allows the creation of environment-friendly municipalities while fostering economic inclusivity.

Keywords: Environmental economics, framework, management decisions, sustainable land use

4.1 Introduction

To address the demands of both the present and upcoming generations, sustainable land use ensures the long-term productivity of natural resources. The approach deals with critical environmental matters resulting from improper land management strategies, which include climate change, biodiversity reduction, soil degradation and water deficiency. Sustainable land use practices unify human wellness with environmental safeguards while promoting financial expansion through their assessment of ecosystem–human interactions [1]. The implementation of sustainable land practices develops many essential benefits that include sustainable economic development, improved access, better ecosystems and reduced poverty levels through strong land resource management. It supports the maintenance of sustainable practices through afforestation along with conservation

Kundan Meshram and Chatrabhuj. A Sustainable Approach to Land Use, (67–84) © 2026 Scrivener Publishing LLC

initiatives and smart agriculture and agroforestry systems [2] that reduce greenhouse gas emissions and offer climate change resistance. The present situation needs urgent simultaneous action on environmental defence and economic expansion, since urbanization and farming development operate alongside diminishing natural resource reserves [3, 4]. Business success in Earth conservation needs a detailed analysis of the economic factors involved in sustainable land management. The knowledge required for sustainable land use economics demands expertise about how land bases its position between economic expansion and natural resources assets. The economic sector operates from agriculture to forestry, real estate development and infrastructure construction because land serves as a fundamental resource base. The utilization of land as a resource normally results in adverse impacts on environmental systems, producing biodiversity loss and soil deterioration with rising carbon emissions [5–7]. Officials in government, as well as stakeholders, gain knowledge about land degradation externalities when they assign monetary worth to ecosystem services. Through Payments for Ecosystem Services (PES) systems, land conservation measures earn financial benefits along with sustainable land practices.

4.2 Foundations of Sustainable Land Use Economics

Sustainable land use economics relies on multiple theoretical structures and concepts to evaluate and research economic sustainability factors related to land management [8]. The theoretical framework presents both theoretical knowledge for understanding economic land use, sustainability elements together with investigation guidance, policy development and implementation directions for generating equitable and environmentally sustainable land management strategies, as depicted in Figure 4.1.

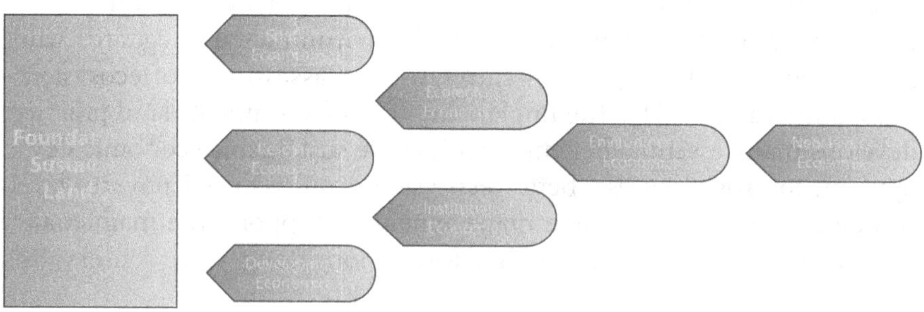

a. Environmental economics:
Environmental economics investigates external effects together with market breakdowns that result from land use management. The operations of land use led to harmful environmental consequences, which include pollution together with biodiversity decline, deforestation and habitat destruction. Economists who specialize in environmental issues use technical tools to measure external effects on nature, after which they create economic solutions such as taxes, subsidies and market-based strategies like carbon markets and pollutant permits. Environmental health benefits from economic activity through responsible land use practices, which the work of these professionals establishes to support environmental sustainability objectives.

b. Neoclassic economics:
As a basis for this method, it is assumed that rational landowners and users act to maximize their profits or utility. The analysis demonstrates how limited supply and market competition affect price levels for agricultural lands about manufacturing establishments and residential properties as well as natural preserves and other preservation zones. Through its framework, the system enables policymakers to create land policies that promote efficiency yet resolve land resource mismanagement problems.

c. Spatial economics:
The land use patterns are influenced by multiple factors including their geographical position together with their mobility options, their closeness to market areas and the extent of urbanization and regional advancements. The field researches analyse the spatial relationships between economic factors, land rental and market concentration economies to analyze urban spread and zoning systems. Urban planners, together with smart growth developers, need this analysis to develop their strategies.

d. Institutional economics:
According to institutional economics, the patterns of land use emerge from institutions together with laws and property rights, as well as governance structures. The framework evaluates how official rules through legislation, as well as policies and social customs regarding community practices,

determine land ownership while influencing land accessing and managing behaviours.

e. Resource economics:
   The study of resource economics focuses on sustainable management of renewable resources with non-renewable natural resources that reside on land territories. Through their models, resource economists evaluate both resource allocation, maximization, resource shortages, impact on future generations and preservation methods needed to sustain resources.
g. Development economics:
   Development economics studies economic growth together with poverty elimination, as well as their connection to land use patterns across both urban and rural areas specifically in developing nations. The researchers explore ways land-based expansion generates work opportunities while minimizing inequality and enhancing living conditions for all groups. Economists focused on development analysis of land transformation programs together with rural growth initiatives alongside policies that grant marginalized groups entry to land resources to achieve balanced national development.
h. Urban economics:
   Urban economics focuses on the study of land utilization patterns in cities to solve problems regarding urban spread, housing price levels, zoning and infrastructure development. The methodology enables urban land market analysis as well as land use optimization, compact city development, mixed-use development and transportation network planning to reduce environmental damage from land use.
i. Behavioural economics
   Behavioural economics is the pattern of psychological elements of human behaviour in the selection of land use. The discipline analyzes both risk assessment beliefs and social group standards together with psychological distortions and choices under uncertain circumstances. The application of behavioural economics principles enables better development of community-level sustainability approaches through improved land policy, incentives and educational programs.

4.3 Economics of Land Use Change

The economics of land use change investigates the intricate interactions of economic factors that cause shifts in land use patterns from rural to urban locations [9–12]. Several significant drivers impact these changes, which are represented in Figure 4.2 below.

Under the economics of land use change framework, multiple complex elements exist that determine land management and utilization throughout regional and country territories. The process of population growth, together with economic development, determines how land is used. Population expansion controls economic development through its impact on labour availability, consumer habits and resource requirements which include homes, infrastructure systems and food system development. Demographic trends involving population age groups, birth rates and death rates with migration patterns create direct effects on the socioeconomic framework that determines land distribution for residential, commercial and agricultural applications.

Social traditions together with cultural elements strongly affect how people utilize land resources. The components originate from cultural beliefs alongside societal structures that influence human relationships and their way of defining identity, alongside their daily choices and behaviours. The protection of cultural heritage emphasizes guarding both physical and abstract cultural elements of monuments alongside archaeological sites

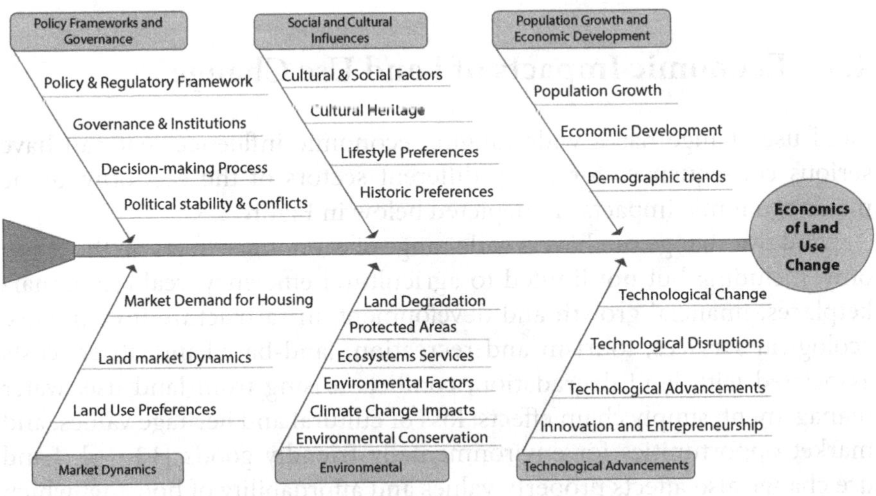

Figure 4.2 Economic factors of land use change.

and traditional practices, languages and cultural customs. Cultural heritage protection enhances social unity, together with the promotion of cultural diversity, and strengthens community identity. The way people live, along with their cultural backgrounds, determines how groups and individuals make land-use decisions because their social values, economic standing and personal experiences affect where they reside, along with their recreational and consumption activities.

Government programs dedicated to creating national parks, wildlife reserves and other conservation areas enable soil conservation as well as prevention of land degradation. The combination of deforestation together with overgrazed land and hasty urban development leads to damaging land and makes farming less productive, while making communities more exposed to climate change effects. The environmental problems become worse when climate change transforms rain patterns, heightens severe weather occurrences and disrupts ecosystems, agricultural production and human residential areas.

Land use modifications result directly from environmental conditions together with natural procedures as well as climate elements. Changes in climatic patterns and temperature variation, known as climate variability, create adverse effects on farming practices, water availability and health outcomes for people. Environmental degradation, such as deforestation, soil erosion, and pollution, undermines ecosystem services and biodiversity, threatening human livelihoods and ecological balance. Natural disasters cause property destruction along with human fatalities while breaking down economic operations.

4.4 Economic Impacts of Land Use Change

Land use change has a wide-ranging economic influence that can have serious consequences for many different sectors of the economy. Some major economic impacts are depicted below in Figure 4.3.

Land use change can have a wide range of economic effects on the economy, including but not limited to agricultural efficiency, real estate marketplaces, financial growth and development, infrastructure investments, ecological services, tourism and recreation, land-based industries, costs associated with land degradation; conflicts arising from land use, water management, supply chain effects; loss of cultural and heritage values, and market opportunities for environmentally friendly goods [13–16]. Land use change also affects property values and affordability of housing, which raises land prices and housing costs in urban areas and causes development,

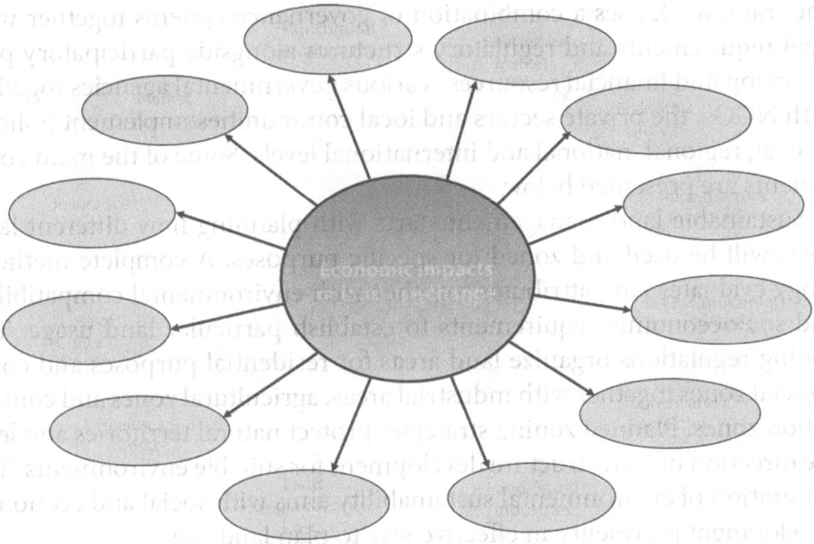

displacement, and housing affordability issues for locals. These factors all have an impact on real estate markets. The modification of land usage brings economic advantages that attract businesses while offering job possibilities and improving regional financial situations, while drawing investment. Regional prosperity and development receive support from these individual influences. However, changes in land use also come with consequences, including those related to the environment, soil erosion, loss of ecosystem services, and disputes over the distribution of resources and land access that have an impact on the economy. Modifications to land use patterns may have an impact on public health, cultural heritage, insurance rates, water availability and quality, risk assessments, and market opportunities. The economic effects of modifications in land use require complete understanding from policymakers, corporations and communities to generate both sustainable management plans and decisions that lead to social development and economic progress with environmental conservation.

4.5 Policy Framework

The establishment of well-organized policies provides the necessary support for sustainable land usage alongside economic growth, the preservation of environmental systems and the establishment of social balance.

The framework uses a combination of governance systems together with legal requirements and regulatory structures alongside participatory participation and financial resources. Various governmental agencies together with NGOs, the private sectors and local communities implement policies at local, regional, national and international levels. Some of the main components are presented below.

Sustainable land management starts with planning how different land areas will be used and zoned for specific purposes. A complete methodology evaluates land attributes together with environmental compatibility and socioeconomic requirements to establish particular land usage. The zoning regulations organize land areas for residential purposes and commercial zones together with industrial areas, agricultural zones and conservation zones. Planned zoning strategies protect natural territories and lead the direction of infrastructure development for suitable environments. The integration of environmental sustainability aims with social and economic development represents an effective way to plan land use.

The legal basis of environmental standards establishes quantitative restrictions for pollution tolerance and resource maintenance, and it preserves the state of natural lands. Land usage operates under policies that enable sustainability along with the protection of the public health system and natural environmental basics. The evaluation of the proposed development project's environmental effects follows a mandatory requirement through environmental impact assessments. Environmental regulations implement standards that control air pollution as well as water pollution together with hazardous waste management practices and protection strategies for ecosystems and conservation measures for soil. Organizations that adhere to environmental standards develop responsible resource usage while simultaneously decreasing their environmental safety hazards.

Planning sustainable land use policies needs community involvement as it establishes their credibility and working efficiency. The process of participatory planning requires stakeholder consultation which includes both local communities, indigenous peoples and additional parties who have a direct interest in the decision-making processes. Such methods keep respect for community traditions, native practices, local wisdom and cultural standards. The process results in social justice standards along with equal resource division and conflict settlement. The implementation of land use policies improves when citizens participate actively because trust develops while transparency remains high, and these policies become socially acceptable and all-inclusive.

4.6 Economics of Sustainable Urban Development

The field of sustainable urban development economics ensures cities exist through planning, management and operation methods that promote environmental safety and economic strength together with social equality. The concept of sustainable cities functions through a strategy that joins urban expansion to environmental ecosystem conservation with resource equality between present and future generations.

The key element in sustainable urban development requires that resources be used efficiently. Proper utilization of natural resources consisting of water, energy, land and resources should be implemented to minimize wastage and environmental pollution. The essential step toward urban sustainability involves supporting renewable power systems based on solar and wind generation. Water conservation methods should include both rainwater collection and wastewater treatment procedures. Construction technology paired with sustainable materials should be used to promote energy-efficient building design. Environmental efficiency relies heavily on waste reduction, which occurs through recycling and responsible choice of products for consumption.

The creation of infrastructure remains central to the development of urban areas sustainably. The development of public transportation based on metro, railways, and electric vehicles, along with sustainable buses, leads to reduced environmental pollution and improved transportation flow. The construction of intelligent waste management platforms is essential for performing waste retrieval and waste elimination operations effectively. The establishment of green infrastructure through parks, gardens and green rooftop construction yields improvements to the quality of living areas. The infrastructure approach to sustainability incorporates both digital management systems for smart cities, water supply, power systems and drainage systems that provide sufficient capacity to serve expanding urban communities.

Urban development leads to inclusive growth by providing social advantages to every segment of society with a particular focus on disadvantaged and at-risk population groups. The delivery of economical housing stands as an essential requirement for vulnerable income groups. The public must have equal opportunities to access healthcare services and education alongside public services in every community. Local employment opportunities, together with skill development programs, should focus specifically on women and minority groups and people with disabilities. The government should implement policies that promote small enterprises and local vendors to build economic growth at the community level.

The creation of resilient plans functions as an essential feature for achieving sustainable urban development. Cities use this approach to develop readiness against climate change, natural disasters and extreme weather occurrences. Urban planning requires infrastructure development which demonstrates resistance against floods, storms and heatwaves. A network of early warning systems needs to be developed for emergency protection of citizens. Wetlands, forests and coastal areas must remain as natural buffers since they minimize disaster perils. The process of including communities in disaster preparedness programs, together with emergency response initiatives, will ultimately create a stronger community resilience.

Sustainable urban development heavily depends on the establishment of public-private partnerships between the government and private sector entities. Cities successfully utilize combined resources and knowledge of government institutions, private entities, civil society groups and local communities to find solutions for urban problems. The relationship between public institutions and private entities enables them to use investments for building infrastructure, renewable energy development, affordable housing and urban development practices. The combination of different sectors creates superior innovative solutions that improve service delivery throughout urban locations.

4.7 Challenges in Balancing Economic Goals with Sustainability

In recent years, many researchers, scientists and lawmakers have been in a dilemma about whether to conserve resources or attend to economic development. To achieve a balance between them, there are many challenges, which are discussed below.

a. Externalities and market failures: The occurrence of market failures, together with externalities proves to be a major challenge. The actual monetary values omitted from market pricing lead to environmental destruction along with social imbalances from economic operations. Companies that generate negative external effects by polluting and resource exploitation do not pay the full costs of it.

b. Inequality and social justice: The problems of inequality and social justice affect numerous geographical areas. Economic growth produces benefits that are not distributed in such a way that would benefit the entire population uniformly. Economic progress has failed to reduce the increasing gap between people's incomes; poverty levels, as well as social isolation, have continued to grow.
c. Behaviour modification and consumer preferences: Building sustainable economic systems requires developing sustainable changes that affect consumer conduct, lifestyle choices and consumption habits as well as methods to reduce resource depletion, waste production and carbon emissions. Achieving economic targets must support sustainability by promoting awareness and environmental education about sustainability and implementing sustainable consumer choice initiatives through pricing mechanisms, labeling programs and public training strategies. Customers face considerable difficulties during the transition because they stick to traditional cultural behaviours.
d. Technological and institutional challenges: Many obstacles related to technology and institutions stop sustainable development from occurring. A green economy needs businesses to work with governments and consumers, implementing new technologies through organizational reforms and behavioural changes. All sectors need to cooperate deeply through joint commitment toward collaboration to address regulatory challenges, coordinate their efforts and counter vested interests for successful institutional obstacle resolution.
e. Measuring progress and accountability: Balancing economic goals with sustainability requires robust monitoring, evaluation, and reporting mechanisms to track progress, measure impacts, and hold stakeholders accountable for their commitments and actions. However, existing metrics and indicators often focus on economic performance and GDP growth, which might not fully convey sustainability's social and environmental aspects. The development of comprehensive and integrated metrics that track progress toward several sustainability goals, encourage responsibility, openness, and continual improvement, is necessary to strike a balance between economic aims and sustainability.

f. Complexity of interconnected systems: Because social, environmental, and economic systems are intricately linked and have intricate feedback loops and unintended consequences, it can be difficult to forecast the full effects of actions and policies meant to achieve sustainability. Systems thinking and holistic approaches that take into consideration the relationships and dependencies among various sectors, stakeholders, and variables, which can be challenging to model and manage effectively, are necessary to strike a balance between economic goals and sustainability.
g. Financial restraints and limited access to capital: Multiple firms, particularly startups and small and medium-sized enterprises (SMEs) face financial challenges while acquiring funding, which blocks their path toward inventive and sustainable business choices. The combination of economic targets with sustainability needs financial challenges to be resolved while establishing green finance mechanisms to support sustainable investments.
h. Transitioning from fossil fuels: Transitioning to renewable energy demands the full replacement of fossil fuels starting from today, which leads to important economic disruptions affecting all industries, the regions they serve, all workers who participate in fossil fuel extraction and production projects. The transition toward sustainability demands organized efforts between stakeholders to handle displacement effects on employment. It also needs the technical education of staff and assistance for impacted communities that are going through this transition.
i. Globalization and interdependence: Economic choices made by countries, together with their corresponding policies, demonstrate that national decisions have worldwide impacts on sustainability as exhibited by cross-border environmental issues like climate change and others.

4.8 Strategies for Mitigating Conflicts between Economic and Environmental Objectives

Innovative strategies that incorporate and aid multiple aspects of economic and environmental preservation are required. techniques to make this happen are described below.

a. Policy integration: Public institutions must include environmental and social elements within their complete policy development procedures. Through this method, environmental targets become integrated with economic planning initiatives. Governments need to establish assessment frameworks that evaluate choices according to economic and environmental factors, alongside social criteria systems. The integration method finds applications in land-use regulations, procurement systems and infrastructure development for creating sustainable results. Multiple development goals can be reached simultaneously when sectors align their policies.
b. Sustainable finance: Financial institutions need to support investments in environmentally sound projects. Investment choices made by banks alongside financial institutions and investors require sustainable factors together with environmental and social concerns according to sustainable finance principles. The money directed towards renewable energy, along with eco-friendly infrastructure and sustainable agricultural projects, enables financial institutions to lead sustainability-focused market development. The procedure combines environmental risk assessment in credit evaluation with institutional backing for businesses that maintain sustainable commitments.
c. Green innovation and technology: The critical step for attaining environmental sustainability involves funding research that develops green technologies. The strategy focuses on technological innovation as a fundamental approach to handle environmental issues. Governments, together with private sector entities, need to fund R&D initiatives that will develop economic solutions with environmental protection as a core objective.
d. Eco-labeling and certification: Marketing eco-labels provides people with tools to support green purchasing decisions. Various eco-labeling and certification systems help users get clear details about products, services and environmental effects. The system provides consumers with the capabilities to support companies that focus on sustainable operations. Businesses earn market advantages through eco-labels which stimulate more organizations to implement sustainable practices. The certification systems promote

sustainable manufacturing together with sustainable consumer practices, which leads to environmental preservation.

e. Community engagement and participation: When local communities take part in development planning and execution, their trust develops together with their sense of ownership. Local populations, marginalized groups and local stakeholders who participate in decision-making ensure that development plans suit local requirements and protect natural elements and cultural values. When communities take part in development projects, their collaboration creates better social bonds that lead to lasting project sustainability.

f. Market-based instruments: The use of economic instruments should motivate businesses towards more sustainable operations. Financial rewards and punishments through market-based instruments help promote environmentally friendly practices. There are many types of economic tools for sustainable business practices such as promotion of pollution taxes, emissions trading systems, green subsidies and payments for ecosystem services. The implementation of internal environmental costs through MBIs makes businesses adopt pollution reduction and resource conservation practices as well as sustainable technologies that synchronize economic activities with sustainability objectives.

g. Capacity building and education: Effective sustainable development depends on the improvement of skills and knowledge acquisition. Capacity building delivers training and education and distributes resources to public servants, corporate entities and non-profit groups. Participation in workshops combined with technical assistance and knowledge-sharing programs enables stakeholders to acquire the required abilities for sustainable practice implementation. Organizational capabilities that receive focused development enable efficient policy execution and generate new ideas for lasting development.

h. Sustainable tourism development: The tourism industry should produce economic gains through methods that preserve the environment. Tourism operators should help and support ecological lodging while protecting traditional customs and active community involvement in tourism programs. The method guarantees both monetary gains for local communities, together with the protection of natural resources, as well as the preservation of cultural essence.

i. Natural capital accounting: All economic decision processes need to include environmental assets. Natural capital accounting requires businesses and policymakers to estimate natural resource worth and ecosystem service value through economic valuation methods. Environmental cost-benefit analysis through this method enables business operations and policy decisions to assess development activities. Decision-makers are able to develop balancing strategies for economic growth through the appreciation of biodiversity and ecosystem value.

j. Circular economy practices: Reducing waste along with resource consumption becomes possible with circular economy promotion. The circular economy designs products with three key features: durability, repairability, and recyclability. Product sharing, together with remanufacturing and materials recovery techniques, reduce waste and expand product existence. The method disconnects economic expansion from resource exhaustion by creating an environment that supports sustainable management of resources.

k. International partnerships and cooperation: A collaborative worldwide effort between nations and international organizations along with non-state entities must tackle environmental problems of the world. The sustainable development of poor nations receives assistance through multilateral agreements in addition to bilateral partnerships and cross-sectoral efforts that supply capacity training with technology transfers and financial aid. Cooperation on mutual objectives enables the international community to achieve beneficial results for human beings and environmental protection.

4.9 Potential for Economic Prosperity through Sustainable Land Use

Sustainable land use can enact economic development, which brings numerous benefits from maximizing land productivity while guarding against environmental degradation. The economic approach to sustainable land use extends past environmental behaviour promotion to achieve broader achievements in social development and equality maintenance. Economic analysis provides tools to assess how land-use policies affect different groups within the population, so disadvantaged populations, including indigenous communities and smallholder farmers, avoid excessive negative impacts from land-use modifications. Through its methods, economics enables decision-makers to use cost-benefit analysis and risk assessments for identifying resilient land-use options that minimize expenses in the presence of uncertainties, as depicted in Figure 4.4. The economic approach helps decrease waste output and optimize resource utilization in agricultural, forestry and mining sectors through circular economy practices and resource management methods that advance environmental sustainability and business performance [17].

Economics lowers waste production and improves resource utilization from land-based industries such as forestry and agriculture together with mining by promoting circular economy principles and resource efficiency, which results in better environmental sustainability together with economic development. Economics enables the use of local knowledge, cultural values and traditional practices during land-use planning procedures to increase community control over natural resources. Through financing instruments, including ecological bonds, impact investments and payments for ecosystem services, economists activate public and private funding to construct sustainable programs for land management, thus obtaining new financial resources to restore and conserve ecosystems and cultivate sustainable agriculture. Land-use zoning and land-use taxation combined with the use of development rights for transfer protection enhances biodiversity by enabling proper ecological land distribution.

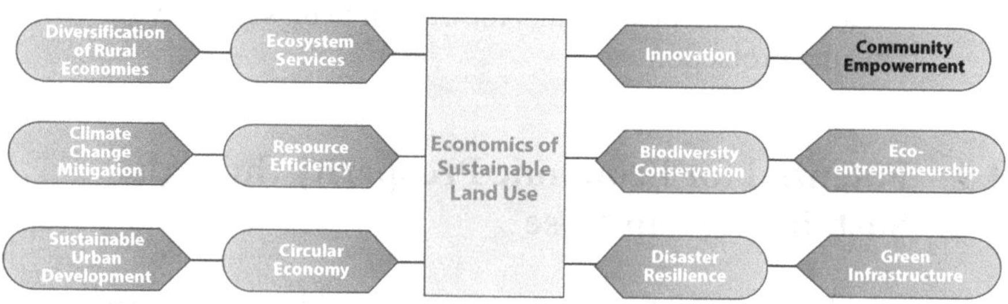

4.10 Conclusion

Contemporary land management systems work on managing sustainability issues caused by poor land preservation by addressing climate change effects alongside biodiversity reduction, soil degradation and water depletion. The combination of afforestation with smart agriculture and agroforestry builds environmental strength while producing endless economic returns and decreasing poverty rates. Environmental stability encounters both opportunities and difficulties stemming from modifications in land usage that happen because of urban development and agricultural domain expansion alongside infrastructure installation activities. Multiple integrated policy frameworks need to exist since these environmental and social forces create a complex situation between economic development, sustainability and equality. Environmental sustainability and economic development create significant obstacles because both objectives persistently fight to be the most important priority. Merging environmental worth assessment into financial decision frameworks requires new approaches to resolve these conflicts, monetary reward systems and joint management structures. Sustainable investments receive evaluation through business ventures and government laws, which demonstrates how sustainability should be integrated throughout business operations and governmental systems. Overcoming existing obstacles in aligning economic targets with sustainability standards requires conflict reduction and the development of sustainable land-based economic growth that incorporates all parties.

References

1. Leng, A., Wang, K., Bai, J., Gu, N., Feng, R., Analyzing sustainable development in Chinese cities: A focus on land use efficiency in production-living-ecological aspects. *J. Cleaner Prod.*, 141461, 2024, https://doi.org/10.1016/j.jclepro.2024.141461.
2. Feleki, E., Vlachokostas, C., Moussiopoulos, N., Characterisation of sustainability in urban areas: An analysis of assessment tools with emphasis on European cities. *Sustainable Cities Soc.*, 43, 563–577, 2018, https://doi.org/10.1016/j.scs.2018.08.025.
3. Chen, D., Duan, Y., Jiang, P., Li, M., Spatial zoning to enhance ecosystem service co-benefits for sustainable land-use management in the Yangtze River economic belt, China. *Ecol. Indic.*, 159, 111753, 2024, https://doi.org/10.1016/j.ecolind.2024.111753.

4. Mirzabaev, A., Nkonya, E., von Braun, J., Economics of sustainable land management. *Curr. Opin. Environ. Sustainability*, 15, 9–19, 2015, https://doi.org/10.1016/j.cosust.2015.07.004.
5. Costanza, R., *et al.*, Changes in the global value of ecosystem services. *Global Environ. Change*, 26, 152–158, 2014, https://doi.org/10.1016/j.gloenvcha.2014.04.002.
6. Wang, J., *et al.*, Exploring the complex trade-offs and synergies of global ecosystem services. *Environ. Sci. Ecotechnol.*, 21, 100391, 2024, https://doi.org/10.1016/j.ese.2024.100391.
7. de Moura, M.S., *et al.*, Beyond land use planning and ecosystem services assessment with the conservation use potential framework: A study in the Upper Rio das Velhas basin, Brazil. *Sci. Total Environ.*, 923, 171437, 2024, https://doi.org/10.1016/j.scitotenv.2024.171437.
8. Heerink, N., van Keulen, H., Kuiper, M. (Eds.), *Economic policy and sustainable land use: Recent advances in quantitative analysis for developing countries*, 2012, https://doi.org/10.1007/978-3-642-57558-7.
9. Wu, T., *et al.*, Triple Bottom Line or Trilemma? Global Tradeoffs Between Prosperity, Inequality, and the Environment. *World Dev.*, 178, 106595, 2024, https://doi.org/10.1016/j.worlddev.2024.106595.
10. Trong, P.T., *et al.*, Analysing Factors Influencing Land Use Planning for Sustainable Land Resource Management in Vietnam: A Case Study of Dan Phuong District in Hanoi City. *Sustainability*, 15, 24, 16557, 2023, https://doi.org/10.3390/su152416557.
11. De Wrachien, D., Land Use Planning: A Key to Sustainable Agriculture, in: *Conservation Agriculture*, L. García-Torres, J. Benites, A. Martínez-Vilela, A. Holgado-Cabrera (Eds.), Springer, Dordrecht, 2003, https://doi.org/10.1007/978-94-017-1143-2_57.
12. Velasco-Muñoz, J.F., Aznar-Sánchez, J.A., López-Felices, B., García-Arca, D., Sustainable land use and management, in: *Sustainable Resource Management*, C.M. Hussain and J.F. Velasco-Muñoz (Eds.), pp. 179–197, Elsevier, 2021, https://doi.org/10.1016/B978-0-12-824342-8.00015-8.
13. UN, World social report 2020, 2020, http://dx.doi.org/10.18356/7f5d0efc-en.
14. Tracey, S. and Anne, A., OECD insights sustainable development linking economy, society, environment: linking economy, society, environment, OECD Publishing, 2008.
15. The World Bank, Poverty and shared prosperity 2022: correcting course, 2022.
16. Figge, L., Oebels, K., Offermans, A., The effects of globalization on ecological footprints: An empirical analysis. *Environ. Dev. Sustainability*, 19, 863–876, 2017.
17. Martens, P., *et al.*, New directions in globalization indices. *Globalizations*, 12, 2, 217–228, 2015, https://doi.org/10.1080/14747731.2014.944336.

5

Social Impacts of Sustainable Land Use Patterns

Abstract

In order to build an equitable, resilient, and successful community it is vital to understand the intricate relationship between social aspects and land use patterns. The framework begins by explaining how land use triggers social consequences and why it is essential to consider social, environmental, and economic factors. Social fairness and justice, along with community well-being and health, are key themes that combine with cultural preservation of identity. Resilient lifestyles and social cohesion form the foundation of social participation, alongside education and awareness development. The chapter concludes by presenting various aspects of socially sustainable land utilisation alongside a summary of present developments. The chapter explores significant barriers and opportunities to establish social sustainability in land use through institutional reform and enhanced cooperation systems that promote community health while building up resilience.

Keywords: Community health, community resilience, land use patterns, social, sustainable

5.1 Introduction

The analysis of land use sustainability presents two main elements, economic sustainability and environmental protection, because they represent fundamental components of sustainable development. The social elements of land use management receive less attention in discussions, even though they are as vitally important as other factors. The goal of sustainable land use is to distribute land use benefits and burdens fairly within social groups and to build communal resilience together with societal togetherness and general well-being [1].

Land use practices that protect the environment cannot work well apart from the social situations in which they take place. For this reason, one must pinpoint and manage the social reasons that affect land use patterns.

Kundan Meshram and Chatrabhuj. A Sustainable Approach to Land Use, (85–104) © 2026 Scrivener Publishing LLC

The social fields considered by these studies are equity, public health, cultural heritage, community resilience, social cohesion, race, and literacy levels [2–5]. Since land use patterns have many effects, the work of social scientists is vital in finding and controlling them. A piece of land should be considered more than an economic or physical item; it actually reflects cultural values, past events and the social setups that build the identity of each community. The way land is put to use influences the health of people living in the community. A connection between public health benefits, quality of life and the availability of green spaces, healthy food and proper environmental upkeep has been well established [6–9]. Environmental damage caused by land use choices often touches the lives of those who are at a disadvantage, adding further to existing inequality. Using sustainable land use approaches can increase community resilience over time by supporting vulnerable people, building stronger bonds and strengthening the community itself. Meeting this vision means bringing local wisdom together with an expert approach and using the existing resources among the community.

5.2 Scope of Social Impact

A wide combination of social effects exists from sustainable land use practices, which link together multiple social elements. By expanding social impacts analysis to cover these newly identified dimensions, people can achieve deeper insights about the interaction of land use patterns and social relationships, which enables better sustainable land use planning [10]. People should focus on establishing fair and equitable communities with resilience to pursue sustainable development targets. This system includes multiple components, as described below.

a. Equity and Justice: People and communities must have equal chances of obtaining land and housing resources alongside employment opportunities and public service access. The procedure forces organisations to identify and resolve problems that cause essential inequalities between ethnic and socioeconomic population groups regarding property distribution. Environmental issues must be properly tackled across communities, while resources must also be distributed fairly so environmental benefits such as parks, open spaces, and green areas are well-balanced and fairly distributed.

b. Community Health and Well-Being: Physical and mental health primarily depend on the proper utilisation of land resources to create health conditions within communities. Appropriately designed parks with open green spaces help reduce people's stress and promote activity, while dysfunctional industrial districts generate environmental pollution that leads to respiratory problems. To reach sustainable community development, both healthy air quality and secure neighbourhoods must exist alongside nutrition-friendly food programs and adequate healthcare facilities that strengthen community immunity. The modification of local employment numbers and economic activity levels emerges from changes in urban development, agricultural rezoning, and industrial operations.
c. Livelihoods and Economic Opportunities: A community creates solid social cohesion through mutual trust and shared values, which leads to assistance among members and collective welfare. Each approach to land use implementation raises or decreases the level of social capital within a community. The visualisation of their direct contributions leads people to boost their trust in the planning system while remaining active in their communities. The long-term economic and social stability requires both belonging and cooperative relationships among people.
d. Social Cohesion and Engagement: Society and culture directly shape the way people feel about land use consequences between different gender groups. Women, along with ethnic minorities, indigenous peoples, and those with disabilities, experience limitations in their ability to participate and access general systems. Conducting these examinations within planning processes creates improved policies which establish fair access for all stakeholders to property ownership, educational opportunities, wealth accumulation and decision-making functions. The process of meaningful engagement allows every member of society to gain development benefits, which strengthens social cohesion and generates mutual respect among people.
e. Ethical: All land use decisions require ethical compliance for their implementation. The process of making land use decisions depends on forming human rights principles through transparent and fair systems that hold responsible

decision-makers. Vulnerable groups find protection from exploitation through ethical planning, while the system operates in the public interest instead of promoting private and political aims. The system brings inclusive governance by creating a platform that enables people to feel their voices matter. Ethical land use development maintains the right conduct both in the present times and across future timelines.

5.3 Importance of Considering Social Impacts

Social consequences need special attention within the sustainable land use development process because they determine community well-being alongside resilience and cohesiveness, as shown in Figure 5.1. Besides managing discrepancies and distributing fair benefits and liabilities, the policy upholds cultural preservation. The promotion of social equity and justice by decision-makers enables them to both resist conflict, perform morally and generate public participation. Health inequities receive attention from

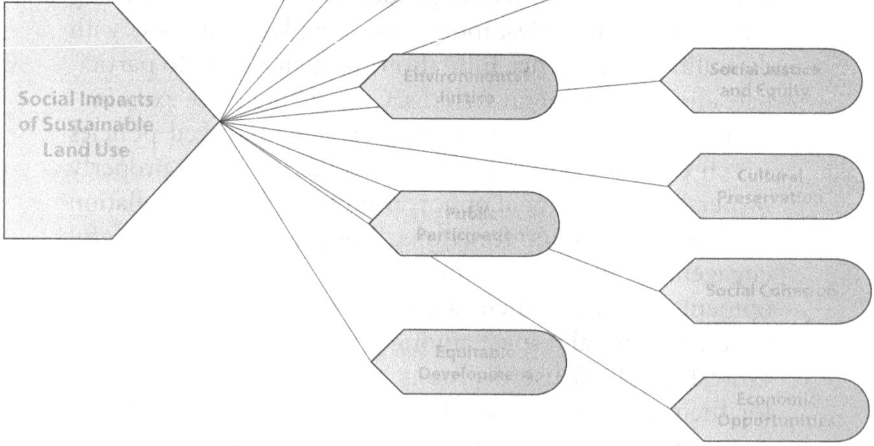

this system, which creates job opportunities and improves community relationships as well [11–13].

Social issues need active inclusion during the planning phases of land use and decision-making operations to ensure development results that are equal, sustainable for all and inclusive to communities. This integration is necessary because it enables effective public involvement and community-based involvement, which ensures various voices, particularly the most disadvantaged groups, participate in decision-making frameworks. Health equity analysis enables solutions to address inequalities concerning clean air, along with safe housing, green spaces and essential services to create better community health outcomes. Both environmental justice initiatives and economic opportunities creation prevent specific groups from facing excessive environmental harm and simultaneously support neighbourhood development through new job markets and safeguard current traditional economies. Through social innovation and empowerment, local communities discover original solutions that adapt to their particular needs while teaming up with different experts from social sciences, ecology, urban planning and public health to produce better land use results. The development of international solidarity between nations enables more effective collaboration to address transnational problems, including migration patterns alongside biodiversity reduction and environmental climate change. Additional factors need attention because they require measures that will support gender equality while addressing specific impacts of land policies on women and marginalised populations, cultural preservation for community identity maintenance, public education and transparent governance for human rights protection.

5.4 Promoting Community Well-Being through Sustainable Land Use

Sustainable land use techniques for community well-being represent a holistic approach that evaluates the three-dimensional links between environment, social and economic systems that influence population health outcomes [14]. Sustainable land use techniques use member health as the top priority to establish conditions that advance the physical, mental and social well-being for all members of the community. These three key approaches exist to support community welfare through the implementation of sustainable land utilisation practices.

a. Access to green spaces: Land use planning that is sustainable focuses on city growth as well as protecting and enhancing green spaces such as parks, community gardens, green corridors and protected reserves. Such places help improve the health of people living in the community. Urban green spaces support the survival of a wide variety of species, while also helping to purify the air, regulate city temperatures and give animals and plants room to live in the city.

b. Environmental quality: The choices made about land utilisation determine most environmental outcomes. Industry, along with tree-clearing operations and expanding cities beyond the sustainable limits of the land, causes ecosystem destruction, damaged soil conditions, air and water pollution. Proper environmental regulations through land management ensure the fusion of conservation zoning initiatives, barrier restrictions and vegetation preservation to achieve sustainability.

c. Mixed-use development: Effective urban neighborhoods are created by including important types of space—residential, commercial, institutional and recreational—all located close to each other. Using this type of planning, residents can reach essential services near them which means travel is less, roads are less crowded and more greenhouse gases are prevented. With this development model, community activities are meshed together in a way that leads to more sustainable and more livable cities. Including business and residential options together in one place brings energy to neighborhoods by letting residents meet and supporting nearby businesses.

d. Active transportation: Planning for sustainable land use tries to design infrastructure that makes it easy for people to walk, cycle and use public transport. Using this approach means installing sidewalks, separate bike lanes, secure spots for people to cross and comfortable transit areas. Such planning helps the public stay healthy by urging them to move more and sit less. As a result, relying less on cars helps to cut air pollution and encourage cleaner, more environmentally friendly conditions. With ramps, textured pavement and spacious access routes, inclusive design helps users with disabilities and improves access for people who are poor. Building active transportation options improves access and encourages community members to interact with each other in real life.

e. Safe and healthy housing: Sustainable housing strategies are intended to provide safe, energy-saving and cost-effective living spaces for everyone. It is essential for such housing to meet health and safety requirements by installing good ventilation, having secure entry points, being structurally sound, having natural light and including environmentally safe building materials. Besides, adding acoustic design features ensures less noise pollution, making it easier for people to have a better, healthier life, regardless of where they live or work.
f. Healthful food systems: Effective care for land supports local food systems by helping to set up urban farms, farmers' markets and food cooperatives. Community gardens and rooftop farms on-site help both residents and institutions obtain fresh produce in places where there are few supermarkets. Helping consumers and local producers work together supports efforts to address food scarcity, enhance food health and lower the environmental costs of delivering food from longer production steps.
g. Social cohesion and community engagement: Community spatial organisation establishes significant effects on social connections between people and group actions. Sustainable land use practices result in public designs that unite communities through interaction and sharing of identity with each other. Through their organisation of public centres and open plazas, along with community involvement in planning decisions, residents can connect and participate with each other. The local environment acquires higher levels of pride and ownership through these interactions, along with the establishment of trust and inclusion among community members. Strong social networks prove essential during emergencies since they allow members to work together and provide support to each other.
h. Disaster resilience and climate adaptation: When communities embrace sustainable land use practices, they create structures that have built-in abilities to survive and adjust to flood disasters, along with storms, heatwaves and wildfires. Effective regulation of floodplains, together with green infrastructure, wetlands, and forest preservation, helps mitigate

climate change and extreme weather events in urban areas. These protective actions lower vulnerability levels while fostering readiness, which jointly protects people from harm, secures damage to properties and accelerates post-crisis restoration phases.

i. Economic opportunities and employment: Economic growth through sustainable land development occurs through support for renewable energy projects and eco-friendly industries, along with green construction, community-based services, local entrepreneurship, and protection of environmental resources. Residents benefit from work opportunities located near mixed-use zones because these areas decrease residential commuting expenses and provide easier access to job opportunities. Such an inclusive economic framework develops diverse local economies, which strengthen community resistance to economic disruptions and raise general living conditions for everyone.

5.5 Cultural Heritage Conservation in Land Use Planning

The process of land use planning heavily depends on cultural heritage conservation because it protects both historical sites and traditional practices, together with culturally important landscapes, during modern development implementations. The conservation process requires communities to honour both physical items and non-material inheritances that illustrate their historical lineage to a particular environment. The inclusion of a properly constructed cultural heritage strategy into land use planning increases social cohesion while creating a sense of belonging, along with cultural continuity and attracting cultural tourists to the region.

a. Identification and documentation: The start of heritage conservation demands proper identification and extensive documentation of all vital cultural resources throughout the planning region. Early identification of heritage sites in danger occurs through established survey methods and inventory work. Planners receive the ability to prioritise conservation efforts and effectively distribute funding by utilizing detailed heritage documentation right at the beginning of land use development.

b. Public participation and engagement: Preserving cultural heritage requires active community participation because residents possess traditional knowledge about their landmarks and authentic emotional relationships with their cultural sites. Public land use planning requires participation through consultation programs, mapping processes and heritage evaluation sessions to assimilate the traditional knowledge of native peoples and regional citizens into planning decisions. Through community involvement, the project promotes both involvement and enduring caretaking responsibilities.
c. Mapping and inventory: Development projects can be protected from damage through planners' use of Geographic Information Systems (GIS) tools, which map both heritage zones and buffering areas. The mapping and inventorying of cultural heritage sites will aid development rulemaking as well as land use selection.
d. Education and outreach: Quality education is very important to serve as a basic understanding of cultural heritage. Organisations need to build outreach programs that use educational resources in both school campuses and public areas, along with museums, to deliver information about preservation and conservation methods. Educational materials combined with displays focusing on historical artefacts will be developed for planner and developer education, as well as sessions for public capacity development.
e. Monitoring and evaluation: The monitoring process consists of component examinations alongside community information, as well as drone imaging, which studies heritage structures together with GIS tracking to monitor archaeological sites and landscapes. Data analysis of predefined standards must be performed during the evaluation process to verify visitor satisfaction patterns, preservation quality levels, degradation rates and community engagement measures. The evaluation identifies what works effectively and what does not work within the implementation of adaptive strategies for handling new environmental situations, social changes and economic adjustments. The documentation generated by Monitoring and Evaluation (M&E) systems serves two

functions by providing essential data for both funding applications and long-term strategic development.

f. Legal protection and advocacy: The protection system mandates officials to create laws coupled with regulations that protect archaeological sites by prohibiting unauthorised changes and containing restrictions against demolition and commercial development. The national laws from each country should abide by the UNESCO World Heritage Convention (1972) and the Convention for the Safeguarding of Intangible Cultural Heritage (2003), as well as additional local agreements. The defence procedures that will stand must integrate heritage listing with site selection and protective historic preservation zones under governing zoning regulations. Active advocacy provides the essential pathway to gain backing from public viewers and member representatives of government institutions. The implementation of campaigns through heritage professionals combined with civil society organisations and local communities should create petitions and lobby for upgraded heritage laws and victimised heritage sites.

g. Heritage tourism and economic development: By backing up local economies with their cultural heritage, we strengthen the process of sustainable development. The combination of cultural heritage, community events, museum displays, artisan businesses and diverse cuisine, together with storytelling, draws visitors both from within the country and from abroad. By supporting tourism, various jobs are opened to local people, small handcraft and tourism businesses are developed and existing infrastructure is improved. Tourism will last over time if community leaders work together to become more eco-friendly and support their people's traditions.

h. Conservation easements and land trusts: Nonprofit organisations operating as land trusts acquire land or easements for the purpose of nature conservation through their nonprofit structure. Protection of cultural land for extended periods is possible through mechanisms that defend traditional land practices along with natural landscapes, sacred religious sites and cultural historical trails. Landowners benefit from money-saving tax incentives and grants that result in financial advantages to voluntarily protect their land.

i. Promoting cross-cultural communication and reconciliation: Building inclusive land use systems in the future demands the creation of strong links to culture. Here, the focus lies on visiting colonial settings and zones affected by war, as past circumstances there have resulted in marginalization and exclusion for indigenous people, displaced groups and minorities. An equitable strategy for land use planning needs to address and correct previous damage such as the loss of indigenous land to colonizers and the forbidding of cultural ways. To ensure justice and sustainability, one must carry out initiatives that support conversations between cultures, lead to better understanding and build up relationships among people from diverse cultures.

j. Collaboration and partnerships: Strong collaboration among several parts of the community strongly supports the preservation of cultural heritage. Partnerships should exist among different sectors, including ministries in the government, universities, research centers, non-governmental groups, agencies such as UNESCO and ICOMOS and local groups. By joining forces, these groups help institutions improve, gain more funding and foster new ideas in saving heritage. Bringing together different areas of work allows for the creation of shared policies, field learning for archaeologists, archaeological research, educational efforts on public heritage and appropriate management of sites. Because they unite under a collaborative structure, different institutions can pool their experience and resources to reach similar goals, making heritage efforts easier to manage on a global scale.

k. International cooperation and exchange: International cooperation plays an essential role in protecting cultural heritage because of emerging globalised world trends. Countries must cooperate for the proper management of transboundary heritage sites and global cultural routes that receive world heritage designations. International exchanges provide various opportunities for scholars and researchers to form academic partnerships and conduct joint research while also allowing professionals to undergo training and participate in global conferences through city and organizational twinning programs. International interactions enable countries to receive and exchange best practices in addition to adopting innovative technologies and international standards in heritage management. National heritage preservation and community development programs gain support from international funding organizations such as the World Bank and Global Heritage Fund, and UNESCO.

5.6 Importance of Social Cohesion in Land Use Planning

Social cohesion is essential in land use planning because it helps to create dynamic, inclusive, and resilient communities in which residents feel linked, empowered, and supported. Integrating social cohesion factors into land use planning procedures promotes a sense of belonging, confidence, and mutual support among community members, which has significant benefits for both people and society as a whole, as depicted in Figure 5.2.

Land use planning benefits strongly from social cohesiveness because of its many positive effects. The establishment of pleasant social connections between residents, together with their relationships and neighbourhood networks, improves community well-being. Community cohesiveness creates stronger systems that resist all kinds of adversities such as environmental, economic and social problems. Through land use planning, communities achieve resistance to hardships and develop team-based choices. These processes drive down tensions between stakeholders because they foster dialogue and collaboration, thus creating fair development results.

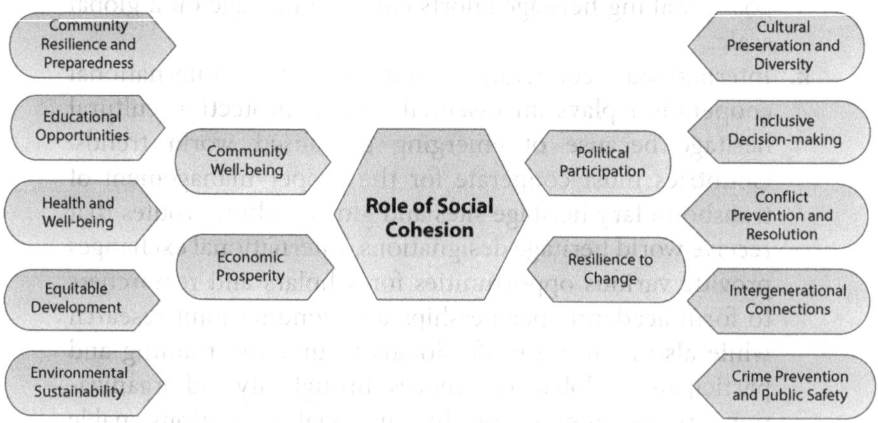

Figure 5.2 Role of social cohesion in sustainable land use pattern.

5.7 Identifying Key Challenges in Addressing Social Impacts

Multiple issues in land use planning form interconnected networks, so complete resolution requires multiple approaches using different concepts to achieve success. Social challenges currently require the most urgent solution because they represent the fundamental issues that need resolution. These are explained below.

a. Lack of genuine community engagement: Community engagement plays a critical role in land use planning since its exclusion from genuine influence in matters affecting the community directly creates current planning difficulties. Local communities lose trust and experience disputes when they are prevented from meaningful consultation about their land. Real participatory planning begins with community engagement throughout its initial development to finally grant complete authority for decisions. Civilised populations that include indigenous tribes endure long-term impacts from European colonial activities that caused them to lose ancestral territories and move forcibly from their homes. The separation of cultural commerce systems from land-based relationships and social organisational frameworks occurred due to historical injustices that caused disruptions.
b. Marginalization in standardized land use plans: Land use plans that apply the same to all regions miss important needs of women, ethnic minorities, those with disabilities, elders and youth. These groups are excluded when public works, services and community amenities are not built or offered in ways they can use. A good example is that women want safe areas outside their homes, disabled individuals need places that are easy to use, and youth prefer schools and activities that are fun and learning-friendly. A lack of consideration of these ideas in land use planning means more people face inequality and become separated from others in communities. Consulting with them during the planning and zoning process helps satisfy a wider range of group needs.
c. Conflicting interests among stakeholders: When planning land, policymakers must think about the needs of officials,

developers, environmental groups and local communities. Problems usually come up when commercial or industry activities endanger the environment or force residents to leave their homes. Without official means of conflict resolution, problems can often end up in stalemates, lawsuits and protests. The creation of multi-stakeholder groups, mediation bodies and easy-to-join decision processes is necessary. Such platforms must encourage people to respect each other, value independent concerns and reach agreements on land conservation.

d. Weak enforcement and governance: Good land use policies may fail unless the institutions involved can enforce them. Such structures, made up of a small team, inadequate training, dishonesty and poor control, allow people to build where they should not, take land unlawfully and spoil the natural resources. Such poor governance of planning upsets the public and damages both fairness and respect for the environment. Better rules for land management, giving the public easier ways to access land details and adding independent bodies to oversee government action are ways to address these issues. The efforts of anti-corruption campaigns, systems for both rural and urban land registration and official and citizen watch groups help improve how countries are run.

e. Barriers to participation for rural and peri-urban communities: There are many substantial obstacles to joining formal land use planning for those in rural and peri-urban areas. It is common for small farmers to not be able to understand how to go through the legal, environmental and regulatory steps. These areas' grassroots groups are often unable to make a strong impact on important decisions. Because of these financial limits, people living in poor communities are unable to go to meetings, hire attorneys or lead campaigns. Besides, some local governments may struggle to employ staff who are experienced in organizing activities for all members of their community. Dealing with these barriers calls for specific educational measures, simple processes for all to join in and useful funds for marginalized people. Cooperation between different organizations and government groups improves the ability to plan inclusively.

f. Insufficient social infrastructure to build community resilience: How well communities manage change and continue to flourish often depends on having access to parks, community centers, marketplaces, schools, clinics and organizations. They act as places for people to come together, support one another, share cultural ideas and take joint action. It is important to ensure that planning for land use gives equal importance to infrastructure to strengthen society and support emergency help. In addition, money spent on social infrastructure helps mental health, reduces loneliness and preserves cultural traditions. Neighborhoods should be involved in planning these spaces so that they are useful and meaningful.
g. Power imbalances and suppression of marginalized groups: Land use is affected by power differences that help political leaders, corporate developers, and rich landowners have the upper hand over zoning, permission for projects and the use of tax money. Because power is concentrated in this way, minorities are overlooked and unequal treatment continues. Gentrification and forced relocation occur when low-income people are no longer able to stay in their community, since they cannot find affordable housing or use community resources. When decisions are not clear or oversight is missing, inequality is made worse. Power imbalances in land care can be reduced by involving everyone, using hearings for the public, letting many groups decide how to allocate resources, protecting at-risk communities and making civil society watch over the process.
h. Gender inequality in land ownership and governance: Women are not given fair opportunities to have land, maintain their heritage and partake in managing land, though they often look after land and people. The restrictions on women's land rights are mainly due to their culture, laws and customs. Often, neither women's points of view nor their specific requirements are considered in land use policies, which leads to poor policy outcomes and continued gender differences. To support gender equality, measures should be put in place that recognize women as owners of land, allocate places for women in planning committees, perform gender

impact evaluations and give women access to money, education and opportunities for leadership. Working for gender equity benefits communities and ensures that land management plans are both sustainable and include everyone.

5.8 Social Factors Affecting Land Use Patterns

Analysis of sustainability in land use patterns requires accurate social evaluation to create resilient communities capable of being socially inclusive and dynamic. Sustainable land use practices produce various social effects that are affected by multiple variables. The set of influencing elements changes significantly from one context to another; several prevalent examples are shown in Figure 5.3.

Community land utilisation patterns experience major changes because of individual elements, which consist of demographic patterns and lifestyle orientation choices alongside traditional influences, financial abilities and individual behaviour choices. Statistics about population numbers, age distribution and household variations determine both housing demand patterns and community characteristics. The way people want to live together with their cultural beliefs shapes both their place of residence, their use of utilities, and their sense of community identity. Land use planning becomes more effective through an individual understanding of specific

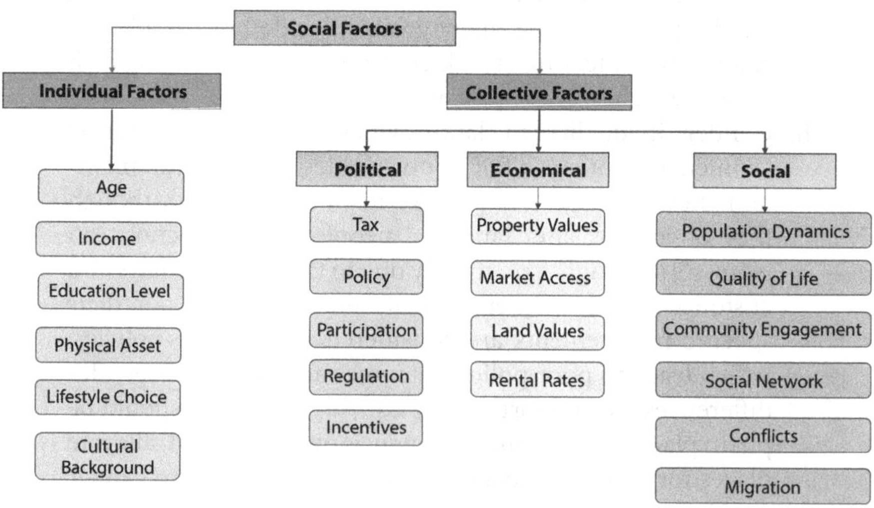

Figure 5.3 Social factors influencing land use pattern.

planning aspects to create environments matching community requirements, preferences and goals. The economic elements that control land use patterns in communities include market fluctuations, together with infrastructural investments, workforce availability and monetary systems. The shifting requirements of the market produce multiple land use needs that direct both property investment choices and property valuation trends. The expenses made for transport systems and utility services impact the availability and attractiveness of locations where economic operations and residential population expansions manifest. Employment opportunities, together with labour markets, determine where job centers and commercial hubs develop, thus affecting both residential travel patterns and housing market dynamics. Business investment choices determine how fast and large land development happens because of money factors and legal framework requirements. Our method of handling economic issues in planning helps create better living conditions for everyone in the area and makes our economy prosper fairly.

5.9 Conclusion

In the present scenario, it is essential to promote social cohesion by developing sustainable land use patterns. It aids in the development of sustainable, resilient and inclusive communities, which is done by policymaking, mutual understanding, collective vigilance, and accountability. Socially cohesive societies provide individuals with a fertile ground for economic development, thereby boosting local and regional development. Sustainable land use also promotes empowered communities, unity, shared development, health, environmental stewardship, long-term sustainability, protection of cultural identities, and effective resource management. This intercultural relationship thrives on opportunities, mentorship, mutual understanding and development in all respective for all age groups. Well-developed land use promotes open space like parks, community centres, green areas, gardens and mixed-use development that develops residential, commercial, and recreational areas, which strengthens social ties among different groups. It also promotes equitable land distribution, which ensures all individuals have access to healthcare, education, and public transportation services. Unequal land leads to social segregation and inequality; it hinders growth, leads to disintegration of the community and contributes to economic disparity.

References

1. Wang, Z., Lin, L., Zhang, B., Xu, H., Xue, J., Fu, Y., Zeng, Y., Li, F., Sustainable urban development based on an adaptive cycle model: A coupled social and ecological land use development model. *Ecol. Indic.*, 154, 110666, 2023, DOI: 10.1016/j.ecolind.2023.110666.
2. Badiu, D.L., Ioja, C.I., Ătroescu, M.P., Breuste, J., Artmann, M., Nița, A.R., Grădinaru, S.R., Hossu, C.A., Onose, D.A., Is urban green space per capita a valuable target to achieve cities' sustainability goals? Romania as a case study. *Ecol. Indic.*, 70, 53–66, 2016, DOI: 10.1016/j.ecolind.2016.05.044.
3. Fang, C., Liu, H., Li, G., International progress and evaluation on interactive coupling effects between urbanization and the eco-environment. *J. Geogr. Sci*, 26, 8, 1081–1116, 2016, DOI: 10.1007/s11442-016-1317-9.
4. Wang, Z., Fu, H., Liu, H., Liao, C., Urban development sustainability, industrial structure adjustment, and land use efficiency in China. *Sustainable Cities Soc.*, 89, 2023, DOI: 10.1016/j.scs.2022.104338.
5. Walker, B., Gunderson, L., Kinzig, A., Folke, C., Carpenter, S., Schultz, L., A Handful of Heuristics and Some Propositions for Understanding Resilience in Social-Ecological Systems. *Ecol. Soc.*, 11, 1, 2006, DOI: 10.5751/es-01530-110113.
6. Uitto, J.I., Sustainable Development Evaluation: Understanding the Nexus of Natural and Human Systems. *New Dir. Eval.*, 2019, 162, 49–67, 2019, DOI: 10.1002/ev.20364.
7. Sharifi, A., A critical review of selected tools for assessing community resilience. *Ecol. Indic.*, 69, 629–647, 2016, DOI: 10.1016/j.ecolind.2016.05.023.
8. McPhearson, T., Cook, E.M., Berbés-Blázquez, M., Cheng, C., Grimm, N.B., Andersson, E., *et al.*, A social-ecological-technological systems framework for urban ecosystem services. *One Earth*, 5, 5, 505–518, 2022, DOI: 10.1016/j.oneear.2022.04.007.
9. O'Byrne, D., Mechiche-Alami, A., Tengberg, A., Olsson, L., The Social Impacts of Sustainable Land Management in Great Green Wall Countries: An Evaluative Framework Based on the Capability Approach. *Land*, 11, 352, 2022, DOI: 10.3390/land11030352.
10. Liu, J., Jin, X., Xu, W., Yang, F., Wang, S., Zhou, Y., Assessing trade-offs and synergies among multiple land use functional efficiencies: Integrating ideal reference and key indicators for sustainable landscape management. *Appl. Geogr.*, 158, 103037, 2023, DOI: 10.1016/j.apgeog.2023.103037.
11. Al Arjani, A., Modibbo, U.M., Ali, I., Sarkar, B., A new framework for the sustainable development goals of Saudi Arabia. *J. King Saud Univ. Sci.*, 33, 6, 2021, DOI: 10.1016/j.jksus.2021.101477.
12. Cao, M., *et al.*, Multi-Scenario Simulation of Land Use for Sustainable Development Goals. *IEEE J. Sel. Top. Appl. Earth Obs. Remote Sens.*, 15, 2119–2127, 2022, DOI: 10.1109/JSTARS.2022.3152904.

13. de Jong, L., De Bruin, S., Knoop, J., van Vliet, J., Understanding land-use change conflict: a systematic review of case studies. *J. Land Use Sci.*, 16, 223–239, 2021, DOI: 10.1080/1747423X.2021.1933226.
14. Riveira, I.S. and Maseda, R.C., A Review of Rural Land-Use Planning Models. *Environ. Plann. B: Plann. Des.*, 33, 165–183, 2006, DOI: 10.1068/b31073.

6
Environmental Impact of Land Use Patterns

Abstract
Unsustainable land use patterns have a multifaceted impact on the environmental aspect. Unsustainable land use is due to unplanned rapid expansion, deforestation, agriculture, and industrial development, which lead to the degradation of natural resources, loss of biodiversity and disruption of ecological processes. These unsustainable patterns contribute to soil, water, air, and natural resource pollution, which increases vulnerability to natural hazards like floods, landslides, and wildfires. Land use patterns directly influence climate regulation by forming carbon sinks and reducing greenhouse gas emissions. Conversion of forest land, green areas and wetlands into urban and agricultural areas reduces the land's capacity to sequester carbon. Sustainable land use prioritises ecology, conservation, and balanced development for all. It can aid preservation of environmental resources, enhance ecosystem services, and promote long-term well-being. Effective land use patterns have more ecosystem service values. It is essential to integrate environmental considerations into land use planning and decision-making for achieving sustainable development goals for future generations.

Keywords: Biodiversity, ecological principles, environmental impact, environmental sustainability, sustainable land management

6.1 Introduction

Environmental legislation often concentrates on land development because it has the potential to impair water quality, change biodiversity and decrease many useful services across the ecosystem. Converting natural lands for urban, industrial, or agricultural areas can result in soil erosion, habitat loss, water being polluted and ecosystem degradation. Such interference with natural systems makes it harder for ecosystems to supply basic services, in particular clean water, a good climate and healthy soil. Intended environmental rules may occasionally cause results that lower

Kundan Meshram and Chatrabhuj. A Sustainable Approach to Land Use, (105–132) © 2026 Scrivener Publishing LLC

their effectiveness [1, 2]. As a result of these effects, environmental policies do not work well, and the environment could end up with even greater damage. Since many land development actions like road building, mining and city expansion are irreversible, their damage to the environment may remain for years, often leading to long-lasting ecological problems.

Preserving nature within fast-expanding cities can be achieved through open space laws, green belts and conservation easements. They help the environment to stay healthy and also help neighbourhoods by preserving green spaces and preparing for climate change. The combination of these tools reveals that experts are analysing how people's actions and the environment are connected [3]. Moreover, planning for land use considers the essential ideals of sustainability. It is now common for planners to place special emphasis on resources, future generations and the environment. They are followed through Environmental Impact Assessments (EIAs) and Strategic Environmental Assessments (SEAs), which look into what land use changes might affect the environment [4, 5]. With these tools, planners, policymakers and stakeholders can locate risks, work to prevent damage and look for ways to foster sustainability. These assessments are strengthened by including science, the opinion of local communities and future projections. In addition, the way land is used helps to guide the progress of climate change.

Using land sustainably, by building cities closer together, planting trees and using sustainable farming, helps reduce climate change and makes the environment more resistant. For this reason, the land use sector is vital for applying climate solutions based on the Paris Agreement and Sustainable Development Goals (SDGs), focusing in particular on Goals 11, 13 and 15. Planning for land use ought to consider changes in population levels, cities growing larger, and people moving between urban and rural areas. The effects of these forces can strain the availability of land, add to unfairness in access to housing and lead to problems with important services. If environmental justice is included in land use decisions, those who are most at risk from environmental harms are no longer exposed more than anyone else. For nature and development to go together sustainably, planning for land use should always give priority to environmental concerns. Meeting government requirements and deciding on proactive, interactive and adjusting approaches that match ecological changes and society's demands is also part of it.

Figure 6.1 depicts the network visualisation of interconnected relationships between key concepts associated with environment and land use patterns. Each node represents a keyword or topic, and the size of the node represents its importance. This network map visualises how environmental

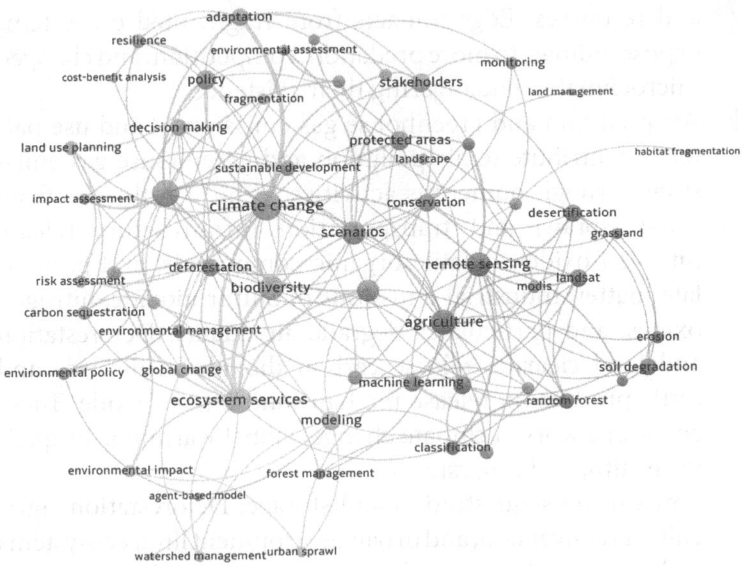

Figure 6.1 Environmental aspect of land use pattern.

and land use topics are conceptually and practically interlinked. It helps researchers with interdisciplinary research approaches for policymakers to analyse interconnections between environmental priorities, technological development, and planning sustainable land use strategies.

6.2 Environmental Aspects of Land Use Patterns

A wide range of environmental elements that affect ecosystems and natural resources belong to the environmental aspects of land use patterns. The specific aspects in land use planning and management work as essential elements because they assist in lowering environmental damage and promoting sustainable development practices [6]. These environmental elements must be recognised from land use patterns, as described below.

a. Habitat loss and fragmentation: Unsustainable land use destroys natural habitat, biodiversity through urbanisation, agricultural growth and massive development. These operations also fragment habitats, undermining ecological connections and wildlife migration. Fragmentation can cause biodiversity loss as species struggle to find habitat and resources. Edge impacts from fragmented ecosystems expose animals to more predation, competition, and changed microclimates, endangering their existence.

b. Air pollution and greenhouse gas emissions: Land use patterns contribute to air pollution and greenhouse gas emissions through numerous activities. Vehicle emissions from urban sprawl and transportation infrastructure release nitrogen oxides, volatile organic compounds, and particulate matter. Industrial emissions of Sulfur dioxide, nitrogen oxides, and particulates degrade air quality. Deforestation and land clearing release carbon dioxide, while rice and cattle production release methane and nitrous oxide. These emissions worsen climate change, global warming, air quality, health, and ecosystems.
c. Low carbon sequestration and storage: Deforestation, agricultural conversion, and urban development limit ecosystems carbon sequestration and storage capability. Photosynthesis in forests, wetlands, and other natural ecosystems stores carbon dioxide from the atmosphere in biomass, soils, and organic matter. Land conversion and habitat loss undermine carbon sequestration, releasing carbon into the atmosphere and raising CO_2 levels. Loss of carbon sequestration worsens climate change and lowers ecosystem resilience.
d. Diversity decline: Land use changes and habitat damage cause species extinction, population decreases, and genetic diversity loss. Natural habitats support varied ecosystems and ecological processes by providing crucial supplies and habitats for many plant and animal species. However, land conversion for agriculture, urbanization, and infrastructure development splits habitats, disrupts ecological connectedness, and reduces species abundance and distribution. Loss of biodiversity decreases ecological resilience, services, human well-being and livelihoods.
e. Natural resource depletion: Natural resource exploitation without sustainability leads to environmental breakdown, besides damaging important ecological processes. Rapid forest clearing occurs to support timber extraction while developers construct new agricultural areas and urban settlements, hence causing considerable forest habitat loss. Forest ecosystems become less capable of carbon sequestration because their ability to absorb carbon dioxide deteriorates,

thus making it difficult to regulate the climate. Causes from overfishing, together with destructive bottom trawling and cyanide fishing practices have resulted in severe losses of fish populations and destruction of specific marine ecosystems. Such practices endanger both fishery-producing communities and the natural aquatic system equilibrium. Mining projects create major water contamination, together with large-scale topographical modifications, forest clearing, and soil degradation. Operations that mine release dangerous heavy metals and toxic substances into water paths while threatening aquatic species and human health simultaneously. Several activities together produce extended damage to ecological services and diminish environmental stability.

f. Risk of natural hazards: The expansion of human populations without planning, along with deforestation and agricultural development of susceptible regions, creates higher natural disaster exposure to floods plus landslides and wildfires and coastal storms. Flood-vulnerable zones and coastal regions have become targets for human development despite the vital role that natural wetlands and mangroves play as weather buffers against high water. Deforestation and agricultural land clearing operations that destabilize hillsides and remove vegetation create conditions that promote landslide occurrences whenever heavy rainfalls occur. The implementation of agriculture in dry surroundings with fire-prone conditions increases the likelihood of devastating wildfires, since climate change produces both greater frequency and intensity of wildfires. Wildfires consume houses, agricultural areas, and forests while releasing harmful greenhouse gases that endanger biodiversity.

g. Loss of wetlands and riparian zone: Wetlands and riparian zones face destruction because people turn them into agricultural fields and construct urban areas and infrastructure projects. Wetlands function as natural water treatment facilities by catching pollution, sediment and mineral substances, which generates better water purity. The riverbank stabilization processes, together with the habitat maintenance function of riparian zones, operate similarly to downstream water filtering processes and wildlife migration corridors. The removal or alteration of natural areas leaves water quality decreasing while erosion rates increase, along with more intense flooding, which brings about severe biodiversity loss. Such ecosystem destruction affects hydrological cycles, thereby reducing environmental resilience to outside impacts such as climate change effects.

h. Noise and light pollution: User cities have rapidly expanded, combined with industrial growth, which triggered major growth in noise and light pollution that seriously disturb wildlife behaviour along with ecological stability. Natural animal communication becomes disrupted when noise from vehicular, construction and heavy machinery emissions disrupts the behaviour patterns of birds, as well as amphibians and marine mammals. The disruption of communication systems leads to problems with mating calls, predator warning signals, and social connection communication, which reduces reproductive success and causes changes in feeding behaviours of affected species. Chronic noise exposure causes animals to experience physiological distress that weakens their body defences as well as decreases their chances of survival. Artificial light pollution originating from streetlights, combined with commercial signs together with urban buildings creates a disruption of natural daily and migratory behavioural cycles in species that are active at night. Bright urban lighting causes migratory birds to lose directionality, which produces fatal crashes against buildings, yet artificial nighttime lighting hampers sea turtle hatching, alongside disrupting the regular activities of nighttime pollinators.

i. Loss of cultural landscapes and heritage sites: Changes in land use patterns like urban expansion and industrial development threaten cultural landscapes, archaeological sites, and historic landmarks. They erode traditional knowledge and practices and diminish cultural heritage, identity, and sense of place. Cultural landscapes reflect human societies' interactions with nature and their inherited values, beliefs, and behaviours. Human history and cultural progress are shown by archaeological sites. Cultural variety, social cohesion, and intergenerational knowledge transmission require protecting cultural landscapes and heritage assets.

j. Loss of ecosystem services: Land use changes can reduce ecosystem services like pollination, pest control, water purification, and carbon sequestration, reducing resilience, vulnerability to environmental change, and human well-being and livelihoods. Ecosystem services like food, water, climate regulation, disease control, recreation, spiritual values, soil formation, and nutrient cycling benefit humans. Deforestation, wetland drainage, and urbanization can disturb ecosystems and reduce their functions. Pollinator habitat is reduced by deforestation, lowering crop yields and agricultural production. Wetland destruction reduces water filtration, increasing waterborne disease and pollution risk. Green places and recreation are lost with urbanization, impacting health. Ecosystem services and human societies depend on preserving and restoring ecosystems.

k. Urban heat island effect: Urbanization and land development change the surface cover, vegetation patterns, and heat absorption capacity, creating urban heat islands that raise temperatures, increase heat-related health risks, increase cooling energy use, and disrupt local climate patterns. Impervious surfaces, less greenery, and anthropogenic heat sources raise urban temperatures. Urban heat islands increase heat waves, heat-related illnesses, and death, especially among vulnerable populations, including the elderly, children, and low-income communities. Demand for air conditioning and cooling increases greenhouse gas emissions and energy prices, worsening climate change and socioeconomic inequality. Green roofs, urban trees, and reflecting surfaces reduce the urban heat island effect, improve air quality, and boost urban liveability and resilience.

l. Land subsidence and groundwater depletion: Overusing groundwater for irrigation, urban water supply, and industrial uses can cause land subsidence, soil compaction, ground instability, compromise infrastructure, water availability, and land degradation. Aquifers and water tables drop due to groundwater extraction exceeding recharge rates. Land subsidence and sinking occur as aquifers deplete and soil pore spaces collapse. Subsidence threatens buildings, roads, and pipelines, raising maintenance costs and safety threats. Sustainable groundwater management strategies include recharge, conservation, aquifer monitoring, protecting groundwater supplies, and reducing land subsidence and soil degradation.

m. Loss of cultural ecosystem services: Changes in land use patterns can reduce cultural ecosystem services like spiritual and recreational values, traditional ecological knowledge, aesthetic appreciation, and connection to nature, affecting cultural identity, well-being, and quality of life. Humans gain from cultural ecosystem services like legacy, locality, and identity. Traditional civilizations and indigenous cultures convey their values, beliefs, and practices through rituals, ceremonies, and customs. Culture, intergenerational knowledge, social cohesion, and community resilience are weakened by the loss of cultural landscapes, sacred sites, and traditional livelihoods. Cultural ecosystem services and sustainability require protecting cultural variety, recognizing indigenous rights, and incorporating traditional knowledge into land management.

n. Natural disturbance regime interference: The natural disturbance patterns of ecosystems are altered as a result of human interventions in land use, such as the suppression of fires, the management of floods, and the restoration of habitats. This has an impact on the ecosystem's dynamics, biodiversity, and resilience. Frequently, these interventions cause disruptions to the equilibrium that ecosystems depend on, which ultimately results in ecological imbalances and degradation.

6.3 Environmental Consideration in Land Use Planning

The vital elements of environmental factors must be included in effective land use planning initiatives. Apart from preserving the natural environment, planners must make sustainable development advancement a primary factor during land use planning operations [7–9]. Multiple important factors drive the vital position of environmental aspects in land use planning operations and administration strategies.

a. Sustainable development: Sustainable development serves as a fundamental framework for delivering present development requirements through methods that sustain future population's capabilities to fulfil their needs. Sustainability principles within planning strategies lead to economic and social development, which does not damage vital environmental systems supporting ecosystems. The environmental considerations mandate equal growth rates with nature protection by supporting urban density and defending natural resources.

b. Managing natural resources: Organizations that make land use choices serve as key players in developing systems to protect vital natural resources, including soil, fresh water resources, forests, minerals and biodiversity. The effective management involves establishing protected areas together with delivering land conservation methods as well as supporting land uses which promote both ecological balance and sustainable livelihoods.
c. Climate change mitigation: The incorporation of environmental factors into land use planning serves as the key method to confront climate change factors during both cause generation and impact management. The way one uses land needs to be sustainable since such measures help combat climate change through support of renewable energy projects together with protection of forests and wetlands, and by minimizing urban growth. Administrative planning needs to recognize climate-sensitive areas to enforce policies which defend the environment through green development and protect flood zones and peril zones from development. Community preparedness together with adaptive capacity, becomes achievable through these measures that decrease climate-related event vulnerability.
d. Legal and regulatory compliance: Incorporating environmental elements into land planning processes helps organizations meet all environmental law requirements and international agreements, and national and regional standards. Environmental plans that obey statutory requirements, such as environmental impact assessments (EIA) requirements, along with biodiversity conservation laws, water protection acts and climate agreements, protect land development projects from legal troubles and negative public image.
e. Economic viability and resilience: Sustainable land use acts powerfully to establish long-term economic sustainability, together with resistance against impacts. Environmental planning brings about job creation through the establishment of green industries, together with renewable energy, sustainable agriculture and eco-tourism sectors.

Environmental hazard reduction benefits governments by reducing expenditures as well as businesses by improving investment stability.

f. Natural hazard management: The management of risk areas vulnerable to natural disasters, including floods, wildfires, earthquakes, landslides and hurricanes, requires environmental assessments as its core basis. The combination of hazard assessments influencing land use planning achieves two goals: protection of high-risk areas from development and the implementation of standards that decrease exposure to danger.

6.4 Environmental Consequences of Land Use Decisions

The evaluation of land use effects on the environment requires teamwork between ecological, social, economic and governance professionals.

a. Environmental Impact Assessment (EIA): The implementation of EIAs determines environmental effects for proposed land use projects and developments. The purpose of EIAs is to identify and predict the environmental results of new projects along with their alternative options. Environmental impact assessments cover air and water quality, biodiversity, soil health, climate change, and socioeconomic implications. Baseline investigations, impact prediction modelling, alternatives analysis, and stakeholder involvement are typical evaluation steps.
b. Ecological and environmental studies: Ecological studies reveal land use–ecosystem linkages. Studies may include habitat assessments, biodiversity surveys, ecosystem service evaluations, and landscape ecology analyses. Land use influences on water resources, air quality, soil health, and vegetation cover are quantified in environmental studies. Remote sensing, GIS, and ecological modelling are used to collect and interpret data.
c. Risk assessment: Such assessments evaluate the possibility and potential repercussions of negative environmental outcomes from land use decisions. This includes exposure pathways,

ecological sensitivity, health concerns, and environmental hazard likelihood. Risk assessments assist decision-makers in prioritizing mitigation measures, allocating resources, and making educated environmental choices.

d. Cost-Benefit Analysis (CBA): A CBA assesses environmental costs and benefits of different land use scenarios. They evaluate the environmental consequences and economic value of development's economic advantages. CBAs help decision-makers weigh short-term economic rewards against long-term environmental sustainability.
e. Stakeholder engagement and public participation: Involving stakeholders and the public in evaluation drives transparency, accountability, and legitimacy in decision-making. Local communities, indigenous groups, government agencies, NGOs, and industry representatives are stakeholders. Public participation incorporates varied viewpoints, local knowledge, and values into land use choices, improving social and environmental results. The policy analysis and planning section evaluates land use policies, regulations, and planning frameworks to determine their effectiveness in resolving environmental concerns. It finds policy implementation and enforcement flaws and potential for change. Planning integrates environmental issues into land use decision-making by creating sustainable development goals, objectives, and strategies.
f. Monitoring and adaptive management: Policymakers monitor environmental indicators to evaluate land use decisions and management approaches. Monitoring data enables repeated land use strategy adjustments in response to new knowledge or changing conditions in adaptive management.

6.5 Integration of Environmental Policies

Environmental policies must be integrated into sustainable land use patterns to minimize environmental consequences and promote long-term sustainability [10, 11]. Sustainable land use patterns can incorporate these environmental policies.

a. Land use zoning and regulation: It balances urbanization with conservation by ensuring proper development in specific locations. Setbacks, environmental impact assessments, and building rules reduce development-related habitat damage, soil erosion, and pollution. Smart growth promotes compact, mixed-use development, decreasing long-distance transportation and preserving nature. Transferable development rights preserve ecologically valuable property while allowing development in less sensitive places. Supporting infill development revitalizes urban cores, improves land usage, and relieves urban peripheral land pressure.

b. Incentive-based conservation programs: The establishment of incentive-based conservation programs, such as payments for ecosystem services (PES) or conservation easements, can serve as a means of encouraging landowners to preserve biodiversity, safeguard natural habitats, and continue to provide ecosystem services. To encourage landowners to participate in these programs, financial incentives, technical help, and legal assurances should be provided.

c. Protected areas and conservation policies: Marine protected areas preserve biodiversity, ecosystems, and fisheries against overexploitation and degradation. Restoration of damaged ecosystems, biodiversity, and resilience must be made through species recovery programs and habitat restoration projects. Regulations against illicit logging, poaching, and wildlife trafficking safeguard fragile species and ecosystems from exploitation and habitat degradation. Supporting indigenous and community-led conservation recognises local communities' responsibility for biodiversity and resource management. Conservation finance tools like ecosystem service payments and biodiversity offsets encourage landowners to preserve and restore natural environments.

d. Sustainable agriculture and forestry: Agroecological farming approaches boost soil health, reduce chemical inputs, and increase biodiversity in agricultural fields. Agroforestry systems give shade, erosion control, and income to farms. Sustainable forest management promotes responsible timber harvesting while preserving forest health, biodiversity, and carbon sequestration. Farmers receive financial incentives to pursue organic farming, crop rotation, and riparian buffer zones through agri-environment initiatives. Research and extension services enable climate-resilient agricultural and forestry practices.

e. Pollution prevention and remediation: The enforcement of legislation to avoid pollution from land-based activities such as industrial operations, agricultural practices, and waste disposal is necessary for safeguarding the quality of the soil, water, and air. Pollution prevention and remediation are also important. To address legacy contamination, policies ought to encourage the implementation of pollution control technologies, pollution preventive measures, and remediation initiatives.
f. Urban green infrastructure and sustainable development: Adding green roofs, rain gardens, and permeable pavements to urban architecture can minimize stormwater runoff, mitigate flooding, and enhance water quality. Compact, walkable areas with green spaces, public transit, and facilities encourage active lives, social cohesiveness, and mental health. Energy-efficient lighting and renewable energy systems reduce resource consumption and greenhouse gas emissions when retrofitted into existing infrastructure. Energy-efficient, resource-efficient, and environmentally friendly buildings are promoted by green building standards and certification schemes. Building community gardens and rooftop farms improves food security, minimizes food miles, and strengthens local food systems.
g. Water resource management: The management of water resources is essential for the preservation of healthy ecosystems and the provision of clean drinking water. This is because the enforcement of legislation to safeguard water quality and ensure sustainable water usage is essential. Both human population and the creatures of the ecosystem need sustainable water management to thrive. Water policies must guarantee fair water access, with clean water supply, protection of water origins and systematic water resource utilization. A framework of regulations needs to exist to stop contamination coming from agricultural runoff, along with industrial effluents and landfills. The process of integrated water resources management should exist inside land use planning to join different sectoral targets while promoting coordinated efforts between government agencies.
h. Urban planning and green infrastructure: Urban designers should implement a network of green corridors as well as permeable flooring and diverse, multifunctional open

spaces. Urban heat islands receive reduced strength from these features, which also perform stormwater absorption and support biodiversity growth. Ecosystem-based planning in urban master plans needs to be implemented for climate risk reduction and mobility improvement, and an increase in urban liveability.

i. Mitigation and adaptation to climate change: The necessary policy solutions include reducing carbon emissions as well as reinforcing infrastructure resistance to changing climate conditions. The appropriate arrangement of land zones during planning allows for the direction of sustainable development through designated areas for renewable power construction, together with designated forested zones and transit-based urban developments.

j. Community engagement and stakeholder participation: Through participatory processes, environmental policies gain representations of the various values and priorities and knowledge systems from every affected group, including local communities, together with marginalized populations and indigenous peoples. Social equity together with enhanced transparency results when stakeholders join early decision-making processes, which then produce contextually right and socially acceptable land use decisions. The implementation of community mapping tools enables residents to determine their most important ecological, cultural, and economic resources for planning decisions. Through public hearings, policymakers can measure community opinions because people get opportunities to provide their input, which helps resolve citizen concerns.

k. Environmental education and outreach: Education enables people to make wise decisions through awareness while it stimulates behavioural adjustments that lead to environmental care across both personal and community levels. Environmental instruction in its formal presentation must become an integral core subject for both schools and universities to teach students about ecosystems and their components, including sustainability together with climate change knowledge and information about land use effects. Non-formal outreach initiatives consisting of public information outreach, environmental festivals, eco-literacy events, and workshops target adults and government representatives together with wider community members. Training programs about sustainable practices plus energy efficiency together with green infrastructure must be provided by workplaces to their employees. Through community-based environmental monitoring networks citizens gain increased capacity for data collection, which builds their participation in environmental governance programs.

l. Assessment of environmental impact: The assessment procedure functions as an anticipatory mechanism for preventing and reducing detrimental effects that affect both the environment and public health, along with local communities. Successful EIA procedures require multiple steps that start with initial background environmental research, followed by the prediction of possible changes and necessary interactions with stakeholders before creating plans to protect and supervise project development. Effective project approvals depend on stakeholder consultation because they combine local intelligence with community concerns to create a more comprehensive project clearance. The implementation of EIA procedures requires strict policies that establish project threshold standards together with explicit time specifications and definite requirements for public involvement. Only approved, independent experts with accreditation can execute EIA procedures effectively. Post-implementation audits, together with compliance monitoring, help establish that mitigation measures are successfully implemented and also detect any effects that were unanticipated.

m. Protection and restoration of wetlands: Natural pairs of lakes provide multiple species with breeding habitats while minimizing effects from severe environmental conditions. Wetlands continue to face severe ecosystem threats worldwide because human activities frequently convert and devalue their lands through pollution and invasive species. The key interventions in restoration programs should include wetland drainage rewatering, invasive species control, native vegetation replanting and the reinstallation of natural hydrological cyclical patterns. No net loss of wetland functions can be achieved through mandatory requirements for mitigation banking and wetland offsets when wetland destruction becomes unavoidable. Wetland management methods will succeed best with community support and public promotion efforts that enhance local capacity to maintain these ecosystems in the long term.

n. Ecosystem-based management (EBM): The main goal of EBM is to protect ecosystem structures as well as functions and productivity so humans can use resources in sustainable ways. The management approach expects leaders to use watersheds along with other ecological boundaries as their basis for decision-making instead of administrative divisions, and include examination of feedback loops combined with climate variation and societal economic factors. EBM encourages stakeholder collaboration as well as knowledge integration between scientific approaches and traditional and local knowledge systems. When EBM becomes integrated with land use planning, it enables developers to unite environmental preservation and economic activities into sustainable goals within changing social-ecological systems.

o. Biodiversity offsetting and compensation: Development projects must pay for biodiversity loss that results from land use changes through conservation measures made in alternative locations. The objective centers on achieving alignment, which sustains or bolsters biodiversity levels while the approach is known as "no net loss" or "net gain." Biodiversity offset programs must implement the mitigation hierarchy to enforce three steps starting from impact avoidance followed by impact minimization before performing offsets to compensate impacts. This will ensure that restored or conserved land fulfills the same ecological purposes and provides the same ecological worth as the destroyed area it replaces.

6.6 Ecosystem Services

Ecosystem services deliver several advantages that ecosystems provide to humans and other living organisms by means of climate regulation and prevention of diseases, alongside safe air and water sources. Social and economic development heavily depends on these services, yet they commonly get neglected when considering traditional land management and economic decisions [12–15].

Ecosystem services provide an exceptional advantage by demonstrating that human communities and natural components exist in a constant mutual relationship. The framework, which explains ecosystem contributions to human welfare and quality of life, separates services into provisioning, regulating, supporting and cultural types. An explanation of the concept is provided below.

1. Provisioning services: The direct benefits humans obtain from ecosystems exist as tangible items available for their use. Examples of provisioning services include food manufacturing from crops, fish, live animals, availability of fresh water and medicinal plant resources. The provisioning services provide necessary vital resources to support human survival, along with sustaining their existence and livelihood.
2. Services of regulation: These services operate to manage natural processes to sustain ecosystem stability and preserve their equilibria. The three main climate regulation examples are water purification, air quality management and flood control practices. Among the examples of climate regulation are carbon sequestration processes, as well as temperature regulation and precipitation pattern regulation. The regulation of environmental services establishes positive outcomes for environmental health, which reduces natural catastrophes and provides climate change resistance.
3. Supporting services: All other ecosystem services rely on supporting services to maintain the proper functionality of ecosystems. Ecological processes occur in five primary ways, namely soil development, nutrient cycling, pollination, seed dispersal and photosynthesis. The supporting services are accountable for establishing both the base for ecological biodiversity as well as the foundation that supports production and ecosystem service delivery.
4. Cultural services: These services take into account the non-material benefits that are derived from ecosystems and that contribute to the well-being of humans, their identities, and their cultural heritage. The following are some examples: experiences that are spiritual and artistic, possibilities for recreation (such as hiking, animal watching, and ecotourism), sources of inspiration for art and literature, and cultural traditions that are deeply connected to nature. Through the promotion of mental and emotional well-being, the preservation of cultural diversity, and the cultivation of connections to nature, cultural services contribute to the enrichment of human life.

Table 6.1 Ecosystem services.

Area	Sub-area	Definition
Resources	Food	A wide range of food resources, including fruits, vegetables, grains, fish, and meat, are provided by ecosystems. These resources are necessary for the maintenance of proper nutrition and subsistence.
	Water	Rivers, lakes, and aquifers are examples of freshwater ecosystems that supply humans with clean water that can be used for drinking, irrigation, and industrial purposes.
	Medicine	A significant number of pharmaceuticals and traditional remedies are obtained from plants, animals, and microorganisms that are abundant in their natural environments.
	Raw material	Ecosystems are the source of raw materials that are utilized in the construction, manufacturing, and energy production industries. These commodities include lumber, fiber, minerals, and fuels.
Regulation of Natural Processes	Climate Regulation	The sequestration of carbon dioxide (CO_2), the maintenance of stable temperatures, and the influence of precipitation are all important functions that ecosystems perform in the process of controlling global and regional climate trends.
	Air and Water Purification	Soil and vegetation act as filters, removing contaminants from the air and water for use in human consumption.

(*Continued*)

Table 6.1 Ecosystem services. (*Continued*)

Area	**Sub-area**	**Definition**
	Flood Control and Erosion Prevention	These ecosystems drain water before it floods, and they hold moisture, which helps decrease coastal erosion while maintaining soil stability.
Support for Livelihoods and Economies	Livelihoods	Ecosystems are directly important to the livelihoods of many groups, such as those that depend on fishing, farming, forestry, and tourism.
	Economic Value	Agriculture, fisheries, forestry, biotechnology, and ecotourism are just some of the industries that use ecosystem services to help the economy grow and improve.
Cultural and Recreational Benefits	Recreation	Ecosystems make it possible to do fun things like camping, hiking, birdwatching, photography, and nature-based tourism.
	Cultural Identity	Environmental landscapes, wildlife, and cultural activities related to ecosystems all contribute to identity, cultural heritage, and spiritual wellness.
	Inspiration and Creativity	Ecosystems inspire art, science, and culture, strengthening human creativity and ingenuity.
Health and Well-being	Physical Health	People who have easy access to green spaces combined with outdoor leisure opportunities will participate in physical activities while their stress levels decrease and their health improves.
	Mental Health	Natural surroundings minimize anxiety, depression, mental tiredness, improving mental health and resilience.

(*Continued*)

Table 6.1 Ecosystem services. (*Continued*)

Area	Sub-area	Definition
	Nutritional Security	Ecosystems provide a wide range of healthy food supplies that people can use to support food security and good health.
Resilience to Environmental Change	Climate Resilience	Healthy ecosystems buffer harsh weather, retain biodiversity, and enable ecosystem-based adaptation mechanisms, making them more resilient to climate change.
	Adaptive Capacity	Ecosystem diversity and functional redundancy help humans adapt to environmental change and unpredictability.

6.7 Benefits of Ecosystem Services

Humans obtain an extensive variety of benefits from ecosystem services, which are provided by healthy and functioning ecosystems. These services are indispensable for the sustainability of the environment, economic growth, and human welfare. Provisioning services, encompassing essential commodities such as food, water, basic materials, and genetic resources, serve as the bedrock for economic endeavours and human subsistence [16]. Pollination, climate regulation, water regulation, and air quality regulation are regulating services that contribute to environmental equilibrium, the mitigation of natural hazards, and agricultural output. Supporting services comprise habitat provision, soil formation, nutrient cycling, and primary production. These services are fundamental to the functioning of ecosystems, the conservation of biodiversity, and ecological resilience. Cultural identity, educational, scientific value, recreation and tourism, aesthetic, spiritual values, and scientific value all contribute to the enhancement of human existence, the promotion of social cohesion, and the nurturing of a connection to the natural world [17]. Economic services play a pivotal role in enhancements to livelihoods, disaster resilience, and overall economic development by providing employment opportunities, economic value, and risk reduction. Informed decision-making, sustainable development, and the preservation of natural resources for the benefit of current and future generations all require the recognition and valuation of ecosystem services [18], which are shown in Table 6.1.

6.8 Land Use Patterns Affect the Delivery of Ecosystem Services

Different factors determine which ecosystem services result from specific land use arrangements. Land use factors that affect ecosystem services include spatial patterns of land cover distribution, together with ecological features of landscapes and levels of human interaction. The following part demonstrates how landscape management practices affect service delivery from ecosystems.

a. Natural ecosystems:
   Natural ecosystems provide numerous ecosystem services through different environments, starting from forest habitats to wetlands and grasslands and ending at marine systems. Different ecological procedures heavily rely on forests since they perform essential functions such as carbon storage, protection of soil and management of water supplies, together with preservation of biodiversity. Wetlands act as vital natural flood reducers, and they function as water purifiers while offering environmental habitat for multiple species.
b. Urban regions:
   Urban land use patterns define ecosystem service delivery because they include dense infrastructure, coupled with hardened surfaces together with dense human populations. Urbanization induces severe ecological consequences that lead to habitat disintegration, biodiversity reduction, natural habitat and green space destruction. Urban zones create several negative water cycle disturbances that lead to increased runoff and water contamination, flooding and erosion problems. Infrastructure coupled with strategically established parks and urban green spaces results in serviceable improvements of ecosystems while presenting recreational benefits and helping to regulate temperatures and clean the air.

c. Agricultural lands:
Agricultural land serves multiple purposes since operations cover extensive monoculture farming and diverse agro-ecosystems, with agroforestry among them. Soil erosion, together with water pollution and destruction of biodiversity emerges as an outcome of intensive agricultural methods, especially when agrochemicals and monoculture farming prevail at large scales. Conversely, sustainable agricultural approaches, including agroforestry, conservation agriculture, and organic farming, can improve ecosystem services, including water infiltration, biodiversity conservation, and soil fertility. Agroecosystems that integrate integrated pest management, cover crops, and diverse crop rotations support pollination and natural pest control services, thereby enhancing ecosystem resilience.

d. Environmental constraints of industrial and extractive activities:
Land use patterns associated with industrial operations, such as mining, manufacturing, and extractive sectors, can exert substantial pressure on ecosystems and the services they provide. Mining operations frequently lead to the degradation of ecosystem functions, including habitat destruction, soil erosion, and water pollution. These negative effects can be mitigated through the implementation of sustainable practices, including habitat restoration, ecosystem-based mine rehabilitation, and reclamation.

e. Transportation corridors and infrastructure:
The presence of transportation infrastructure, encompassing roads, railways, and waterways, has the potential to fragment habitats, disrupt wildlife corridors, and exacerbate the decline in biodiversity. By integrating ecological factors into transportation planning, such as the implementation of habitat restoration initiatives and wildlife crossings, green bridges, and green bridges along transportation corridors, these adverse effects can be alleviated, and ecosystem connectivity can be improved.

6.9 Sustainable Land Use Enhances Ecosystem Services

The enhancement of ecosystem services, which are the advantages that ecosystems offer to human beings, is significantly aided by the implementation of sustainable land utilisation practices. Through the use of sustainable land management practices, one can guarantee the uninterrupted delivery of essential services while simultaneously preserving the robustness and health of the ecosystem. There are several ways in which sustainable land use might improve ecosystem services.

a. Land restoration and rehabilitation: The restoration of degraded lands through the use of reforestation, revegetation, and soil conservation methods improves ecosystem services such as the retention of water, storage of carbon, and supply of habitat. Climate change mitigation, biodiversity conservation, and soil stabilization are all areas that can be improved by the implementation of afforestation projects. These projects include agroforestry, forest restoration, and tree planting programs.
b. Green infrastructure for climate resilience: The implementation of green infrastructure projects, such as urban forests, permeable pavements, and built wetlands, improves the ecosystem services that are provided in urban and peri-urban settings. The reduction of heat island effects, the management of stormwater runoff, and the provision of habitat connectivity for animals are all forms of green infrastructure that contribute to the mitigation of the effects of climate change.
c. Agroforestry and sustainable agriculture: Agroforestry systems integrate trees and crops on the same ground, offering many benefits such as better soil fertility, higher biodiversity, and enhanced carbon sequestration. Agroforestry systems also provide several other sustainable agricultural benefits. The prevention of soil erosion, the regulation of water flow, and the provision of habitat for animals are all accomplished through the use of practices such as alley cropping, windbreaks, and riparian buffer zones. Sustainable agricultural practices, such as organic farming, conservation agriculture, and integrated pest management, lessen the need for

agrochemicals, improve the health of the soil, and provide support for pollination and natural pest control services.

d. Conservation of forests and sustainable logging: Sustainable forest management approaches, such as selective logging, reduced-impact logging, and certification systems (for example, the Forest Stewardship Council), encourage the preservation of biodiversity, the sequestration of carbon, and the protection of watersheds. The protection of high conservation value areas and the preservation of completely intact forests contribute to the supply of ecosystem services such as regulation of climate, purification of water, and provision of habitat for species.

e. Restoration and management of riparian zones: Restoring riparian zones along rivers and streams brings about improvements in water quality, decreases the amount of erosion that occurs, and improves aquatic habitat. In addition to providing habitat for fish and other aquatic species, planting of natural vegetation, installation of buffer strips, and implementation of streambank stabilization measures all contribute to the filtering of pollutants and the reduction of sedimentation.

f. Wetland ecosystem: Restoring degraded wetland ecosystems and maintaining them are ways to improve ecosystem services. These services include the purification of water, the regulation of floods, and the provision of habitat for migrating birds and aquatic species. Water quality in urban and agricultural regions can be improved by the use of constructed wetland areas for treatment of wastewater, management of stormwater, and other purposes.

g. Urban green infrastructure: The incorporation of green areas, parks, urban forests, and vegetated corridors into urban development is a way to improve the ecological services that cities provide. By reducing the effects of urban heat islands, improving air quality, lowering the amount of stormwater runoff, and providing homes for urban wildlife, green roofs and green walls play a significant role. City gardens, community orchards, and rooftop farms all contribute to the production of locally grown food, the preservation of biodiversity, and the overall well-being of the community.

h. Protected area networks and collaborative land management: Through participatory decision-making and stewardship, establishment of collaborative land management arrangements, such as community-based conservation projects and co-management agreements, improves the quality of ecosystem services. At both the regional and landscape scales, the development of protected area networks, ecological corridors, and biodiversity hotspot conservation initiatives helps the connectivity of habitats, the conservation of species, and the resilience of ecosystems.

i. Ecosystem-based coastal management: The implementation of ecosystem-based approaches to coastal management, such as restoration of mangroves, stabilization of dune systems, and protection of coral reefs, increases the resilience of coastal areas and provides essential services such as the protection of fisheries habitats, opportunities for tourism, and protection from catastrophic storms. Both the conservation of biodiversity and the sustainable utilization of marine resources are achieved through the establishment of marine protected zones and the management of sustainable fisheries.

j. Rangeland management: Practices such as rotational grazing, rest times, and livestock exclusion zones are examples of sustainable grazing management practices that contribute to the preservation of rangeland health, the prevention of soil erosion, and the promotion of biodiversity. The soil fertility, water infiltration, and carbon sequestration in grassland ecosystems are all improved by the implementation of holistic land management strategies. sustainable practices that improve grassland environments are managed grazing programs alongside native grassland restoration efforts.

6.10 Conclusion

Land use patterns greatly affect both the environment and the well-being of people, both today and in the future. This study has revealed that modifying the environment for agriculture, industry or city building can have significant effects on ecosystems. Problems such as the loss of soil, habitats, biodiversity, polluted water sources and damaged ecosystems occur because of unsustainable land management. Many of these impacts are lasting and, for some cases, permanent, for example, when land is changed through activities like mining, expanding infrastructure and deforestation.

The problems with the land use–environment connection happen because the consequences to nature can develop slowly, so by the time they become noticeable, much harm has already been done. It is crucial for sustainable development to know that land use decisions directly affect ecosystem services, including climate regulation, water cleaning, flood control and food supply. To promote sustainability in land use, environmental issues should be closely aligned with policies, urban planning and economic development. Among these measures is working with environmental impact assessments, planning for the environment, and zoning regulations that support environmental needs. In addition, taking care of the land should focus on restoring damaged environments, guarding sensitive habitats and ensuring everyone gets to use natural resources.

References

1. Zhang, X., Lu, X., Chen, D., Zhang, C., Ge, K., Kuang, B., Liu, S., Is environmental regulation a blessing or a curse for China's urban land use efficiency? Evidence from a threshold effect model. *Growth Change*, 52, 1, 265–282, 2021, https://doi.org/10.1111/grow.12465.
2. Woodward, R.T. and Wui, Y.S., The economic value of wetland services: a meta-analysis. *Ecol. Econ.*, 37, 2, 257–270, 2001, https://doi.org/10.1016/S0921-8009(00)00276-7.
3. Saint-Macary, C., Keil, A., Zeller, M., Heidhues, F., Dung, P.T.M., Land titling policy and soil conservation in the northern uplands of Vietnam. *Land Use Policy*,27,2,617–627,2010,https://doi.org/10.1016/j.landusepol.2009.08.004.
4. Nickerson, C.J. and Lynch, L., The effect of farmland preservation programs on farmland prices. *Am. J. Agric. Econ.*, 83, 2, 341–351, 2001, https://doi.org/10.1111/0002-9092.00160.
5. Irwin, E.G. and Bockstael, N.E., Land use externalities, open space preservation, and urban sprawl. *Reg. Sci. Urban Econ.*, 34, 6, 705–725, 2004, https://doi.org/10.1016/j.regsciurbeco.2004.03.002.
6. Arnott, R.J. and Lewis, F.D., The transition of land to urban use. *J. Polit. Econ.*, 87, 1, 161–169, 1979, https://doi.org/10.1086/260744.
7. Irwin, E.G., Bell, K.P., Geoghegan, J., Modeling and managing urban growth at the rural-urban fringe: a parcel-level model of residential land use change. *Agric. Resour. Econ. Rev.*, 32, 1, 83–102, 2003, https://doi.org/10.1017/S1068280500002525.
8. Nickerson, C.J. and Bockstael, N.E., Preservation or development: competing uses over the future of farmland in urbanizing areas, 2001.
9. Cen, X. and Zhang, H., Impacts of multi-scenario land use change on ecosystem services and ecological security pattern: A case study of the Yellow

River Delta. *Res. Cold Arid Reg.*, 16, 1, 30–44, 2024, https://doi.org/10.1016/j.rcar.2024.03.003.

10. Kang, P., Chen, W., Hou, Y., Li, Y., Linking ecosystem services and ecosystem health to ecological risk assessment: A case study of the Beijing-Tianjin-Hebei urban agglomeration. *Sci. Total Environ.*, 636, 1442–1454, 2018, https://doi.org/10.1016/j.scitotenv.2018.04.427.
11. Baró, F., Haase, D., Gómez-Baggethun, E., Frantzeskaki, N., Mismatches between ecosystem services supply and demand in urban areas: A quantitative assessment in five European cities. *Ecol. Indic.*, 55, 146–158, 2015, https://doi.org/10.1016/j.ecolind.2015.03.013.
12. Barbosa, O., Tratalos, J.A., Armsworth, P.R., Davies, R.G., Fuller, R.A., Johnson, P., Gaston, K.J., Who benefits from access to green space? A case study from Sheffield, UK. *Landscape Urban Plann.*, 83, 2–3, 187–195, 2007, https://doi.org/10.1016/j.landurbplan.2007.04.004.
13. Caspersen, O.H., Konijnendijk, C.C., Olafsson, A.S., Green space planning and land use: An assessment of urban regional and green structure planning in Greater Copenhagen. *Geogr. Tidsskr.-Dan. J. Geogr.*, 106, 2, 7–20, 2006, https://doi.org/10.1080/00167223.2006.10649553.
14. Pauleit, S., Ennos, R., Golding, Y., Modeling the environmental impacts of urban land use and land cover change—a study in Merseyside, UK. *Landscape Urban Plann.*, 71, 2–4, 295–310, 2005, https://doi.org/10.1016/j.landurbplan.2004.03.009.
15. Bowler, D.E., Buyung-Ali, L., Knight, T.M., Pullin, A.S., Urban greening to cool towns and cities: A systematic review of the empirical evidence. *Landscape Urban Plann.*, 97, 3, 147–155, 2010, https://doi.org/10.1016/j.landurbplan.2010.05.006.
16. Breuste, J., Haase, D., Elmqvist, T., Urban landscapes and ecosystem services, in: *Ecosystem Services in Agricultural and Urban Landscapes*, pp. 83–104, 2013, https://doi.org/10.1002/9781118506271.ch6.
17. Colding, J., 'Ecological land-use complementation' for building resilience in urban ecosystems. *Landscape Urban Plann.*, 81, 1–2, 46–55, 2007, https://doi.org/10.1016/j.landurbplan.2006.10.016.
18. Qiu, J., Nassauer, J.I., Ahern, J., *et al.*, Advancing landscape sustainability science: key challenges and strategies for integration with landscape design and planning. *Landsc. Ecol.*, 40, 25, 2025, https://doi.org/10.1007/s10980-024-02042-4.

7
Technology in Land Use Planning

Abstract
Sustainable development can be achieved through land use planning, as it determines the distribution of land resources and natural habitat. Advanced technologies like Geographic Information Systems (GIS), Remote Sensing, and Blockchain have transformed the efficiency, mapping, monitoring, and transparency in determining land use patterns. This has aided the development of more knowledgeable and fair resource management techniques. This chapter emphasises the role and advantages of technological advancements in information development, openness, mapping, managing and involving communities in decision-making. In current scenarios, there is a need for a well-balanced approach that uses technology to its potential and understands its limits. Citizens, researchers, society, government and stakeholders can utilise technologies in social, environmental and economic dimensions of land use patterns. There is a need to combine technology and participatory tools to develop a resilient, fair, and habitable environment for upcoming generations.

Keywords**:** Environmental sustainability, land use planning, management, remote sensing, sustainable development, techniques

7.1 Introduction

Technology is essential in shaping how land is used by offering useful tools and intelligent systems that support all phases of the land use planning process for planners, policymakers, researchers and local communities. The connection between human actions and nature is mainly influenced by how land is handled and divided. Making decisions about how land is used has wide-ranging effects on the environment, the economy and society [1, 2]. As people move to cities, populations rise, climate and resources become scarce, and it is getting more important to rely on technology to guide us in how land can be used efficiently.

A major role of today's technology lies in gathering and examining spatial and environmental data. The use of GIS and remote sensing allows

Kundan Meshram and Chatrabhuj. A Sustainable Approach to Land Use, (133–162) © 2026 Scrivener Publishing LLC

planners to analyze images of the Earth and its data to map current land uses, track ongoing changes and identify trends of land use changes, urban encroachment, deforestation or increased farming [3]. They also simplify the study of landforms, floodplains and soils, making it easier to recommend locations good for housing, farming, conservation or industrial development. Using artificial intelligence, planners can prepare for changes in the environment or society based on different ways of managing land resources. Long-term planning benefits most from understanding the risks that might arise, especially those from climate change, loss of various species and overuse of infrastructure [4].

Technology improves the way people take part in and monitor land use planning. Through interactive systems, 3D models, augmented reality tools and dashboards accessible online, different stakeholders are able to look at land development ideas and study their effects. This technology encourages all interested parties to participate, speak openly and reach an agreement on sustainable land management. Tools for community GIS and joint mapping support residents by involving them and helping them share their local views and ideas, which results in fairer and better-suited choices. Technology is important for supervising and enforcing land use policies and their rules [5–7]. Drone surveys, remote satellite observation and sensor technology are used to oversee land developments and discover unauthorized use of land, while also protecting zoning and environmental guidelines. To take care of protected zones, prevent unplanned urban growth and protect sensitive habitats, these abilities are necessary.

When planners include technology in land use planning, it becomes possible to plan effectively and reach balanced development while caring for the environment. Technology helps by allowing us to plan for the future rather than wait for events to occur, so risks and opportunities can be noticed well in advance. It helps cities become smarter, manage the effects of climate better in the countryside and smooth rural-urban changes [8]. Land management can be made easier and more sustainable by technology which guarantees land resources benefit us today and tomorrow. A complete study of technology and land use interactions produces ideas to improve future environmental conservation approaches, as seen in Figure 7.1.

The systematic examination of comprehensive data linked to social and economic variables and environmental factors becomes possible through technology in land use pattern development. The analysis includes essential elements such as soil properties, terrain aspects, climatic variations, water management patterns, vegetation specimens and population development patterns. The most transformative technology in land use identification is

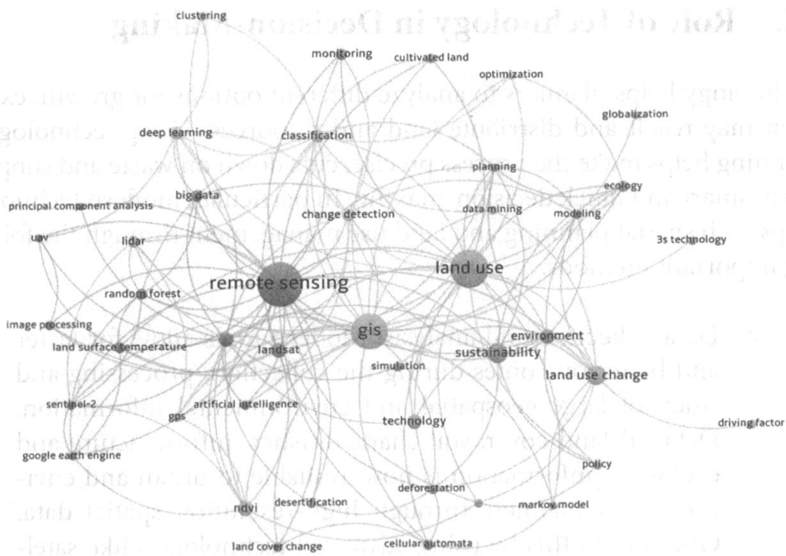

Figure 7.1 Relationship between technology and land use pattern.

Geographic Information Systems (GIS) since it enables developers to analyze multiple layers of spatial data to find the best placement spots. The combination of GIS systems with remote sensing features offers complete and immediate geographical information after it collects satellite and aerial images [9]. These useful analytical tools enable users to pinpoint locations which suit farming activities as well as urban growth areas, environmental protection zones, infrastructure construction sites and disaster reduction programs. Time-based monitoring technologies assess both nutrient conditions and plant health, as well as soil dampness and pest detection in the field. Survey-based management enables farmers to apply water and traditional farming products more efficiently by optimizing their use, which produces better crop yields and enhances environmental protection and preserves agricultural resources to sustain the land health [10]. Modern mapping technology helps organizations execute their administrative work consisting of land registration tasks and zoning regulations, along with permitting operations. The tools promote transparent processes by handling land ownership matters effectively while helping administrators to enforce land use rules. Land use policy implementation, along with sustainable development objectives, can be better monitored through process streamlining by governments.

7.2 Role of Technology in Decision-Making

Technology helps planners to analyze different options for growth, expect what may result and distribute land and resources. Using technology in planning helps make the process precise, cuts down on waste and supports open, smart and quick decision-making. In particular, modern technology helps with spatial planning and land use management through the following important methods.

a. Data collection and analysis: Modern tools allow for faster and better outcomes during the collection, processing and study of large geospatial and environmental information. Detailed land cover, soil characteristics, infrastructure and ecological information is now available to urban and environmental planners through high-resolution spatial data. One can do this by using advanced technologies like satellite remote sensing, LiDAR, survey work from drones and field sensors. Such software helps in overall assessment of the environment, thorough inspections of the infrastructure and detailed examination of land use practices. Due to modern technology, planners now use a lot of geospatial data such as satellite images, air pictures and ground surveys, to update their land use and environmental choices.
b. Remote sensing and monitoring: Planners can evaluate development initiative impacts through the real-time data provided from satellite imaging as well as unmanned aerial vehicles (UAVs) about land cover and land use, and environmental conditions. Through their capabilities, these tools enable planners to track changing rates of urbanization along with observing deforestation, vegetation health and water body status. Remote sensing data enables the identification of land cover modifications, along with vegetation health assessment and urban growth measurement, which assists organizations in creating better adaptive management schemes and policies for land use.
c. Visualization: Through technological applications, stakeholders and decision-makers receive useful explanations about complex planning ideas through maps and virtual simulation models. A combination of visualization tools allows stakeholders to view future development forecasts

while they evaluate appearance-based impacts and the broader outcomes of planning decisions, which helps obtain shared agreement regarding planning decisions. The tools help planners achieve transparent and inclusive planning through their applications during public hearings, stakeholder meetings and urban design development.

d. GIS technology: Using geographic information technology, planners can combine geographical data for analysis and map generation to build detailed decision-based maps and data systems. GIS functions as a central system to process, visualize and analyse various spatial datasets. One GIS interface integrates several data sets, which include topography as well as zoning regulations alongside environmental constraints. It creates suitability maps as well as pinpointing essential development locations and executing risk assessments by following ongoing spatial changes over time.
e. Scenario modelling and simulation: Planners use scenario modelling tools that link to GIS platforms for evaluating various development outcomes of land use by trying different policy choices and environmental factors combined with population projections. Such tools provide a systematic method to evaluate both short- and long-term effects that arise from multiple planning strategies.
f. Real-time monitoring: Planning sector units gain access to real-time environmental data and real-time updates about atmospheric quality, river conditions, land use, acoustic and climatic data. Instant data accessibility enables organizations to stop upcoming threats while fulfilling regulatory needs and serves as a fast response force to pollution incidents and undocumented construction.
g. Stakeholder participation: Various digital platforms aid in participatory planning for different stakeholders. The adoption of inclusive planning methods establishes trust with the community while enhancing legitimacy, which ultimately leads to outcomes that represent multiple community requirements. Stakeholders can use accessible mapping platforms such as ArcGIS Online and OpenStreetMap to share data while collaborating with peers as well as the community for local knowledge input, development feedback and decision-making processes.

h. Decision support systems: Planners should use GIS with databases combined with analytical modelling tools to assess different land use options and decide which strategy meets specified evaluation requirements. The decision-making systems enable planners to conduct evaluations of different planning strategies through economic, social and environmental criteria assessment.
i. Risk assessment and resilience planning: Technology enhances risk assessment and resilience planning by identifying and minimizing potential hazards and vulnerabilities in land use development. Geographic risk assessment methods, such as flood mapping, wildfire risk modelling, and seismic hazard analysis, help planners identify high-risk locations and execute mitigation measures to decrease the impact of natural disasters on communities and infrastructure.
j. Performance monitoring and evaluation: Technology enables performance monitoring and evaluation of land use policies and planning projects through metrics, indicators, and dashboards that track progress toward goals and objectives. Performance monitoring systems allow planners to evaluate the efficacy of land use interventions, measure outcomes, and make data-driven changes to planning methods as needed.
k. Policy analysis and implementation: Planners utilize technological tools to determine the effects that emerge from policy decisions, along with regulations and incentive systems. Technology enables experts to study the economic and social effects, together with environmental effects, that result from land use policies, regulations, laws and incentives. Policy analysis frameworks, which include cost-benefit analysis and multi-criteria decision analysis, provide planners with the capability to measure competing objectives alongside policy gap detection and create methods to reach desired planned outcomes.
l. Community engagement and empowerment: Technology use enables local communities to take action in planning decisions through its ability to provide innovative capabilities. Community members gain access to mapping tools that let them track problems along with cultural landmarks and affect which planning requirements receive priority attention. Through social media channels, government agencies

receive additional feedback instruments that enhance their accountability and their ability to be responsive. People who participate in mapping activities and data collection become key contributors to transparent and accountable decision-making during the planning period by showing local data while defending personal interests.

7.3 Evolution of Technology

Mapping of land use and land cover has evolved with time. As technology rises, it has helped society to easily map different land use patterns, as with time, more and more sophisticated tools are available. The tools and techniques which have evolved over time are described below.

7.3.1 Mapping Techniques Before 20th Century

- Hand-drawn maps: In the past, land use maps were made by hand using manual techniques such as drawing and sketching, which relied on observations and field surveys.
- Topographic maps: These maps emerged in the 19th century; they offered intricate depictions of topographical characteristics, encompassing information about land usage and land cover.

7.3.2 Aerial Photography (Early to Mid-20th Century)

- Aerial photography: This technique was developed in the early 20th century; it brought a significant transformation in land use mapping over vast areas. Through aerial photography, precise images could be obtained of natural terrains and their features from above ground.
- Photogrammetry: This procedure generates three-dimensional data from aerial images through its established process. This aids in the development of precise maps and effective land use classification.

7.3.3 Satellite Remote Sensing (Late 20th Century)

- Earth observation satellites became a critical development in remote sensing technology when they appeared during the

1960s. Major advancements in satellite technology became prominent in the later decades with Landsat 7 and Landsat 8 satellites, as well as the SPOT (Satellite Pour l'Observation de la Terre) series. These satellites introduced advanced sensors to monitor and capture top-quality multispectral Earth surface images throughout various wavelengths. The enhanced imaging technology permitted researchers to deliver precise examinations of land use patterns, besides detailed evaluations of living plant conditions and aquatic resources in combination with urban expansion and environmental transformation patterns throughout time. The multispectral imaging technology helped experts observe minimal surface variations that standard human vision would miss. Satellite technology has achieved such advancements that scientists, environmentalists and decision-makers now use it to handle natural resources while measuring climate change effects and building sustainable development plans.

- Digital image processing methods have improved the capacity to examine satellite data because they aid effective Land Use and Land Cover (LULC) and satellite image analysis. These analytic methods, consisting of image classification, enhancement, filtering and change detection algorithms, enable scientists to retrieve valuable details from unprocessed space-based data. Supervised and unsupervised methods empower analysts to classify land use categories, including agricultural fields, urban areas, forests, vegetation, bare land, open areas, and water bodies with better speed and precision. Through image enhancement tools, analysts gain a simpler identification of essential landscape features because the tools adjust brightness along with contrast and spectral features.

7.3.4 Geographic Information Systems (GIS) in the Late 20th Century

- Before GIS was introduced to the world, land use analysis was analysed manually, which involved time-intensive, manual mapping activities, producing static representations of basic geographic data that was vulnerable to errors. GIS established a digital structure that allowed perfect integration and data management of geospatial and attribute data across numerous data sources like satellite images, aerial pictures, censuses and ground surveys. Through this system, researchers and other experts gained better control of geographic information visualizations to conduct interactive analyses that significantly enhanced spatial analysis efficiency.

- The established tools allowed users to perform advanced land use classification as well as track changes over time, along with a sophisticated analysis of spatial trends and patterns. Remote sensing, with GPS and real-time data sources and GIS, forms an integrated system that continues to expand its applications throughout different fields of study.

7.3.5 Machine Learning (ML) and Artificial Intelligence (21st Century)

- The manual interpretation of satellite data by experts, together with large amounts of human-operated input, restricted scalability while introducing uncertainties into classification results. The image classification process uses ML algorithms, including decision trees and support vector machines (SVM), together with random forest models to analyse labelled data patterns, then apply them for unseen image classification. Through automated processes, the mapping workflow becomes faster while achieving better accuracy levels and reducing operator mistakes. Through algorithmic application, urban planners, together with environmental agencies and researchers, can now perform extensive land use and land cover assessments, which enable them to make superior and expedited decisions.
- Deep learning methods, such as convolutional neural networks CNNs perform excellent feature extraction and learning of abstraction-based patterns and spatial relationships from imagery data throughout multiple layers of the network. Through their processing of vast amounts of high-resolution satellite imagery without extensive preprocessing, they enable better and more accurate identification of residential, commercial, agricultural, industrial and

forested areas. The CNN-based modelling system exhibits an outstanding ability to execute different tasks, like semantic segmentation together with object detection and change detection.

7.3.6 Unmanned Aerial Vehicles (UAVs) or Drones (21st Century)

- UAV technology, referred to as drones, fundamentally changed how land use mapping operates during the 21st century. UAVs function differently from aerial and satellite systems by providing adaptable, economic and fast access to high-definition imaging capabilities targeting specific, exact areas. These devices achieve better data collection through their capacity to operate at low flight heights, traverse challenging terrain and thus suit the requirements of detailed mapping for projects including agricultural field studies, along with urban development and environmental research. The effortless deployment method and accessible nature of UAV technology have removed data collection barriers so researchers, land managers and local governments can execute immediate surveys through UAV surveys.
- UAV-based mapping methods produce better representations of the land's condition that upgrade its precision and authenticity for analyzing land use and land cover (LULC) maps. High-resolution cameras mounted on UAVs, along with multispectral to hyperspectral, thermal and LiDAR sensors, provide the capability to obtain precise images of different elements within land cover areas. Because of their low altitude operations, drones obtain greater spatial resolution than conventional satellite imagery and keep track of small objects like tree individuality, crops, small water sources and built-up characteristics. UAV-based remote sensing has emerged as a fundamental technology for modern land use mapping because it provides essential support to decisions in urban planning, together with agricultural and forestry applications and environmental conservation initiatives.

7.3.7 Utilization of Open Data (21st Century)

- Open data utilization during the 21st century established a major role in distributing geographic information to the public while improving land use mapping techniques. NASA's Earth Data, together with USGS Earth Explorer, Copernicus Open Access Hub and OpenStreetMap, allow the general public, government bodies, research professionals and planning agencies, along with non-profit organizations, access to high-quality spatial data through open-access satellite imagery, aerial photographs, and geographic datasets. The lowering of entry barriers through this accessibility improves land use analysis and supports more stakeholders who become involved with increased innovation and collaborative partnerships. The wider involvement attained through open data leads to transparent processes along with community-driven planning strategies, while building decisions based on data throughout environmental monitoring, urban development tasks, disaster situations and resource allocation activities.

7.4 Technology in Determining Land Use Patterns

Modern technology has meaningful effects on land use through its provision of innovative solutions for environmental management, sustainable development and resource stewardship. Through technological integration in land use processes, stakeholders enhance their efficiency alongside their decision-making abilities and effectiveness, which leads to superior results for human beings and the environment [11–14]. Technology plays an indispensable role throughout the entire span of land use planning, together with management and decision processes. Through advanced technology, one can collect data while conducting analysis, making models and generating visuals which enable sustainable land use decision-making. An extensive study of the way technology modifies land utilization patterns is shown below.

7.4.1 Data Collection and Monitoring

Technology allows scientists to monitor environmental change across extensive spaces through time. GIS enhances various technologies through its ability to integrate, analyze, display geospatial data and create spatial maps for land utilization pattern recognition and environmental assessment, as well as land use conflict detection. Environmental conditions, biodiversity, soil moisture, air quality and temperature data can be collected through a combination of ground-based sensors, UAVs and IoT devices.

7.4.2 Spatial Analysis and Modelling

Through land suitability analysis, habitat modelling and land use change modelling methods based on GIS suites, the planning sector can foresee and measure the environmental effects of land utilization choices regarding ecosystems, hydrology and biodiversity. The models assist in selecting ideal land use choices by means of reducing environmental hazards and enhancing sustainable practices. The evaluation of socioeconomic factors with environmental implications from multiple land use possibilities is made possible through scenario planning and decision support systems (DSS). The simulated models provide stakeholders the opportunity to see different trade-off combinations, which leads them towards adaptive planning and evidence-based policymaking for reaching sustainable land development goals.

7.4.3 Natural Resource Management

The utilization of advanced geospatial and sensing technologies has aided in the management and monitoring of natural resources. Irrigation, fertility levels and moisture content, together with erosion risks, remain accessible through advanced tools designed to assess soil health and composition. Soil information can be used for developing sustainable land management approaches, which include crop rotation combined with contour farming and conservation tillage systems. Forest cover assessments, biomass evaluations and deforestation detection occur through forestry sector applications of inventory systems along with remote sensor technologies to achieve sustainable forest management goals.

7.4.4 Urban Planning and Infrastructure Development

The combination of tools enables stakeholders to perform land use arrangement before designing sustainable urban plans through environmental assessment procedures. Smart system implementations lead to diminished resource requirements while reducing operational costs and increasing overall urban service delivery quality throughout municipal areas. Urban analytics provide cities with predictive power that enables them to perform scenario-based analysis of traffic congestion, pollution problems and urban sprawl. Technology enables experts to detect vulnerable areas so they can develop protection strategies and engineer buildings that endure severe weather events. City operators employ resident platforms together with planning tools to include citizens in their local administrative decisions. Modern urban development underwent changes because these advanced systems unite three main elements of data-driven planning with resident participation and environmental sustainability. Digital technologies help cities secure sustainable development through balanced growth by providing better city life quality, yet making them resilient against forthcoming uncertainties.

7.4.5 Community Engagement and Decision-Making

Through their access to participatory mapping tools, mobile applications and citizen science platforms, community members both gather geospatial information and indicate their area's influential sites as well as report problems of land degradation, unauthorized land use activities and ecological threats to local planners. The planners receive location-specific information in real time through crowdsourcing, as a result of which locals obtain the power to combine their professional expertise with personal experiences and their priorities. Inclusive methods protect land use decisions when they come from community-based perspectives and social equity principles. Modern planning tools provide policymakers with planners, ordinary people, and interactive capabilities to examine zoning rules together with infrastructure plans, natural and social limitation layers, thus increasing knowledge access for everyone. The combination of sophisticated technologies provides clear and simplified information that helps parties collaborate through scenario development for making evidence-based decisions. Through digital communication platforms, including social media platforms and web-based engagement tools and online forums, the public can maintain essential dialogue with stakeholders by using participatory feedback mechanisms. Through these platforms, people receive immediate information while they maintain channels to share their comments and grade recommendations to authorities.

7.5 Various Technologies Used for Land Use Patterns

The collection and analysis of spatial data through technological methods are vital for land planning and management because they let researchers handle and interpret spatial information. Through these technologies, users gain important geographical data about Earth's surface, which includes data related to land cover and land use patterns, with additional details on topography and environmental condition information. Multiple procedures operate in combination for this purpose, are described in detail below.

7.5.1 Remote Sensing

Remote sensing technology involves collecting Earth's surface information by using sensors that operate from the atmosphere at remote distances. The implementation of sensors takes place on satellites, together with aeroplanes, drones, ground-based platforms and satellites for remote sensing. The technology uses electromagnetic spectrum wavelengths starting from visible to infrared to microwave radiation for both collecting data and generating images and measurements of Earth surface features, along with land conditions and environmental details [15–18], which is shown in Figure 7.2.

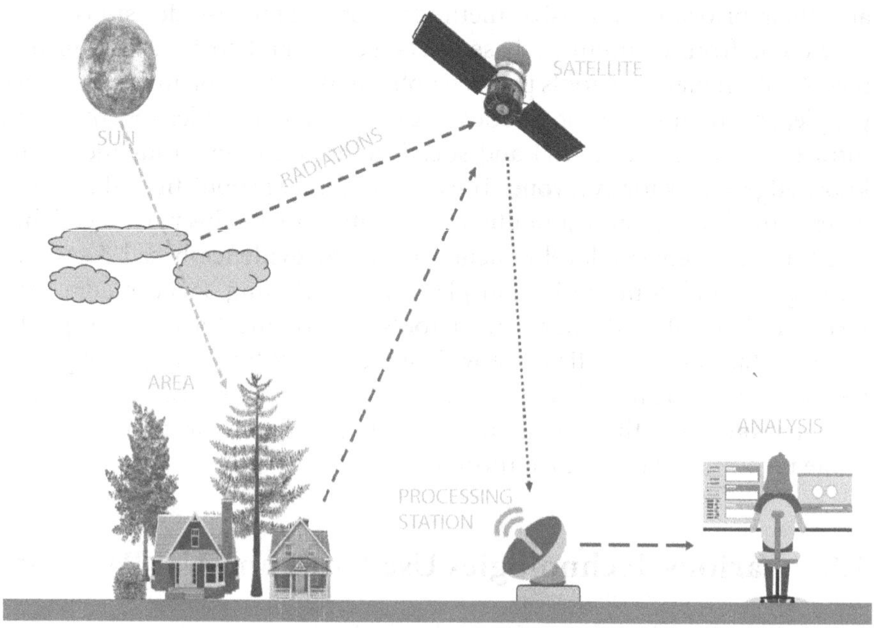

Remote sensing technology functions through capturing and recording electromagnetic signals that arise from surface objects. How diverse materials and surfaces reflect radiation and how they absorb or emit such energy enables remote sensing instruments to identify multiple attributes and events. Useful environmental insights about land cover, vegetation vitality, water bodies, urban regions and additional environmental factors can be obtained through processing the acquired data, followed by analysis and interpretation [19–22]. The different sectors, like agriculture, forestry, environmental monitoring, urban planning and disaster response, utilize remote sensing technology for their operational needs. This tool enables researchers, along with policymakers and land managers, to detect environmental changes by analyzing physical attributes of land territory and conduct assessments across the ecosystem for making strategic decisions regarding sustainable natural resource management and conservation efforts.

a. Satellite Imagery
Advanced sensors mounted on satellites create satellite imagery that displays the Earth surface appearance through orbiting space vehicles. The detectors on satellites measure electromagnetic radiation extending from visible to infrared and microwave spectrums to provide extensive information about Earth surface features and atmospheric conditions irrespective of environmental conditions. The produced imagery serves as a vital tool for tracking land cover modification while allowing researchers to inspect vegetation health patterns, observe urban area growth, and understand various natural and human-generated environmental processes. Satellite data enables the support of multiple applications such as precision agriculture with crop monitoring, tracking deforestation, health status, urban planning of spatial development and infrastructure needs. The most beneficial quality of satellite imagery consists of its systematic high-resolution spatial monitoring features, which enable scientists to observe climate patterns chronologically.

Satellite imaging systems include multiple options for obtaining high-definition images which support land use analysis, among other purposes. These organizations operate among the most well-known satellite image delivery services:

- Landsat Program: Since 1972, the NASA-US Geological Survey (USGS) joint operation of Landsat has continuously collected Earth-based data through its satellites. Spin-off satellites from Landsat use multiple wavelength spectrals, including visible range, together with infrared and thermal range data to capture scenes with 15 to 30 meter spatial precision.

- Sentinel Satellites: Under the Copernicus program run by the European Union the Sentinel satellites function as part of this system. The Sentinel-2 optical sensors use multispectral technology to create images at a precise 10 to 20 meter scale through 13 different spectral bands that include visible light, near-infrared and shortwave infrared bands. Sentinel-1 takes radar pictures with different polarization modes, so it can be used in all kinds of weather and at all times of the day and night for things like recording sea ice, tracking land deformation, and logging trees.
- WorldView Satellites: The WorldView satellites (WorldView-1, WorldView-2, WorldView-3, and WorldView-4) are run by Maxar Technologies and provide some of the best spatial resolution images that can be bought. The resolutions range from 30 centimeters to 1 meter. The panchromatic, multispectral, and shortwave infrared images that the WorldView satellites take make it possible to study land use, plan cities, respond to disasters, and keep an eye on infrastructure.
- Planet Labs: Planet runs a group of small satellites called Doves that take pictures with high return rates all over the world. Planet's pictures, which usually have a spatial precision of 3 to 5 meters, are used to keep an eye on things that are changing on Earth's surface, like the growth of cities, farming, changes in forest cover, and natural disasters.
- Google Earth Engine: Google Earth Engine is a tool in the cloud that gives access to a huge collection of satellite images and 3D maps. Earth Engine collects data from many satellite missions, such as Landsat, Sentinel, MODIS, and others. This lets researchers and analysts do time-series analysis, large-scale land use analysis, and classification jobs based on machine learning.
- Moderate Resolution Imaging Spectroradiometer (MODIS): NASA's Terra and Aqua satellites use the MODIS instrument to obtain images of Earth surface areas at intermediate spatial resolution levels from 250 to 500 meters. MODIS gathers information from various spectral bands, which combine visible and near-infrared and thermal infrared wavelengths for extensive observations of surface and atmospheric

operations. Through its critical functions, MODIS contributes to monitoring worldwide vegetation patterns while detecting changes in surface temperatures, wildfire detection and atmospheric observations of clouds water vapour and aerosols.

- Advanced Spaceborne Thermal Emission and Reflection Radiometer (ASTER): NASA's Terra satellite operates ASTER as its high-resolution imaging instrument that gathers multiband data from visible and near-infrared to thermal infrared wavelengths for detailed observations. ASTER offers a collection of spatial resolutions between 15 to 90 meters, which produces superior detail than moderate-resolution sensors and well suits applications with strict accuracy needs. ASTER satellite data find extensive use in terrain mapping because they enable the production of Digital Elevation Models (DEMs) as well as provide land cover classification capabilities, mineralogical, geological analysis functions and environmental monitoring systems.

b. Aerial Photography

Specialized aerial equipment on aircraft performs the task of capturing detailed Earth surface images through elevated photography operations. Architects and urban planners rely on aerial imagery, and land managers need it to analyse terrain. The practice extends to cartographic work and ecological testing as well as infrastructure planning and terrain assessments. The data set reveals specific details regarding land topography, along with vegetation mapping and hydrological characteristics and land utilization distributions to support exact spatial analysis and mapping operations. Aerial photographs provide better resolution and clearer images than satellite imagery, thus enabling precise feature identification of individual structures, roads and vegetation units as well as geological elements. Aerial photography provides a high degree of acquisition flexibility because operators can adjust flight paths together with altitudes and camera settings to fulfil project demands or pinpoint specific monitoring regions. The scientific community, along with applied providers of spatial information, continue to rely on aerial photography as an essential instrument.

c. LiDAR

The remote sensing technology known as LiDAR or Light Detection and Ranging performs accurate Earth surface distance evaluations through laser pulse operations. This system utilizes laser light bursts that it fires towards targets to determine reflection time for sensor detection. The usage of time-of-flight measurements from LiDAR permits system operators to generate precise three-dimensional surface representations of Earth through GPS locations and inertial measurement unit orientation data. The generated 3D models from LiDAR technology embrace all types of natural and urban environmental elements, specifically including elevation features, plant cover, and built structures and hence demonstrate immense utility across these settings. The penetration capabilities of LiDAR create accurate ground surface models across forested areas since the technology can perceive beyond vegetation. The obtained point cloud data enables production of Digital Elevation Models (DEM) and Digital Surface Models that serve multiple applications. The technology delivers highly precise topographic information to support land use planning and performing safety assessments on construction sites as well as geological hazard measurement. LiDAR significantly contributes to the monitoring activities related to both forestry management and vegetation maintenance. Scientists utilize LiDAR to generate building maps alongside utility, road mapping for urban substance development and engineering design needs. The technology finds widespread usage in engineering projects because it enables measurements for road alignment, bridge construction, volume calculation and earthworks monitoring. The robust elevation and structural capabilities of LiDAR elevate it to a significant positioning in scientific research and environmental analysis, along with engineering functions.

7.5.2 Geographic Information Systems (GIS)

GIS functions as an advanced technological instrument used to collect spatial or geographic information, which subsequently gets stored before undergoing analytical operations, and serves as a manager for this information along with visual presentation capabilities. Through its capabilities, users can recognize interactions among data points in geographic areas, thus acquiring a better understanding of natural landscapes and constructed environments. GIS brings various types of data, like maps combined with imagery, photographic data and numerical datasets, under one system for spatial analysis to support decision-making processes. The system integration lets users position diverse information layers on top of each other while examining spatial relationships through processes that would demand traditional analysis tools. All the elements required to establish a complete GIS infrastructure include hardware devices together with software programs, datasets, and expert human analysts. The system's hardware embraces computers with GPS functionality in combination with servers, together with digitizing tablets to process information from data entry and output data. The specialized GIS programs, such as ArcGIS, QGIS and Google Earth, make up software that enables users to map areas alongside conducting spatial analysis and visualization. The main components of GIS data consist of geographic data elements that involve vector and raster layers together with topographic maps, remote sensing images and databases containing feature description attributes. The human presence within GIS enables the creation of systems along with data interpretation to develop applications that help professionals solve real-world problems. The wide diversity of fields enables GIS to find useful applications. The functionality of GIS aids urban planners in areas of zoning, together with infrastructure design and land use analysis. GIS proves vital for environmental management teams because they use it to analyze ecosystem dynamics alongside pollution tracking and climate modelling activities. The tracking of deforestation along with watershed management and agricultural land evaluation receive assistance through GIS applications for natural resource initiatives. GIS serves as a key tool during emergencies because it enables risk evaluation, route planning for evacuations and emergency operation coordination. The implementation of GIS in public health allows health authorities to track epidemic trends, also helps locate unmet service requirements and create health intervention strategies. The application of GIS creates improved decisions through visual spatial representations which enable stakeholders to understand complex information more easily. GIS strengthens planning operations through tools that enable simulations as well as predictions and scenarios for testing different potential outcomes. The combination of interactive maps with dashboards and geospatial visualizations creates platforms that enable better interaction between government, business organizations, researchers and community members. GIS operates as an essential tool that helps people understand environmental and social spatial interplay while facilitating management of these interactions.

7.5.3 Global Positioning System (GPS)

A satellite-based platform named GPS offers precise position data and speed information to users through any weather along with spot-on time updates on Earth's surface. The GPS operates through interconnected satellite arrangements linked to control base stations and GPS receivers used by users. The satellites operate by sending continuous signals that a GPS device receives to create precise spatial positioning through triangulation calculation methods. The system functions 24 hours a day and achieves widespread recognition as a highly precise and dependable system that various industries and disciplines require. GPS satellites consistently emit signals that are received by GPS receivers on the ground, enabling them to accurately determine their exact position, speed, and time. GPS receivers accurately establish the user's location by calculating the time it takes for signals from numerous GPS satellites to reach the receiver. Through the process of triangulation, the receiver can accurately determine its precise geographic coordinates, including latitude, longitude, and altitude.

GPS technology is extensively utilized in multiple industries, such as transportation, agriculture, surveying, and land use planning, to support navigation, tracking, and monitoring activities. GPS receivers are incorporated into automobiles, smartphones, and portable devices to offer step-by-step guidance, optimal route planning, and live monitoring of resources, vehicles, and personnel.

GPS technology at present serves as a vital element for land use planning operations through its precise methods of collecting spatial data effectively. GPS serves various applications in field surveys together with digital mapping and land parcel boundary definition. The accurate property boundary definition, along with infrastructure network mapping and documentation of natural features, exists through handheld and integrated GPS receiver devices used by surveyors, planners, and land managers. Modern land records depend on this data for maintenance, while property disputes receive resolution and zoning decisions draw from the information obtained. The geographical positioning system supports thorough spatial evaluation through its ability to supply detailed location-based data which GIS technology accepts to construct sophisticated analysis frameworks. GPS technology provides numerous advantages when used for land use planning. Accurate and precise data collection occurs because GPS technology reduces measurement errors, thus improving spatial information quality. Operational efficiency, along with productivity increases through GPS technology because it delivers faster and more efficient fieldwork processes, particularly in regions that are hard to access using traditional survey methods. Through real-time location tracking, GPS enables field teams to achieve simple tasks while obtaining convenient access to targeted areas. Data obtained from GPS systems facilitates sound choices throughout activities related to land usage for infrastructure growth, land resource management, urban development and environmental protection. Environmental management professionals deploy GPS systems to monitor the protected regions, identify ecological areas and record the shifts of animals along with environmental patterns through time.

7.5.4 Unmanned Aerial Vehicles (UAVs)/Drones

Aircraft drones, also known as Unmanned Aerial Vehicles, function either through remote human control or autonomous operations under predefined flight instructions and internal navigation capabilities. These air-based tools are installed with different sensors and camera systems in addition to LiDAR and other data acquisition instruments, which help capture detailed aerial information through their sensors. UAVs comprise numerous types that span from tiny quadcopter models that customers buy to advanced professional drones utilized for scientific research and industrial operations. Drones continue to favour land use planning thanks to their functional design, which produces sophisticated data collection. UAVs demonstrate their highest capability as they simultaneously gather topographic details and visible images as data sources.

UAV systems enable environmental scientists, together with planners, to develop and perform extensive high-definition digital elevation models to run detailed terrain examinations. Multiple land use planning and management tasks rely on drone technology for their operations. The systems gather images which enable processing for land type identification between urban regions, agricultural zones, forest areas and aquatic surfaces. UAVs enable ecological monitoring and environmental management through their assessment of plant health as well as biomass measurement and time-based vegetation change observation. Through infrastructure inspection UAVs allow operators to check roads bridges, buildings and utility networks safely, thus both enhancing maintenance operations and upgrade planning. UAVs deliver various direct benefits for land use planning activities. The use of drones proves to be better than manned aircraft flying operations by delivering time-saving performance and economic efficiency. Such systems demand reduced resources yet offer quick deployment, which makes them appropriate for both unexpected survey tasks and consistent monitoring activities. Drone survey capabilities enable detection operations within inaccessible or dangerous terrain coverage, such as mountains, forests and disaster recovery zones, because of their compact shape and nimble movement. The video streaming feature of drones enables stakeholders to actively observe developments in projects, agricultural processes and environmental alterations in real time. Current data delivery to stakeholders makes possible both prompt management decisions, prompt interventions and open communication.

7.5.5 Machine Learning and Artificial Intelligence

The combination of Earth observation systems along with ML and AI algorithms enables them to generate useful data points from enormous information sources that include satellite imagery and GIS database collections. The algorithms can spot important data patterns that standard analytical techniques cannot detect reliably, including urban growth patterns, transformations in land usage, forest cover mapping, wetland deterioration and vegetation health changes. The supervised learning models within the ML domain help organizations achieve land use categorization and change detection through classification and regression operations. Spatial data analysis through unsupervised learning algorithms enables users to identify hidden patterns without input labels because this approach works best when studying unknown patterns and in areas lacking extensive recorded data. Reinforcement learning enables researchers to develop optimized land management strategies through simulated environments because it is becoming more prevalent for this purpose. The predictive features of these models transform what planners and policymakers can do through proactive strategy development for sustainable development, sustainable urban planning, disaster prevention programs and agricultural success enhancement. The precise classification system enables better planning decisions for land use, conservation efforts, infrastructure design and environmental impact assessment operations.

Machine learning provides the capability to unite many different types of data sources, which include sensor data, geospatial datasets, satellite imagery, demographic records, and social media feeds for thorough analysis. The multi-dimensional data fusion enables complete land use analysis because ML algorithms unite diverse inputs to understand socioeconomic, environmental, and cultural factors. ML-based models that join high-resolution satellite data with demographic statistics create effective, validated methods to detect and display areas undergoing high urban expansion, fast population growth and land use disputes that aid spatial planners in their work and infrastructure development project planning. The main advantage of using ML approaches for land use decision-making involves their ability to generate precise measurements of uncertainty and risk assessments. The methods enable administrative officials and decision-makers to evaluate scenario confidence levels, together with understanding multiple probable outcomes. The decision-making process becomes more resilient and data-based because these methods help planners forecast variable outcomes, establish intervention priorities, minimize negative impacts from environmental damage, resource allocation inefficiencies and urban growth disorders. The blend of these tools provides both enhanced availability of data-driven information and upgrading the ability of decision-makers to involve a wider audience in their decisions despite complex datasets. This approach includes diverse stakeholders to develop sound collaborations that enhance trust between them and lead to better, sustainable, and equitable land use results. The automated detection together with monitoring of land use changes through machine learning substantially decreases the time and workforce needed for manual analysis.

7.5.6 Blockchain Technology

Blockchain serves as a digital disruptive technology that maintains transactions securely through distributed yet decentralized ledgers across a computer network to create an impervious and transparent transaction record. All transactions undergo verification and chronological connection into blocks, which forms an imperishable and tamper-resistant series of data. The core elements of blockchain are transparency, security and decentralization that create favourable conditions for modernizing land use planning and land administration systems. Land use applications using blockchain technology create dependable systems that improve land database quality and accessibility and eliminate traditional central authorities from managing ownership records and related information. A digital registry based on blockchain functions as a trusted information hub that offers essential, valid data that supports planning decisions for governance. Through self-executing digital agreement systems blockchain enables automatic property transaction security because of its self-executing nature. Through its digital registry system, blockchain avoids fraud because it reduces theft opportunities and simplifies both property transfer processes and transaction systems to support efficient land development operations.

Blockchains enable local community planners and policymakers to collaborate with consensus voting toward zoning permissions, along with land development permits and land use regulation decisions. The unalterable character of blockchain records protects land data by validating its authenticity, which reduces all opportunities for data tampering and consistent unethical changes and false record creation. The financial gain in such locations remains high since inadequate institutions create numerous disputes about land property rights. The land use planning benefits from blockchain technology through its improved permitting systems. Planning departments maximize the efficiency of their development permit operation through blockchain-enabled digitization of workflows that include application records and documentation of approval processes and regulatory data on blockchain platforms. The process of land asset tokenization creates divisible fractional shares thanks to blockchain which produces digital tokens. The tokenization system enhances real estate investment access, which lets more people participate in the market and strengthens property market liquidity alongside its power to integrate funding streams with development schemes. Through tokenization the planning system becomes more inclusive as everyone has equal access to acquired property and land investment participation.

The blockchain technology enables users to develop digital tokens of biodiversity credits and carbon offsets. Blockchain-based platforms let landowners receive financial benefits for preserving ecosystems, carbon sequestration and natural habitat protection activities, thus connecting payments to conservation targets. Land management schemes include these assessment methods to measure quantitative improvement of stewardship practices while maintaining complete transparency. The secure data storage and access system on blockchain helps stakeholders enhance their information sharing potential effectively. The tamper-proof shared database allows all groups such as urban planners, researchers, developers, policymakers and community members to access this single collective source for better coordinated decisions through data-driven methods. The decentralized nature of the system enables participants to find immediate equal access to verified data, which establishes it as vital for team-oriented adaptive planning methods.

7.6 Equitable Benefits from Technological Innovations

Technology serves its desired purposes when it targets underserved communities because such approaches lead to equal land use practices that benefit everyone. The implementation of superior technological systems within land use management provides uniform chances for all communities across all sectors. Multiple important aspects need consideration.

The achievement of fair distribution requires financial support together with technological assistance and leading-edge innovations to reach every community, but especially those from disadvantaged socioeconomic groups. Marginalized and rural communities do not gain access to technology deployment because they lack both adequate funding and proper infrastructure. Organizations and governmental bodies should develop core innovation ecosystems in underserved areas through funding mechanisms, along with incubators and technology-based centers under their programs. Every voice needs to be encouraged into innovation networks, which helps define land use technology development together with its deployment plans.

Sustainable community development needs community members to develop technology capabilities so they can use it effectively in their role. The national investment pool should create training courses that integrate technical education with digital literacy training and entrepreneurship foundations. Teaching combined with digital literacy and vocational instruction makes it possible for more people to take part in the digital economy through educational programs. Skilled local actors obtain sustainable developmental capabilities by enhancing their skills, resulting in long-term projects of community-driven innovation together with social and economic progress. The core driver involves developing and executing policies for advancing digital equity with inclusion. The government must establish legislation for digital rights protection, yet it needs to ban discriminatory practices which block people from using technological services and should create data privacy laws to defend personal information.

7.7 Disadvantages of Technology

The modern development of technology improves both the efficiency, accuracy and transparency of land use planning while presenting new management requirements for handling emerging problems. The negative effects of technological use in planning result in improved organizational resources and reduced stakeholder access to planned outcomes. Each main problem resulting from technical implementation in land use planning is examined below.

a. Expenses and accessibility: Organizations with limited financial resources need to establish major capital expenditures to create technical solutions. Makeup costs for organizations that need to acquire servers, computers, software, storage systems and networking equipment prove excessively expensive. Higher costs emerge from modern Geographic Information Systems program acquisition through remote sensing technology and spatial analysis module implementation, along with permanent software fees and service maintenance charges. Planned organizations that invest in advanced technology break away from limited resource organizations because of their technology expenditure costs.

b. Outsourcing expenses: Organizations must pay elevated prices to external providers who manage software development alongside integration work, together with data analytics processes, thus developing a need for substantial dependence on outside services. Widespread vendor dependencies by organizations create multiple operational control and privacy concerns, which threaten sustainability and service continuity.
c. Absence of diverse data: Planning algorithm accuracy depends entirely on the quality of training data, which developers provide to the system. When an ecosystem uses limited datasets, the existing social inequalities become more pronounced in the models' output generation. The mathematical expertise of algorithms helps produce plans that lead to vulnerable population victimization and negative social results that were not intended. Ethical AI operations, together with data representativeness, provide the necessary basis to avoid these results.
d. Over-dependence on technology: The strong capabilities of technology need careful usage since deep reliance on digital tools causes essential knowledge derived from community expertise, public engagement and individual evaluation capabilities to fade away. Social and cultural engagement combined with community participation needs to be time-sensitive because all stakeholders require collaborative interaction to work together with planning experts. Planning experts should employ digital technology to supplement participatory engagement methods instead of completely replacing traditional participatory engagement practices.
e. Challenges in fragmentation and interoperability: Each governmental department follows its own system by using separate networking software while keeping different data formats. Separated systems block proper data exchange and organization between different groups. When different information systems function poorly with one another, the result becomes duplicated work alongside inconsistent data that produces flawed decision-making processes.
f. Data quality and accuracy: A land use planning system requires precise and proper documentation of base data

for achieving accurate decisions. Multiple errors arise from sampling problems, along with sensor malfunctioning and inaccurate data entry through manual procedures. Weak planning results stem from unacknowledged data inaccuracies that generate incorrect analyses. Verifying data and validating methods need to be established strongly because these measures safeguard data reliability and credibility.

g. Cultural and societal impacts: Modern technology integration usually fails to align with the indigenous cultural guidelines as well as social customs that exist in specific local areas. Access to technology proves insufficient because it leads to implementations that do not align with community needs and thus disrupt traditional knowledge practices. In order to achieve successful land use planning, there should be cultural appreciation through supportive technologies that work with local practices to adapt to traditional ways of doing things. Sustainable outcomes require communities to receive equal treatment between technological progress and knowledge of their cultural and social dimensions.

7.8 Conclusion

Land use planning decisions underwent a reform thanks to technological implementation, which delivers quantitative assessment methods to address challenging topics about urban development and ecosystem management. The system ensures fair opportunities to obtain land and related natural resources for both individuals and groups residing in places having low institutional trust and prior land ownership difficulties. Social equity emerges through technological integration because it enables communities to become more involved in planning processes to demand their needs and redirect policy outputs affecting their daily lives. When implementing technological systems for land use planning, organizations must recognize the accompanying issues they create for the process at the same time. The technological issues that planning systems face do not reduce the substantial chances for equitable results that these procedures offer. By strategically investing in capacity building, data governance, and community engagement, stakeholders may effectively utilize the transformative potential of technology to advance social fairness, environmental sustainability, and economic success for both current and future generations. Strategically weighing the benefits and drawbacks of technology, and giving importance to inclusive and participatory methods, may effectively utilize the complete capabilities of technological advancements to create more resilient, fair, and habitable environments for everyone.

References

1. Qi, W., Huang, C., Wang, Y., Zhang, X., Sun, W., Zhang, L., Global-Local Three-Dimensional Convolutional Transformer Network for Hyperspectral Image Classification. *IEEE Trans. Geosci. Remote Sens.*, 61, 1–20, 2023, doi: 10.1109/TGRS.2023.3272885.
2. Haroun, F.M.E., Deros, S.N.M., Din, N.M., Detection and monitoring of power line corridor from satellite imagery using RetinaNet and K-mean clustering. *IEEE Access*, 9, 116720–116730, 2021, doi: 10.1109/ACCESS.2021.3106550.
3. Pourghasemi, H.R., Amiri, M., Edalat, M., Ahrari, A.H., Panahi, M., Sadhasivam, N., Lee, S., Assessment of urban infrastructures exposed to flood using susceptibility map and Google Earth Engine. *IEEE J. Sel. Top. Appl. Earth Obs. Remote Sens.*, 14, 1923–1937, 2020, doi: 10.1109/JSTARS.2020.3045278.
4. Meng, Q. and Hsu, L.T., Integrity monitoring for all-source navigation enhanced by Kalman filter-based solution separation. *IEEE Sens. J.*, 21, 14, 15469–15484, 2020, doi: 10.1109/JSEN.2020.3026081.
5. Ni, Y., Li, X., Ye, Y., Li, Y., Li, C., Chu, D., An investigation on deep learning approaches to combining nighttime and daytime satellite imagery for poverty prediction. *IEEE Geosci. Remote Sens. Lett.*, 18, 9, 1545–1549, 2020, doi: 10.1109/LGRS.2020.3006019.
6. Zhang, G., Xu, P., Xu, H., Hsu, L.T., Prediction on the urban GNSS measurement uncertainty based on deep learning networks with long short-term memory. *IEEE Sens. J.*, 21, 18, 20563–20577, 2021, doi: 10.1109/JSEN.2021.3098006.
7. Gamal, A., Mohamed, R., Abdel-Basset, M., Hezam, I.M., Smarandache, F., Consideration of disruptive technologies and supply chain sustainability through α-discounting AHP–VIKOR: calibration, validation, analysis, and methods. *Soft Comput.*, 1–27, 2023, doi: 10.1007/s00500-023-08819-0.
8. Orlandi, F.C., Dos Anjos, J.C., Santana, J.F.D.P., Leithardt, V.R., Geyer, C.F., Entropy to mitigate non-IID data problem on Federated Learning for the Edge Intelligence environment. *IEEE Access*, 11, 78845–78857, 2023, doi: 10.1109/ACCESS.2023.3298704.

9. Li, T., Wang, C., Meng, M.Q.H., de Silva, C.W., Attention-driven active sensing with hybrid neural network for environmental field mapping. *IEEE Trans. Autom. Sci. Eng.*, 19, 3, 2135–2152, 2021, doi: 10.1109/TASE.2021.3077689.
10. Bengana, N. and Heikkilä, J., Improving land cover segmentation across satellites using domain adaptation. *IEEE J. Sel. Top. Appl. Earth Obs. Remote Sens.*, 14, 1399–1410, 2020, doi: 10.1109/JSTARS.2020.3042887.
11. Zhang, Y., Huang, Y., Han, S., Yang, Z., Hong, D., Meng, W., Machine learning methods for spaceborne GNSS-R sea surface height measurement from TDS-1. *IEEE J. Sel. Top. Appl. Earth Obs. Remote Sens.*, 15, 1079–1088, 2021, doi: 10.1109/JSTARS.2021.3139376.
12. Zhang, Y., Lu, Q., Jin, Q., Meng, W., Yang, S., Huang, S., Han, Y., Hong, Z., Chen, Z., Liu, W., Global Sea Surface Height Measurement from CYGNSS Based on Machine Learning. *IEEE J. Sel. Top. Appl. Earth Obs. Remote Sens.*, 16, 841–852, 2022, doi: 10.1109/JSTARS.2022.3231916.
13. Cao, M., *et al.*, Multi-Scenario Simulation of Land Use for Sustainable Development Goals. *IEEE J. Sel. Top. Appl. Earth Obs. Remote Sens.*, 15, 2119–2127, 2022, doi: 10.1109/JSTARS.2022.3152904.
14. Yusuf, Y.Y., An assessment of land use land cover change detection in Misau local government area, Bauchi state, Nigeria, in: *Proceedings of the 5th International New York Academic Research Congress*, April 23-24, 2022.
15. Reith, J., Ghazaryan, G., Muthoni, F., Dubovyk, O., Assessment of Land Degradation in Semiarid Tanzania—Using Multiscale Remote Sensing Datasets to Support Sustainable Development Goal 15.3. *Remote Sens.*, 13, 1754, 2021, doi: 10.3390/rs13091754.
16. López-Vargas, A., Ledezma, A., Bott, J., Sanchis, A., IoT for Global Development to Achieve the United Nations Sustainable Development Goals: The New Scenario After the COVID-19 Pandemic. *IEEE Access*, 9, 124711–124726, 2021, doi: 10.1109/ACCESS.2021.3109338.
17. Alyami, S.H., Almutlaqa, A.A., Alqahtany, A.M., Ashraf, N., Likelihood of Reaching Zero Energy Building Design in Hot Dry Climate: Saudi Arabia. *IEEE Access*, 9, 167054–167066, 2021, doi: 10.1109/ACCESS.2021.3134365.
18. Parvin, K., *et al.*, Intelligent Controllers and Optimization Algorithms for Building Energy Management Towards Achieving Sustainable Development: Challenges and Prospects. *IEEE Access*, 9, 41577–41602, 2021, doi: 10.1109/ACCESS.2021.3065087.
19. Kumar, V., Panchal, V.K., Goyal, D., Singh, V., Land Features Extraction from Image by the Probability Concept of Honey Bees: A Remote Sensing Perspective. *Int. J. Syst. Algorithms Appl.*, 3, ICEEC13, June 2013.
20. O.S.N., M.K.M., Mudigere, D., Muley, D., Urban Satellite Image Classification using Biologically Inspired Techniques, IEEE, pp. 1767–1772, 2007.

21. Vali, A., Comai, S., Matteucci, M., Deep Learning for Land Use and Land Cover Classification based on Hyperspectral and Multispectral Earth Observation Data: A Review. *Remote Sens.*, 12, 2495, 2020, doi: 10.3390/rs12152495.
22. Digra, M., Dhir, R., Sharma, N., Land use land cover classification of remote sensing images based on the deep learning approaches: a statistical analysis and review. *Arabian J. Geosci.*, 15, 1003, 2022, doi: 10.1007/s12517-022-10246-8.

8

Policy Framework for Sustainable Land Use

Abstract

A policy framework is necessary to govern sustainable land use. Efficient policy formulation is crucial as it informs the control of land resources, melds conflicting interests, and enables conservation of biodiversity. Effective land use policies at the national and international level contribute impressive results for government institutions, communities, enterprises, and ecosystems. Through it, legislation teaches about sustainability, preserving practices and biodiversity preservation, as well as social and economic growth. Effective policy requires institutional infrastructures, land ownership, public participation, and law bodies for successful policy on land use. Sustainability-enabled options are one of the must-have aspects in land use management in the policy of land zoning. In this chapter, the methods of zoning, smart growth, mixed use, and ecology zoning are discussed, which help in maximising the use of land, minimising the hazards to the environment and promoting the development of strong neighbourhoods. Through the coordination of land use policies, socioeconomic development, policymakers in synergy ensure equality, resilience and adaptive system for long-term advantage to future generations.

Keywords: Environmental integrity, decision-making processes, policy framework, socioeconomic development, sustainable land use practices

8.1 Introduction

Land is the starting point for both ecological well-being and for growth of people, so its management needs to be done responsibly so it can be sustained. As the world experiences rapid growth in population, city development, high industrial activity and climate change, it is now more important than ever to use land in a sensible way. Due to these factors, challenges like expanding cities, deforestation, degradation of land, and exploitation of resources endanger the environment and human health. Because of this

Kundan Meshram and Chatrabhuj. A Sustainable Approach to Land Use, (163–192) © 2026 Scrivener Publishing LLC

crisis, countries around the world are agreeing to use land resources sustainably by balancing nature, fairness and the environment [1–3].

Taking good care of land requires bringing ecology, resources and the economy together since they all depend on the land. For this to happen, governments need detailed policies. Sustainable land use depends on policymaking which aid in supporting the alignment of environmental, social and economic goals. Properly formulated policies help governments and planners to set guidelines to conserve natural resources, prevent land from becoming damaged and promote responsible land management. Laws can be put in place to improve farming methods, handle expansion in cities, carry out zoning rules and ensure all have access to land [4]. Effective policymaking makes it easier to use sustainability in land use planning by engaging stakeholders, supporting new solutions and backing policies for protection of the environment. Good policy structures ensure that land development in a region aligns with climate protection, protection of biodiversity and meeting community needs.

Well-defined policies overcome challenges between different land users. They also support inclusive ways of governing that require various government levels, charity groups, businesses and local people to cooperate. They cooperate to carry out strategies that align activities on the land with actions to fight climate change and to protect nature [5].

Relying on evidence helps make land use policies more sustainable, because it guarantees planning and management choices are supported by accurate, objective and scientific information. Much of this approach depends on powerful data gathered from monitoring environment, careful spatial analysis, assessing communities and years of ecology research. Using scientific knowledge and real-life incidents, policymakers can notice trends, plan for impact on land use and develop approaches that are relevant and result-focused for the place. As a result, policymakers depend less on their own opinions or politics, making policies more open, reliable and effective [6]. Furthermore, using evidence in land use planning lets planners keep adjusting and reviewing policies as the environment, population and challenges evolve. It promotes the use of GIS, remote sensing, modeling tools and participatory research to provide useful details for suitability analysis, risk assessments and zoning choices. Because of this strong approach, it becomes possible to arrange policies that maintain care for the environment, evenly share resources and promote development [7, 8]. As a result of taking this approach, policies for sustainable land use both direct existing land practices and shape a future where the environment is healthy, everyone is included, and the economy is strong. They are flexible plans that respond to new observations and the opinions of stakeholders, encouraging creativity and continuing sustainability [9]. All things considered, using evidence to support policies is a necessary method to take on big challenges facing the earth's land, like degradation, less diversity and issues with development. By creating sustainable habitats, boosting communities and managing land products responsibly, it improves life for people throughout generations.

8.2 Why is Policy Framework Necessary?

It is important to develop a sustainable land use policy framework that helps us to handle unusual challenges like planning, goal setting, establishing important legal, government roles, and promoting collaboration and incentives for sustainable practices [10–13].

a. Establishing incentives and mechanisms: It sets up recommendations to guarantee people can manage their land for a long time by raising the idea of economic sustainability through market-based resources, tax benefits and grants from the government. It allows groups such as communities, developers and conservationists to support positive changes for the environment and improve damaged land. When these approaches are employed, stakeholders have a greater ability to sustain both the economy and the environment for a long time.
b. Guidance and direction: Many stakeholders in land use management, such as government agencies, planning organizations and institutions, benefit from the evaluation frameworks provided by policies. The frameworks explain important principles, priorities and aims to unify the effort towards sustainable growth. Adhering to this policy, government departments are able to make decisions that are in line with each other and actively promote similar interests in conservation, developing infrastructure, housing and agriculture for sustainable land development.
c. Fostering collaboration and coordination: The goal of sustainable land use requires all parties to work collectively. Active engagement with a variety of stakeholders consists of public authorities together with private sector stakeholders, civil society organizations and indigenous communities alongside local residents to complete coordinated action.

Policy frameworks build collaboration through partnerships between institutions and multi-sectoral platforms, as well as by establishing well-connected relationships. Through these mechanisms, stakeholders achieve better communication and can jointly decide and resolve challenges while all strategies gain wide acceptance because they follow principles of inclusion and fairness.

d. Facing intricate obstacles: An effective policy framework enables institutions to handle several complex problems through structured strategies. Policymakers use this framework to analyze multiple challenges so they can establish strategic plans by combining different interventions in a manner suitable for the connected relationship between environmental sustainability and economic growth, along with social equity.

e. Encourage long-term planning: What makes sustainable land use possible is forward- thinking and strategic planning that helps deal effectively with upcoming potential risks alongside new possibilities. Policy frameworks enable present and upcoming population needs to be considered through strategic assessment. The framework boosts community and institutional adaptive capacity by promoting extended strategic planning, which allows them to handle environmental as well as socioeconomic shocks, including climate change, resource shortage and population expansion.

f. Legal and regulatory structure: Policy frameworks provide the start for forming complete laws and rules that control how land is used. They set the principles that shape laws, regulations for zoning, land ownership, rights to property and how the law addresses environmental issues. Land use operations are performed in a way that supports national growth, is sustainable and supports the public. Policies allow abstract ideas to become laws by outlining the key principles, ethical codes and usual procedures. Also, they require evidence of land use and development to be kept, which allows for acceptance or rejection by guidelines and lawsuits, increases transparency and backs the rule of law.

g. Juggling divergent interests and priorities: Land use management usually means managing different and sometimes clashing interests such as more development, protecting the environment, creating social fairness, encouraging urbanization and farming. Growth in industry and building more roads and other structures can endanger the environment and age-old land usage habits. A strong policy system helps describe ways inclusive processes can be used to manage such conflicts. It has processes for consulting, involving stakeholders and assessing impact, so that multiple groups like local communities, developers, environmentalists and government are included. Using structured negotiation and including everyone in decisions, the framework works to reconcile different priorities and leads to fair, balanced and both short- and long-term environmentally sustainable land use.

8.3 Beneficiary

The framework of policies designed for sustainable land use brings advantages to numerous actors, comprising government bodies together with academic researchers, private firms and civil society entities. Through these frameworks, both current and upcoming populations receive protection for their environment, together with financial stability along with social acceptance and enhanced ecological strength [14, 15]. Different population segments derive advantages from these frameworks, which are analyzed below.

a. Government agencies: The main group that gains advantages from sustainable land use policy frameworks consists of government agencies. The institutions that perform urban planning alongside environmental regulation, together with agricultural development and natural resource management, need specific, transparent guidelines to achieve operational effectiveness. The policy frameworks supply authorities with necessary legal instruments that enable them to implement regulations, develop zoning plans and coordinate actions across multiple governmental entities at all jurisdictional levels throughout different geographic regions. Through frameworks, governments achieve productive administrative operations and diminish the ambiguity, which leads to better governance efficiency.
b. Local communities: Communities located near the land stand as essential stakeholders for planning how developers

use land since they obtain their means of living from these natural resources while using them to preserve their culture and personal identity. Current policy guides communities to actively contribute to decision-making and protect their rights when involved in planning activities. Governing frameworks support community involvement through meetings, mapping exercises and local governance practices to enable citizens in their environmental development. Platformed land-use participation leads to better community involvement, thus forming coalitions and greater sustainable planning outcomes. The local population gains better access to their land and resources as well as public services including water systems, fundamental sanitation facilities, transportation infrastructure and farming assistance.

c. Private sector: Policy frameworks deliver advantages to the private sector businesses, developers, investors and entrepreneurs because they create clear guidelines and predictable circumstances. The established guidelines and incentives accompanying these frameworks reduce investment uncertainty and drive the industry to plan ahead in the land-based sectors. Businesses receive higher odds to pursue green practices when they receive established support frameworks that push technological adoption and sustainable development investment. Framework systems establish public-private partnership tools that equip businesses with collaborative abilities to access resources and generate new ideas regarding land management. Credit policy frameworks establish paths for both efficient business profitability and ethically sound operations, which build economic development without harming nature or social justice frameworks.

d. Non-Governmental Organizations (NGOs): The establishment of policy frameworks enables NGOs to participate in public meetings while accessing transparent, accountable processes that manage land governance. Through policy frameworks, NGOs acquire the ability to track land use policies while participating in dialogues about policy decisions and making an impact on decision-making procedures. The framework empowers NGOs to defend marginalized communities while revealing bad practices to push for reforms, either by promoting environmental sustainability, human rights, gender equality or rural progress. The policy frameworks provide organizations with essential starting points to work together with public and private entities in coordinated restorations of ecosystems, development programs and climate adaptation endeavours. The organization's ability to function as a watchdog organization, facilitator and implementer of sustainable land use programs increases.

e. Benefit academic and research institutions: Policy frameworks support the generation of scientific evidence, facilitate interdisciplinary research, and inform the development of policies based on that evidence. Policy frameworks facilitate collaboration among academia, government, and civil society to tackle intricate land use challenges, propel knowledge progression, and devise inventive resolutions to sustainability concerns.
f. International organizations: Policy frameworks benefit international organizations such as the World Bank, United Nations, and regional development institutions by establishing a unified framework for harmonizing and coordinating land use practices and policies across regions and countries. In pursuit of attaining worldwide sustainability objectives, such as the Sustainable Development Goals (SDGs) and the Paris Agreement on climate change, policy frameworks serve to bolster international cooperation, capacity-building, and knowledge-sharing endeavours.

Policy frameworks that promote sustainable land use guarantee the prudent management of land resources and ecosystems, thereby promoting responsible stewardship. Policy frameworks safeguard the natural capital and ecosystem services on which the prosperity and well-being of future generations depend through the promotion of sustainable development, conservation, and long-term planning.

8.4 National and International Land Use Policies

National and international land use policies create fundamental conditions that determine how land resources are managed and conserved while developing new areas for infrastructure. These policies determine decision-making through sustainable practices by establishing bases for maintaining environmental, social and economic balance. A study into the implications of the policy of land use in various geographical scales, ranging from the neighbourhood to global levels, requires a complete understanding of the national and international policy alliance [16–19]. This section provides primary elements that should be addressed in the process of policy evaluation.

a. Scope and objectives: A country formulates land use policies within their territorial boundaries based on guidelines that express its unique socioeconomic features, coupled with cultural issues and needs for the environment. The formulated policies aim to achieve balance in terms of domestic requirement formulations, for instance urban expansion and enhancement of infrastructure, agriculture productivity and preservation of resources and land distribution equilibrium. In the execution of the national policy, the government passes the zoning restrictions to regulate urban expansion as well as industrial development in agricultural regions. Such policies focus on the control of climate change, conservation of biodiversity, control of desolation and governance of land rights across various countries. The United Nations Convention to Combat Desertification (UNCCD) and the United Nations framework convention on climate change REDD+ program establish international accords that tie national practices to sustainability agreements and develop mutual nation-to-nation operational frameworks.
b. Legal framework: The legal instruments for national land use are the domestic legal instruments, including land tenure laws, which apply with zoning law, environmental protection acts and urban planning ordinances. Such legal structures make it possible for land use plans to be resorted to and formalized through governmental corrections at each power level. International land use policies draw their basis from treaties, protocols, conventions and multilateral agreements to ratify ways and possibilities for managing common environmental resources. The Convention on Biological Diversity (CBD) and the Paris Agreement are international guidelines that would guide countries while formulating suitable ways of using land to address global environmental issues. The international legal instruments provide standards and duties, which encourage cooperation and impose obligations on the member states that sign the agreements.

c. Stakeholder engagement: All efficient land use policies require taking the initiative of engaging stakeholders in land use policymaking at any level, whether local or international. National governments employ different forms of mechanisms, such as public hearings combined with community consultations and stakeholder workshops to enable affected land users to air their views with regard to planned changes. Local residents, together with indigenous communities, join landowners, businesses and non-governmental organizations in active participation. The international stakeholder engagement process depends on dialogue with multiple countries together with international agencies (including FAO, UNEP, and UN-Habitat) in addition to non-state actors, academic institutions and global civil society organizations. The platforms support building consensus while sharing knowledge and developing essential partnerships, which are essential for creating transnational strategies of sustainable land use. Different perspectives from stakeholders create policies that maintain fairness toward all people and are adapted for their specific environment as well as the wider world issues they need to resolve.
d. Monitoring and evaluation: Nation states typically create departments that collect information about land use changes, policy enforcement, environmental indicators and socioeconomic effect data. Monitoring and Evaluation (M&E) systems enable both continued policy effectiveness assessment and detection of weaknesses that need policy modifications. International M&E operates with sophisticated reporting through country reports, continuing evaluations and standard global performance assessments. The Intergovernmental Panel on Climate Change (IPCC), together with the United Nations Development Programme (UNDP), perform global evaluations to assess country achievements regarding SDG land-related objectives. These evaluations serve two essential functions by providing equal visibility of country performance and enforcing mutual responsibility among different nations.
e. Adaptation to evolving circumstances: National as well as global land policies require flexibility since the pressures on land use do not exhibit constant conditions. Combination of climate change with the growth of city populations,

migration trends, and technological progress shifts the basic requirements of land use and predominant targets. National policies call for periodic examinations for updating as the conditions are changing, to the effect that new measures are required and new best practices are incorporated. The rationale for policy changes surfaces when governments want to incorporate smart city development, as well as climate-resilient agriculture along with nature-based solutions, in their frameworks. The implementation of flexibility within policy frameworks enables both proactive and responsive actions to take place instead of requiring changes in response to existing conditions.

f. Capacity building and implementation support: Land use policy success directly stems from institutions, together with individuals who can properly implement their procedures. For the operationalization of land use policies, national governments typically need technical support, together with experienced staff and institutional backing, along with financial instruments. The success of land use operations depends on providing planners and policymakers with training sessions, investing in geospatial technologies and establishing communities responsible for local land management. International development organizations provide support through training opportunities, together with monetary backing and technological resources, to many nations in the developing world. Various global groups combine forces with donor agencies to back local institutions and national institutions and support policy harmonization and best practice promotion.

8.5 Role of Policies in Guiding Sustainable Land Use

The formation of policies stands as the key element that determines the outcome of sustainable land management, its equal distribution and resilience. Through regulation policies, an environment is established which harmonizes development initiatives with conservation measures in the allocation, development and preservation of land [4, 8]. Government policies create the enabling framework that sustains equal and adaptable land management practices that fulfil the needs of present and future people.

The next section outlines how different policies influence sustainable land utilization.

a. Establishing goals and objectives: Clear strategic goals emerge from policies to support sustainable land use planning, which follows national environmental policies and Sustainable Development Goals (SDGs). Policies establish specific targets to direct how land gets utilized, so areas with environmentally stable potential receive priority along with lowered economic disparity and responsible construction practices. The national policy seeks to regulate urban expansion by endorsing dense city development and protecting ecological areas by dedicating 30% of the national land.
b. Implementation of regulatory frameworks: The identification of effective land governance requires the creation of complete legal and regulatory frameworks based on policies that are well designed. The set legal boundaries guide all choices relating to land usage. Policies that enforce land use categories through law create organized sustainable land use patterns through their enforcement of residential, industrial, agricultural and conservation zones.
c. Advancement of integrated planning methodologies: Governments support planning strategies that examine how environmental components unite with economic factors, together with social elements, when deciding land use applications. Policies follow a comprehensive framework that prevents planning activities from operating in separate compartments because they recognize systemic interrelations. Organizational decisions regarding urban development require integration with water resource management practices, along with biodiversity protection and rural sustainability objectives.
d. Promoting sustainable practices: Sustainable land management adoption is induced by policies that introduce various forms of incentives and disincentives for positive changes in behaviour. Financial programs promote sustainable land management through supporting organic agriculture with subsidies, including eco-development taxes, ecosystem service payments, and forest conservation grants, among many others. Environmentally responsible actions are rewarded

in policies which incentivize land users to take on soil-preservation and biodiversity-enhancement and carbon-reduction practices.

e. Enabling stakeholder engagement: Stakeholder-engagement policies make planning of land uses more collective while achieving complete involvement from multiple groups of stakeholders. With public hearings alongside community consultations, in addition to participatory mapping exercises, policies create forums for civil society bodies, indigenous groups, local parties and private parties to add their voices to decision processes. People's active involvement in land governance strengthens it by creating trust amongst all groups, reducing conflicts at the same time as generating land use patterns that cater to cultural, economic and social interests of the villagers.
f. Capacity development and knowledge sharing: Effective land use management involves policies that consequently foster the development of institutional enablers and human professional development. These include policy support that requires stakeholders acquire the technical skills with tools and instructional materials that offer them the ability to put into practice sustainable land management. In extension and knowledge-sharing projects, it also means research creation, which ensures that services are active and keep coming up with new ideas.
g. Monitoring and evaluation: Metrics include embedded Monitoring and Evaluation (M&E) systems that capture the achievement, compliance and consequences of the land use plans over time. The combination of spatial data collection with socioeconomic metrics and land cover analysis through performance indicators makes M&E systems confirm policy effectiveness. The management process enables both the discovery of implementation flaws and emerging concerns, which lets adaptive adjustments take place.
h. Promoting innovation and adaptation: Policies establish environments of experimentation to perform research and implement experimental projects, which drive innovation and sustainable land use adaptations. The policy should allocate assets for research activities while enabling partnerships

between organizations to test new solutions to land management matters and create communication channels to enable stakeholders to exchange knowledge in identifying innovative practices.

8.6 Key Elements of the Policy Framework for Land Use

Sustainable, equitable and effective land management depends on a strong interlinked set of policy elements which form the backbone of land use governance. The components work together to direct land decisions and protect human and natural systems, which establishes present development practices toward extended community objectives. The essential components of such a policy framework are described below.

a. Goals and objectives: Every land use policy framework requires clear, established goals and objectives to act as its foundation. These strategic viewpoints maintain alignment with principles of sustainable development combined with environmental sustainability, social equity and economic progress. The framework contains defined objectives which consist of maintaining forested areas alongside protective measures for agricultural production and limitation of city expansion. The set of objectives serves to convert the vision into concrete targets that include land protection percentage growth and housing affordability enhancement. The framework creates a shared national direction for decision-makers, planners and stakeholders through explicit priority declaration which supports uniform land use planning initiative coordination.
b. Legal and regulatory framework: The framework consists of the laws and regulations, as well as statutory bodies, that lay down the guidelines for the conservation and management of land utilization. In addition, a robust corporate structure helps in compliance drive, minimises disputes between land owners, and has the organization and necessary muscle to deal with unlawful invasion for development. People require security in their land rights, and law offers this foundation as

a way of attracting investments and putting order in the use of the land.

c. Land use planning process: For land use planning to be effective, the process must be wide-ranging and accessible to all; it must help people to make informed and democratic choices. It works by dividing the process into specific steps, starting with obtaining necessary data about environmental, social and economic issues. Analyzing this data helps us understand how land is being used now, highlight problems and predict future needs. It is very important to consult the public at the planning stage to make sure that community views are available before scenarios for land use are crafted. Land management policy frameworks should clearly spell out how people like local residents, indigenous people, businesses and non-governmental organizations will become involved in setting up and changing land use policies and plans. Adding these participatory parts to the planning process means a wider range of knowledge systems, cultural values and priorities are shared. Taking everyone's needs into account helps make the strategies better and more widely accepted as reasonable, fair and appropriate by local communities. Because of this, the planning of land use is used to arrange space, encourage teamwork and ensure sustainable results over time.

d. Spatial planning and zoning: Spatial planning and zoning act as two crucial factors in the practice. They help in planning of land, subdividing the land and development activities. They also prevent the systematic development of land conflicts. New regulations categorize regions into residential regions, business areas, and industrial regions, together with agricultural and preservation zones. They set requirements about the allowed building heights, densities, setbacks and land coverage.

e. Environmental protection and conservation: Environmental protection and conservation are required in all land use planning and policy creation. For the purpose of maintaining the environment, one should prioritize and protect places such as wetlands, forests, biodiversity hotspots and water reservoirs. Major actions include setting up legally mandated protected areas, carrying out habitat conservation plans and applying environmental regulations like

Environmental Impact Assessments (EIA) and Agricultural Impact Mitigation (AIM). Moreover, use of nature-focused measures—strategies such as green infrastructure, restoring nature and responsible land management—helps tackle environmental concerns in a budget-friendly way and strengthens the world's natural resistance. Laws related to land use have to be crafted in a way that supports sustainable use of natural resources, bolsters pollution control and focuses on soil conservation. All together, these steps take care that land use practices do not harm the environment but play a positive role in maintaining ecosystems and the services they supply for a long time. When we include environmental care in land use laws, development is kept under control, and nature is safe for both future and present generations.

f. Social equity and inclusive development: A successful land use policy framework needs an equalized approach that meets the needs of all demographics, particularly the disadvantaged communities. The policy framework must ensure equal access to ground resources, housing, public facilities and service infrastructure. The framework should also specify actions for affordable housing aims, including reservation of the land and protection of both informal communities/ native areas, as well as formal recognition of community ownership claims. The procedure of inclusive development requires participatory contributing procedures that allow marginalised populations to contribute to decision-making regarding land use. Social equity inclusion in land use policies ensures that developments contribute towards lowering poverty and ensuring social justice and human well-being.
g. Economic development: The relation of sustainable economic development relies to a great extent on the land use policies that facilitate the correct linking of the market investments, land resources and economic activity sectors. Dominant economic diversification comes about when the authorities have put in place policies that give special functions to industrial parks and agricultural hubs, combined with business districts and eco-tourism zones. Measures on the establishment of infrastructure with tax benefits and easy property access encourage innovation by promoting entrepreneurship.

h. Planning and development of infrastructure: Land development and economic operations require infrastructure, which serves as their fundamental enabling factor. Land use policy requires complete alignment between infrastructure development planning and capital investments, which include road, energy grid, water supply system, telecommunications network and sanitation service infrastructure. These systems affect settlement directions and establish the feasibility criteria for development operations. When land use policy includes infrastructure planning, there is proper resource management that prevents infrastructure deficits and decreases environmental impacts of development projects. Through strategic planning, it becomes possible to anticipate urban expansion zones and direct urban development into controlled, sustainable areas.
i. Climate change adaptation and mitigation: The rising threats to land systems from climate change require built-in adaptive and mitigation strategies to be included in land use planning policies. Residential community planning includes making districts resistant to flooding, extreme heat and sea-level rise impacts while carbon sink conservation joins efforts with energy-saving planning and low-carbon transport programs. National climate targets receive support through zoning codes, building regulations and green infrastructure projects when climate considerations become part of their development process. The application of climate-sensitive policies both strengthens community resistance to environmental hazards and limits their exposure to risks from environmental variations.
j. Monitoring, evaluation, and adaptive management: A policy framework can retain its effectiveness by using highly structured procedures for checking, adaptation and evaluation systems. The methods that have been put in place help to track the level of achievement as one way of looking at performance against initial targets and at the same time allowing for immediate policy changes. These systems are necessary for reviewing work, making suggestions for betterment and keeping policy goals up to date as circumstances shift. As part of monitoring, data is systematically gathered to check if the business is reaching its targets and goals. Some activities used include taking satellite pictures from space, visiting

local areas to check what is happening and evaluating the main environmental, social and economic factors. It gives up-to-the-minute information on the use of land, problems in the environment or social impacts, so people know if something is changing. Evaluation works together with monitoring by thoroughly assessing how implemented policies have worked. Reviewing both the things that went right and wrong allows people to find out why they happened and make better plans for what is next. Checking on achievements provides clarity over outcomes and charges officials to be accountable for what happens. Adaptive management looks ahead and boosts how flexible and responsive policies are.

8.7 Zoning

Zoning also serves as a public sector tool for planning land management in use to regulate the way land develops within its limits. As a result of segregating geographic areas, divisions are made into pieces that receive particular regulatory regulations regarding the requirements for construction and restrictions on building height along with population density rules and building separation restrictions.

Purpose of zoning

a. Orderly development: Zoning serves as an important instrument for introducing systematic urban development *via* systematic land management, placing the urban areas into residential, commercial, industrial, agricultural, recreational functions and so on. By doing so, it prevents incompatible land uses from sharing proximity, such as heavy industrial facilities being near residential houses. By implementing the zoning idea, incompatible land usage is severed to reduce the hazards brought about by noise pollution, air contaminate, traffic congestions and also security dangers. Industrial facilities producing noise pollution and emissions should be located as far from designated spaces for schools, residential districts and medical facilities as possible. With the help of realistic land use planning strategies, zoning gives permission to planners to draft essential infrastructure of the future, which includes water supply, roadways and waste disposal systems. Zoning gives rise to an efficient society with an effective living environment.

b. Safeguarding property values: Zoning regulations significantly maintain property values by means of neighborhood maintenance in terms of quality while preserving the character, aesthetic of land and residential areas. The rules set within residential areas prevent the developers from constructing yet larger commercial buildings or tall constructions that may overshadow houses and create issues with the infrastructure. These controls' implementation protects community visual unity, operational integrity and supports the functioning of a stable and well-regulated property market. By controlling undesirable changes in residential areas close to homes, the risk of property values falling is decreased and real estate market health is maintained and improved. Having zoning rules protects the property values and appeal of houses in both the short and long runs, supporting even development and confidence among buyers.

c. Public health and safety: Zoning helps protect communities by controlling how land is used, which in turn lowers risks and supports the well-being of the public. It puts industrial and residential areas apart to reduce the risk of people facing pollutants, loud noises and traffic dangers. By imposing zoning, development is limited in hazardous areas such as floodplains and zones where landslides happen. It promotes safety in buildings by requiring fire protection and establishing minimum distance for construction projects, and it also requires drainage and waste disposal systems to ensure proper hygiene. In addition, zoning creates opportunities to enjoy green spaces, manages how populated areas are and keeps services like public transport or schools running well, which enhances safety and comfort in living areas.

d. Resource management: Land zoning becomes crucial to sustainable resource management as it provides partitions in which there are areas reserved for conservation purposes, and areas that are under prescribed development. With environmental and conservation zoning rules, these regions with key natural resources are protected from development that is not environmentally friendly. Forests are often included in protective zones to stop deforestation and help the environment by storing carbon and limiting soil erosion. Wetlands are also preserved since they purify water and control flooding. Recreational zoning makes it possible for the public to visit natural areas while still protecting their environment and all living creatures inside them.

e. Economic development promotion: Zoning creates a forward-thinking approach to stimulate economic growth by the creation of zones, which ensure there is business investment and opportunities for entrepreneurship. Commercial and industrial areas that also include mixed-use properties lead to the development of business districts, innovation centers and employment centers for zones. Offering zoning benefits such as extra density and tax relief helps convince developers to spend their money in areas that need extra care. Incentives encourage redevelopment, making it profitable and more attractive for use where space is either not fully utilized or important for the city. Often, zoning laws also have flexible rules that support the introduction of businesses in the green energy, logistics, agri-business and tourism sectors. Because of this, the economy can grow sustainably and fit well into broader plans and environmental concerns.
f. Character and identity of the community: Zoning policies serve an important function of organizing the development of the urban environment and influencing the appearance and organization of buildings in districts. These regulations are therefore an expression of a given community's socio-cultural values, present needs, and aspirations towards the prospects of urbanization or a countryside setting. Zoning codes are a means of adopting certain rules and regulations that include the heights of buildings, facade design, architectural design, and the management of sites. In historic towns, zoning usually concentrates on protecting building and architectural image through the establishment of historical areas while in suburban areas, densities are kept low and emphasis is laid on the protection of greens.

The public sector employs zoning as a planning tool for land management purposes to supervise the way land develops within its boundaries. Proper zones emerge through dividing geographic spaces into segments, each of which receives special regulatory measures that control building

Table 8.1 Zoning function.

Zoning function	Subcategory	Description
Regulation of Development Intensity	Building Height, Density, Setbacks	Controls maximum building heights, lot sizes, and setbacks to manage population density, ensure adequate spacing, and preserve community scale and open spaces.
Geographic Area Segmentation	Functional Zoning, Special Zones	Divides a municipality into zones like residential, commercial, industrial, agricultural, and conservation based on infrastructure, demographics, and community needs.
Land Use Regulation	Use Restrictions, Permitted Uses	Assigns specific uses to each zone—e.g., housing in residential zones, manufacturing in industrial zones—ensuring order, compatibility, and appropriate infrastructure allocation.
Urban Planning Integration	Smart Growth, Compact Development	Guides cities towards sustainable growth patterns such as compact neighborhoods, infill development, and reduced urban sprawl to optimize infrastructure and land use.

(*Continued*)

Table 8.1 Zoning function. (*Continued*)

Zoning function	Subcategory	Description
Compatibility Management	Buffer Zones, Transitional Areas	Introduces transitional zones or buffers to separate conflicting land uses like residential and industrial, reducing disturbances and promoting harmony.
Health, Safety, and Environmental Protection	Hazard Mitigation, Environmental Buffers	Protects public health by enforcing setbacks from floodplains, pollution sources, and fire-prone areas; mandates green spaces and open areas for safety and ecological balance.
Affordable and Inclusive Housing	Inclusionary Zoning, Density Bonuses	Requires or incentivizes developers to include affordable housing in projects or contribute to housing funds, ensuring diverse and equitable communities.
Flexibility and Adaptability in Zoning	Overlay Districts, Conditional Use Permits	Introduces flexible mechanisms to adapt zoning laws to changing needs *via* tools like mixed-use overlays, special permits, and planned unit developments (PUDs).

(*Continued*)

Table 8.1 Zoning function. (*Continued*)

Zoning function	Subcategory	Description
Natural and Cultural Resource Conservation	Historic Districts, Conservation Easements	Protects natural landscapes and cultural heritage through special zoning for historic sites, ecological reserves, scenic areas, and environmentally sensitive lands.
Transit-Oriented Development (TOD)	High-Density Nodes, Walkable Communities	Promotes mixed-use, high-density development near transit stations to reduce reliance on private vehicles, support public transit, and increase accessibility.
Integration with Broader Planning Instruments	Comprehensive Plans, Infrastructure Investment	Aligns zoning with long-term municipal plans, transportation networks, and public service provision to support coherent and strategic urban growth.
Community Participation and Governance	Public Hearings, Community Forums	Encourages civic engagement in zoning decisions through open meetings, workshops, and digital platforms, promoting transparency and accountability.

(*Continued*)

Table 8.1 Zoning function. (*Continued*)

Zoning function	Subcategory	Description
Sustainable Economic Development	Commercial Zoning, Industrial Clusters	Supports business growth by designating commercial/industrial zones, encouraging innovation districts and clusters to stimulate local economies.
Environmental Sustainability Regulations	Green Infrastructure, Low-Impact Development	Mandates or encourages energy-efficient buildings, green roofs, permeable surfaces, and rainwater harvesting in zoning ordinances to combat climate change impacts.
Rural and Agro-Residential Zoning	Urban-Rural Interface, Agricultural Preservation	Allows compatible residential use in agricultural zones, promotes community farming, and protects farmland from urban encroachment through agro-residential policies.

requirements, building height restrictions, population density guidelines and building separation rules. Local property standards receive protection by zoning laws while land uses are kept compatible; development becomes orderly and community life improves for residents, as depicted in Table 8.1 above.

The zoning system splits regions into various districts through community needs and infrastructure evaluations; it implements restrictive boundaries to specify what type of development is allowed in residential, commercial and industrial areas. The establishment of buffer zones through zoning creates compatible relationships among land uses while resolving potential problems between them. Through zoning, public health and safety are ensured through environmental hazard control. This provides space for alterations in community demands and allows the delivery of affordable housing. Zones play multiple roles in resource conservation and transit-oriented projects, and they work effectively with planning frameworks to create full land management goals. The zoning system includes participation from public members so they can shape decisions about their neighborhoods.

8.8 Zoning Approaches That Promote Sustainability

To promote effective land use patterns, various dimensions of sustainability must be considered. It therefore becomes necessary to explore all options, including zoning, to satisfy the need for sustainability techniques. Multiple innovative zoning methods exist for organization of land planning, the purpose of which is to link planning practices with sustainable development goals. These different types of zoning are discussed below.

a. Form-based zoning: The planning technique of form-based zoning puts forward physical structure and community space planning as main focuses instead of using the standard zoning categories. The method favors visual and functional connections between buildings and streets and public spaces over rigid functional area separation (e.g., residential, commercial, industrial). The fundamental objective of this zoning system focuses on developing communities which combine various activities within walking distance of each other. The core components are street-level interaction with architecture, public spaces and design consistency that helps form-based zoning to create sustainable communities that engage social connectivity. The strategy employs pedestrian-friendly environments and combined property usage while implementing green systems and energy-saving building requirements for constructing dense, sustainable cities.
b. Transit-Oriented Development (TOD) zoning: The purpose of TOD zoning is to pack densely built residential and mixed-use facilities within attainable walking distance from railway stations, subway stations and primary bus routes. The main objective of TOD zoning is to make transportation systems more accessible, thus reducing automobile usage and supporting environmentally friendly transportation modes.

TOD areas require transportation-oriented residential areas and businesses located near each other, utilizing pedestrian-friendly systems with integrated road systems and minimal constraints on parking and multiple property types. The blueprint comprises features that assist in reducing pollution and traffic jams and improving health outcomes. The adoption of TOD enhances economic performance through the growth of the walker traffic base among local businesses and environmental responsibility in maximization of property usage in high-density urban areas.

c. Inclusive zoning: Inclusive zoning means developers are asked or rewarded for including some affordable housing elements in their projects or for joining affordable housing efforts. It is an important method for dealing with the increased problem of affordable housing, mainly in places where urbanization is rapidly taking place. Mandating low-cost units in market-rate projects encourages different income groups to live together and thwarts the tight grouping of people living in poverty. Integrating these areas helps people from different socioeconomic classes enjoy easy access to important services, education and jobs, which helps reduce social inequality and displacement. Inclusive zoning helps to grow communities where social diversity is valued and where both economic diversity and equal rights are present.

d. Green zoning: Under this strategy, one observes some regulations which could require or recommend eco-friendly building materials that involve renewable energy systems, energy-efficient technology, green roofs and stormwater management systems. Development of urban areas is encouraged by the approval of measures that interlink natural features using buffer protection measures, as well as native vegetation and boosted tree covers. Green zoning is a tool that enables cities to adapt to the changes in the climate through its carbon sequestration properties and its ability to reduce urban heat islands, along with its contribution to the enhancement of water quality and air quality level. Through the sustainable practice approach, green zoning realises its goal of ecological sustainability.

e. Conservation zoning: Conservation zones are designated to prohibit damaging activities to ecosystems and the natural state of landscape. Zoning for conservation is absolutely imperative because it helps in protecting wildlife areas, water bodies systems and reservoirs while providing protection to threatened species. Heritage protection is upheld by conservation zoning due to the fact that the approach defends cultural sites, which protect archaeological and historical sites.

f. Performance-based zoning: In this framework, developers have various options that help them realise zoning targets; this leads to better design innovations towards sustainable development. Such a zoning system requires developers to comply with energy-related objectives using renewable resources and construct stormwater management practices at their sites. Through performance-based zoning, the developers work towards public sustainability objectives through goal achievement rather than adherence to traditional compliance measures, making this a powerful tool for balancing public goals for sustainability and private development.

g. Overlay zoning districts: This flexible zoning mechanism comes in handy to planners when creating specialized zone regulations that do not affect the base zoning fundamentals. An overlay district exists to promote greening building practices, preserve scenic views, protect water-affected sites, and facilitate architectural heritage zones. The overlay zoning-based plans enable the strategic sustainability and intervention measures with the implementation of context-based policies on the basis of zoning categories. The tool has two main goals: to manage the bounded process that offers the developing regions peace of mind to manage that urban expansion and to provide special social, environmental and cultural programs while ensuring the overall organizational development frameworks.

h. Agro-residential zoning: Agro-residential zoning permits the combination of residential dwellings with farming areas, which serves to benefit local food supply networks and ensures farming sustainability in peri-urban and rural terrain. The zoning scheme permits people to actively participate in farming operations and garden maintenance as well as agro-tourism activities from a rural residential base. The zoning structure stops agricultural land from succumbing to urban development through residential area protection of farming activities. Community sustenance through agro-residential zoning develops through the combination of local food production systems, shorter transportation distances and rural area upholding. Agro-residential zoning strengthens community ties to their land base, which in turn leads to better practices of land stewardship and sustainable management strategies.

8.9 Conclusion

Policies are essential for the development of sustainable land use patterns, along with other social, environmental and economic parameters of the region. Policymaking includes regulations and laws and also incorporates land use strategies in a balanced approach. With careful planning, policymakers can formulate regulations on how land resources can be used effectively, decide on environmental policy, and distribute resources. Integrated responsibilities to fuse factual information, public feedback and economic projections when constructing policies allow sustainable land use to thrive. The policymakers maintain a mediator role between different sectors like agriculture, industry, housing, conservation and infrastructure to guarantee equitable land allocation, which prevents one group from receiving excessive benefits. Through their approach of adaptive management, land use policies acquire flexibility to react properly to developing issues that include urban expansion and climate change along with population increases. Policymakers who are effective establish connections between various sectors to align their plans, such as land use with transportation and other policies that lead to beneficial results.

References

1. Turetta, D., Paula, A., Kuyper, T., Malheiros, T.F., da Costa Coutinho, H.L., A framework proposal for sustainability assessment of sugarcane in Brazil (vol. 68, p. 597, 2017). *Land Use Policy*, 72, 578–585, 2018, https://doi.org/10.1016/j.landusepol.2017.11.044.
2. Carrilho, J., Dgedge, G., dos Santos, P.M.P., Trindade, J., Sustainable land use: Policy implications of systematic land regularization in Mozambique. *Land Use Policy*, 138, 107046, 2024, https://doi.org/10.1016/j.landusepol.2023.107046.
3. African Union, Land policy in Africa: A framework to strengthen land rights, enhance productivity and secure livelihoods. Addis Ababa: African Union and Economic Commission for Africa, 2009.

4. Bizoza, A.R. and Opio-Omoding, J., Assessing the impacts of land tenure regularization: Evidence from Rwanda and Ethiopia. *Land Use Policy*, 100, 104904, 2021, https://doi.org/10.1016/j.landusepol.2020.104904.
5. Ali, D.A., Deininger, K., Mahofa, G., Nyakulama, R., Sustaining land registration benefits by addressing the challenges of reversion to informality in Rwanda. *Land Use Policy*, 110, 104317, 2021, https://doi.org/10.1016/j.landusepol.2019.104317.
6. Carrilho, J. and Trindade, J., Sustainability in peri-urban informal settlements: A review. *Sustainability*, 14, 13, 7591, 2022, https://doi.org/10.3390/su14137591.
7. Shkaruba, A., Skryhan, H., Likhacheva, O., Kireyeu, V., Katona, A., Shyrokostup, S., Sepp, K., Environmental drivers and sustainable transition of dachas in Eastern Europe: An analytical overview. *Land Use Policy*, 100, 104887, 2021, https://doi.org/10.1016/j.landusepol.2020.104887.
8. Childers, D.L., Pickett, S.T.A., Grove, J.M., Ogden, L., Whitmer, A., Advancing urban sustainability theory and action: Challenges and opportunities. *Landscape Urban Plann.*, 125, 320–328, 2014, https://doi.org/10.1016/j.landurbplan.2014.01.022.
9. Chirisa, I., Peri-urban dynamics and regional planning in Africa: Implications for building healthy cities. 2010.
10. Chigbu, U.E., Schopf, A., de Vries, W.T., Masum, F., Mabikke, S., Antonio, D., Espinoza, J., Combining land-use planning and tenure security: A tenure responsive land-use planning approach for developing countries. *J. Environ. Plan. Manag.*, 60, 9, 1622–1639, 2017, https://doi.org/10.1080/09640568.2016.1245655.
11. Chigbu, U.E. and Kalashyan, V., Land-use planning and public administration in Bavaria, Germany: Towards a public administration approach to land-use planning. *Geomat. Landmanag. Landsc.*, 2015. 10.15576/GLL/2015.1.
12. Bao, H.X.H., Jiang, Y., Wang, Z., Feng, L., Social capital and the effectiveness of land use policies: Evidence from rural China. *Land Use Policy*, 139, 107069, 2024, https://doi.org/10.1016/j.landusepol.2024.107069.
13. Bao, H.X.H. and Robinson, G.M., Behavioural land use policy studies: Past, present, and future. *Land Use Policy*, 115, 106013, 2022, https://doi.org/10.1016/j.landusepol.2022.106013.
14. Feng, L., Zhang, M., Li, Y., Jiang, Y., Satisfaction principle or efficiency principle? Decision-making behavior of peasant households in China's rural land market. *Land Use Policy*, 99, 104943, 2020, https://doi.org/10.1016/j.landusepol.2020.104943.
15. Feng, L., Bao, H.X.H., Jiang, Y., Land reallocation reform in rural China: A behavioral economics perspective. *Land Use Policy*, 41, 246–259, 2014, https://doi.org/10.1016/j.landusepol.2014.05.006.
16. Tschopp, M., Ceddia, M.G., Inguaggiato, C., Bardsley, N.O., Hernández, H., Understanding the adoption of sustainable silvopastoral practices in

Northern Argentina: What is the role of land tenure? *Land Use Policy*, 99, 105092, 2020, https://doi.org/10.1016/j.landusepol.2020.105092.
17. Bayoumi, N., Tsurusaki, N., Xu, G., Managing the urban-rural transition: A review of approaches and policies for peri-urban land use. *J. Urban Manag.*, 2025, https://doi.org/10.1016/j.jum.2025.04.006.
18. Chung, Y.F., Maron, M., Drielsma, M., Rhodes, J.R., Do private land conservation policies and programs adequately consider climate change? *Biol. Conserv.*, Art. no. 111187, 2025, https://doi.org/10.1016/j.biocon.2025.111187.
19. Ramírez-Mejía, D., Zinngrebe, Y., Ellis, E.C., Verburg, P.H., Land-use spillovers from environmental policy interventions. *Global Environ. Change*, 92, Art. no. 103013, 2025, https://doi.org/10.1016/j.gloenvcha.2025.103013.

9
Challenges in Implementing Sustainable Land Use Patterns

Abstract
Sustainable patterns of land use give rise to complicated and interdependent challenges related to land use. This chapter goes into the complicated spiderweb of problems that impact the long-term viability of land ecosystems and quality of life globally. These challenges include environmental degradation, economic pressures, challenges of governance and challenges related to culture. The chapter addresses these problems by referring to the concepts of complexity theory and by insisting on the global interconnectivity of land use practices. Apart from pointing out the revolutionary power of technology associated with the term digital divide and innovation stimulus, it insists on the ethical imperatives and cultural sensitivities imprinted on sustainable land management. There is a need to address the importance of adopting collaborative, participatory, and adaptable strategies to successfully negotiate the difficulties of sustainable land use. These strategies should aim to achieve resilience, equity, and prosperity for both people and the environment.

***Keywords*:** Adaptable strategies, challenges, collaborative, environment, sustainable land management, resilience

9.1 Introduction

It is important to create and promote sustainable land use patterns to handle the growing problems the world is facing now. They are crucial for the well-being of ecosystems, income sources, ensuring there is enough food and dealing with climate change difficulties. Sustainable land use is needed, not just wished for, because our current and future generations depend on it. Sustainable land use tries to supply natural resources and space to everyone without using more of them than is available. Examples are ensuring biodiversity, managing the supply of water and soil, reducing greenhouse gas emissions and boosting the resilience of agriculture. At the same

Kundan Meshram and Chatrabhuj. A Sustainable Approach to Land Use, (193–222)

moment, it needs to think about the community's economic and social well-being, especially in exposed and vulnerable regions. Because there are many aims and sometimes conflicts, careful planning, using research-based decisions and including all key actors at all stages, is important.

Still, there are many difficulties in shifting to sustainable land use. A major issue is when decisions about agriculture, cities and conservation conflict with each other. Often, these battles end up causing land degradation and deforestation and lead to the loss of critical habitats. In addition, if there are weak rules for governance, ineffective use of land regulations and different departments not working together, the problem becomes even bigger. Sustainable practices are often not well guided in various regions because their governance does not provide enough transparency, coordination or accountability. The use of technology is often constrained by a number of problems. Developing nations find it hard to obtain modern applications like Geographic Information Systems (GIS), remote sensing and advanced data analytics for better land use decisions. There are also big cultural and social gaps that alter how people feel about and use land. Traditional methods for managing land, the absence of official land rights and earlier unequal treatment may stop fair and sustainable use of land. Since they are tied together, these issues build on one another and create an interconnected set of problems [1–4].

To address these challenges well, it is important to see their many sides first. A comprehensive strategy should consider the environment, the economy, culture and politics to work correctly. Policy tools need to consist of designs for land use zones, benefits for following sustainable practices, assessing the impact on the environment and actions to inform and empower land users. However, looking for one solution that fits all learners is bound to fail. Because land use situations vary greatly, it is important to use solutions that suit each particular area. Sustainable land use planning has to be adaptable and involve everyone so that it can keep up with new situations and benefits. Having new ideas and teamwork between all sectors is vital, helping members of government, academia, industry and civil society add ideas, support and inventive solutions.

Making sure everyone in the community and those who are affected can become active in decision-making matters greatly. Reaching sustainability calls for people from all levels to play a part, along with using and sharing local traditions. Besides having the right information and tools, communities should also have the power to decide about their environment and how to live sustainably. Now, seeking perfect outcomes is an approach that does not work anymore. Working together to create stable government structures allows communities to reach sustainable and fair solutions in land management. Besides helping the environment, making this transformation also supports human well-being, keeps economies stable and helps the planet stay healthy for the long run [5–7].

Agricultural lands must aim to produce a lot while at the same time preserving their land. Even so, it is particularly tough to achieve sustainable growth here in developing countries because of not enough knowledge, poor infrastructure and a shortage of resources. The ability of landowners to manage their finances is the main factor that decides if sustainable land use practices start and last. Often, smallholder farmers and owners of small properties do not have enough money to try sustainable farming methods. To address these problems, partnering private landowners with government institutions is very important. Support from the government such as providing funds, subsidies, advice and favorable policies, is important for maintaining profits and making sure farms continue protecting the soil and ecosystem. Although this could lead to many good results, it is difficult to use these systems in areas of Africa that are poor, as their main problems are getting enough food and reducing poverty. The lack of financial investment chances also adds to the difficulty, cutting off access to lasting sustainable land usage.

Also, differences in culture and society add another dimension. The gap between older and younger people often affects views on using sustainable methods. A group of people who practice agroecology might not be willing to accept new farming techniques because they are confident that their traditional ways are sufficient. Conversely, young people who own land are more likely to try new and greener land management methods because of their experience with modern education, technology and understanding of environmental matters. Since each group has different ideas, there is more friction which limits how well sustainability is achieved in society. Disparities in education hold back the general adoption of sustainable ways to use land. There is not enough access to sustainability education in many rural or underserved places, which results in a big group of people not understanding key sustainability concerns. Lack of information supports reverting back to old ways of using land that lead to more land degradation and harm to ecosystems.

Climate change makes things worse for agriculture by causing unpredictable changes. Increasingly unpredictable weather and more frequent extreme weather conditions like droughts, floods and storms have damaged how land is used. This leads to faster soil erosion, a reduction in the quality of soil and an upset to agricultural output. Therefore, existing ways of making farming more sustainable need to be changed to meet these new landscape conditions. The uncertainty about these climate-related events makes it hard to design strong and dependable methods of protecting the Earth. It is very important to develop new techniques for forecasting and management to address these ongoing changes. Innovative tools may improve forecasts, risk management and planning, helping landowners and policymakers become more proactive. As Figure 9.1 shows, success in sustainable use of land requires uniting financial, educational, technological and intergenerational strategies [8–10].

The sustainable utilization of land in urban settings relies on government planning. The recruitment of multiple stakeholders is necessary in the process, and both local and national planning are required to ensure long-term viability. Inadequate planning leads to a lack of sustainability. Despite development usually being directed by agreed-upon rules, ignoring these rules may cause growth to be unpredictable and not meant for long-term success [11–13]. Hence, it is imperative to emphasize the necessity of incorporating pre- and post-evaluation into the planning process. Moreover, education and the allocation of resources are crucial for effective land use management.

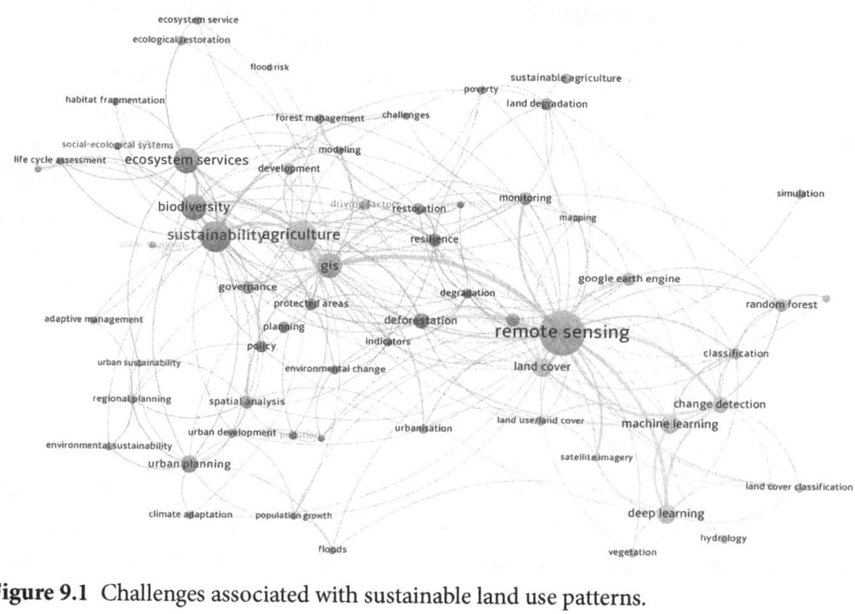

Figure 9.1 Challenges associated with sustainable land use patterns.

9.2 Challenges in Implementing Sustainable Land Use Patterns

It is vital to address various difficulties by using integrated approaches that considers environmental, social, economic, and governance aspects [14, 15]. Effective cooperation between governments, civil society organizations, commercial sector, and local communities is crucial for advancing sustainable land use practices that have positive impacts on both people and the environment.

a. Complexity: Land use sustainability entails several complexities because of joint governance of governments, private actors, indigenous communities, environmental activists and locals with various strategic objectives. Different key stakeholders have varied core objectives, such as economic development, environmental conservation and maintenance of cultural heritages. Adoption of unity among conflicting goals calls for a joint decision-making process that will lead to transparency, accompanied by inclusiveness and trust building. It is important to note that the successful implementation of plans requires a complete understanding of individual needs from the stakeholders as well as clear approaches to resolving disagreements and making common decisions to protect the land.
b. Policy and regulation: Non-integrated policies and outdated laws that do not change with time lead to poor environmental and social fairness. Combining concerns about the environment with poor enforcement makes the already present problems worse, as people are not penalized for harming the environment. This problem can be answered by introducing new rules that coordinate policies from several sectors and have well-defined bodies to implement and enforce them. Organisations adhere to effective governance structures that encourage innovative solutions in concert with flexibility to accommodate changing land use needs.
c. Competing interests: Land development by economic and environmental groups has often caused a conflict between supporting business and conserving nature. Because land use is so important for both industry and conservation, decisions made in favour of economy often hurt the

environment. For these interests to be reconciled, governments must develop unified rules that combine improving the economy with saving the environment. Steps to increase development must pay close attention to the environment, supporting responsible use of land and preventing damage to the environment.

d. Limited resources: Successful application of sustainable land use strategies internationally is possible only when there is enough funding, trained experts and strong institutions. Yet, countries considered developing nations tend to face great obstacles in finding the important resources needed to build environmental sustainability. It is especially clear in the fields of financing, technology and people's skills that these countries face problems. Because there is not enough funding, advanced tools or trained workers, it becomes very hard to use sustainable land management and keep close watch over land use changes. The main purpose of bridging these gaps is ensuring that progress toward sustainable land management is fair and effective for all nations.
e. Land degradation: Sustainable land use is still threatened by land degradation caused by deforestation accompanied by soil erosion, pollution and desertification causes. The solution to stop land degradation needs methods that unite conservation work with restoration projects and sustainable management practices of the lands. To prevent environmental deterioration along with restoration of degraded areas, the essential practices include reforestation together with agroecology, erosion control and soil fertility management.
f. Urbanization: Urban areas expanding very fast and without planning is a major way that leads to the change of important farmland, forests and natural habitats to places for buildings and infrastructure. Because cities keep growing fast, there is an increased risk of habitat loss, more pollution and resource depletion. In addition such changes tend to worsen social inequalities and difficulties, mainly in cities that do not have enough basic amenities and services like clean water, sanitation and affordable housing. For sustainable development in cities and to fight against the negatives, both strategic planning and stricter land use laws are required. It helps manage urban growth so development, environmental safety and social care are all considered.

g. Socioeconomic equity: People and communities need to share land development advantages and costs fairly which depends on equitable and inclusive planning for use. Although environmentally and economically, sustainable land management is necessary for a bright future, it might cause more difficulties for indigenous people, rural farmers and those living in towns with low income. Many times, these marginalized groups deal with problems such as being forced out of their homes, losing their main source of income and accessing fewer of the resources and services they depend on. Also, they are generally excluded from decisions about land use policies that affect them, which makes their situation worse. Including socioeconomic equity in land development requires planning to involve these communities, ensuring fair compensation and resettlement when they have to move and setting policies that values their land, cultural background and economic well-being.
h. Technological limitations: Many advanced technologies like GIS, Remote Sensing, AI and ML techniques have yet to gain widespread use across rural areas, underdeveloped and resource-constrained regions. Modern equipment acquisition, equipment maintenance expenses and requirements for specialized experts through technical professions present significant operational obstacles. Digital implementation faces severe challenges because of insufficient digital infrastructure that consists of poor connectivity, limited software access and inadequate electricity supply. Planning authorities face difficulties in integrating these tools because their staff lack the necessary training and human resources to handle the implementation process. To bridge digital gaps, several initiatives are essential based on local needs, which should combine governmental monetary support with international funding for digital structures and open-source promotion combined with local training sessions and private-public joint ventures. The accessibility and price affordability, together with easy use of technology, will enable local communities and planners to make informed decisions about adaptive land management strategies.
i. Globalization: Globalization has strongly affected how land is used locally by linking local markets with those at a global scale. By doing this, land is reshaped to meet increasing

world demand for agricultural commodities, timber, minerals and biofuels. When countries try to take advantage of international export markets, they often abandon traditional, mixed and rural land practices for mass farming, cutting down many forests and extractive activities. Usually, these changes disregard community and nature's value, changing nature mainly to benefit the economy, without caring about sustainability and the well-being of people. Globalization is also a major driver of land use changes which are often sparked by international trade deals, growth in foreign investment and global subsidies that favour large production. These pressures often cause local land administration to fail in developing and protecting land, which then results in land damage, less biodiversity and the marginalization of both small farmers and indigenous communities that depend on traditional land use.

j. Enhancing institutional capacity: In numerous developing regions and vulnerable places, public authorities addressing land use concerns do not usually have the basic ability to tackle or respond well to issues affecting land. Their inadequacy covers a gap in finance, technology, expertise and good organizational structure to be able to carry out their operations effectively. For this reason, problems such as ineffective planning for land development, problems carrying out policies and common corruption in institutions arise. Conflicting and uncertain roles for government authorities cause items to be handled more than once which leads to inefficiency and delays. Many populations that are marginalized in society such as indigenous peoples, those living in rural places and the urban poor, are rarely included in making policies and choosing projects, which only adds to social inequality and makes fair land governance harder. Upgrading institution capacity calls for professional training, changes in administration, improvements to existing processes and better working relationships between the planning, environmental and agricultural departments. Developing these organizations supports clear, responsible and inclusive ways to oversee land, which is necessary for sustainable development.

k. Climate change: Land sustainability becomes more challenging when climate change causes damage to ecosystems and changes the weather. It makes hazards such as storms, droughts and floods more likely to happen and more destructive. Swings in temperatures, irregular rainfall, long dry spells and flooding all contribute to less availability and productivity of land for farming, forestry, houses and conservation. As a result, productive farmlands are losing their fertility and many coastal areas experience more flooding and saltwater problems. To deal with all these climate effects, using climate-friendly methods to use the land more sustainably is important. Also, strategic land use planning within national climate adaptation strategies should be built as a connected system. This way of dealing with land is important to address negative impacts on the climate and also to use land for reducing global climate change.

l. Market forces: Consumer demand, price fluctuations, and the behaviour of the investment sector act as forces in the market that shape how various territories are utilized. Such forces are more focused on short-term profits at the expense of long-term environmental sustainability. Real estate speculation and speculative land investments increases land value, thereby displacing residents as well as promoting hoarding tendencies, mainly in semi-urban areas. Market forces create significant incentives for land destruction, and these involvements must be minimalized by strict policies. The ability of businesses to prevent such negative consequences involves the demonstration of responsible behaviour in assisting customers to choose eco-friendly supply chains. This includes receiving payments for sustainable land management and implementing a framework for sustainable consumer perception with ethical purchasing habits that will influence the markets' choices for responsibly produced products, thereby having an upstream impact on the decision-making.

m. Cultural and behavioural factors: The ways land is dealt with, valued and controlled are deeply connected to cultural traditions, typical ways people behave and social norms. Across cultures, land is regarded differently and it is seen as important not only for business but also for cultural customs, religious practices and safeguarding the pasts of their ancestors. Not all communities protect land in a sustainable way; some still depend on harmful practices and change only when needed while facing environmental problems.

The best responses should therefore blend cultural understanding and caring for the environment. Public programs, coupled with learning about the environment in schools and extra workshops, may encourage the community to act together for land conservation and sustainability.

n. Land tenure and property rights: Property and land tenure laws give people the power to protect their land year after year. When job security is lacking for farmers, their future plans are uncertain and land is managed poorly, which can result in careless land use and practices damaging for nature due to resource wars and issues of displacement. Many indigenous people, smallholders and city and rural citizens who have little access to land paperwork are the ones mainly affected by this issue. Handling these matters needs both legal changes and participatory governance models that approve and include the traditions of communal land ownership. They help provide better security of land and advance practices meant to save land.

o. Insufficient data and knowledge gaps: The planning and decision-making processes for active land use management require precise and up-to-date information about land features together with detailed resource records, population projections, and environmental danger assessments. Many distant regions, along with regions facing financial limitations, experience problems accessing recent reliable data relating to land characteristics, resource availability, demographic patterns and environmental threats. The lack of knowledge about soil quality together with water resources, biodiversity and land degradation capabilities prevents the implementation of evidence-based policies while preventing adequate monitoring and evaluation processes. Often, policies miss their mark because researchers do not properly involve themselves in what happens locally. This gap can be filled by giving people access to open data, running nearby training courses and respecting traditional ways of learning. When stakeholders at all levels personality work together, it results in easier exchanging of knowledge, which leads to better-informed and locally appropriate land governance practices. They ensure the development of policies that combine certainty with respect for culture which supports the long-term care of farmland.

9.3 Economic Interests vs. Environmental Conservation

Environmental conservation forms an opposition with economic interests, yet these two goals remain deeply linked. The pace and magnitude of economic development usually hinges on businesses obtaining natural resources to construct industrial sectors, which produce employment opportunities for national development. Environmental conservation establishes plans to safeguard both natural resources, ecosystems and biodiversity for maintaining continuous ecological stability [5, 8]. The situation creates a constant conflict in political decisions and land use management because choosing fast economic growth typically destroys ecosystems, but applying strict environmental regulations blocks immediate business opportunities.

9.3.1 Economic Interest

a. Profit motive: The economy depends on natural resources as a basis to achieve financial returns for several economic sectors. Agriculture, together with forestry, mining and manufacturing industries, needs land, water, minerals and additional natural resources to develop products and deliver services. Multiple financial benefits steer people to modify land practices even though these changes result in short-term gain rather than lasting environmental stability. Corporate profit optimization leads to environmental degradation through forest clearing, unsustainable grazing and unsustainable mining activities. To balance profit-making with sustainable resource management, the establishment of laws and financial motivators is essential for achieving economic sustainability.
b. Job creation: Extraction industry, i.e. mining, agriculture and logging, creates employment opportunities that are beneficial to certain rural regions where there is an employment scarcity. These jobs help these poor regions escape their plight as they develop fundamental infrastructure and assist the economy to grow. Such industries, in most cases, provide new employments but their success oftentimes translates into making sacrifices to the environment and local populations. Mining and farming done in unsustainable ways

can be a major cause of water pollution and soil damage. All of this damage leads to several health issues for local people, like facing higher levels of polluted air and water, and reduced farming success.

c. Resource extraction: Past generations prefer resource extraction as a major economic booster. Raw materials, including timber, fossil fuels and minerals, are shipped to other countries for converting them into production, which will result in economic incomes and industrial development in the economy. Countries rich in resources continue to subject their economics to high dependency mainly because royalties are taxed and exclusive exports are earned. Resource extraction operations have significant negative impacts on the environment through destruction of the ecosystem and pollution, having impacts on displaced human communities.
d. Infrastructure development: Constructing roads, along with constructing energy plants, urban settlements, and dams is an important aspect of national development plans. The establishment of these projects causes economic growth because it integrates markets as it ensures improved access to essential service facilities. The land buys required by the construction efforts spur four major environmental challenges, namely: forest destruction, destruction of wetlands, destruction of a creature habitat and an increase level of carbon emissions.

9.3.2 Environmental Conservation

a. Preservation of biodiversity: The core strategies of land use planning must include biodiversity conservation since these threats exist. The conservation strategies include setting up protected areas together with ecological wildlife corridor networks that connect broken natural habitats and keep biodiversity hotspots that host significant endangered species. Sustainable land use demands that development projects carry out minimal ecological impacts through environmental impact assessments while also implementing regulations that transform land and promoting land-use planning to achieve a balance between conservation and human development obligations. Societies that link land use with biodiversity conservation efforts will guarantee effective natural system functioning combined with sustainable economic activities and balanced development of human needs and natural ecological well-being for current and future populations.

b. Climate change mitigation: The methods humans use regarding land have a direct link to climate change mitigation because they determine both greenhouse gas releases and carbon storage capabilities of ecosystems. The natural ecosystem protects forests, wetlands, peatlands, and grasslands, and succeeds in capturing atmospheric carbon dioxide because these systems function as effective carbon storage pools, which stabilize the worldwide climate situation. Efforts should always focus on conserving forests, wetlands and similar regions and exclude the rebuilding of natural landscapes. They help lower the emission of greenhouse gases and help plants store more carbon. Strategies such as replanting forests and planting new trees should be used first in endemic areas, helped by full-scale ecosystem restoration. They help reduce climate change and also make the environment more resilient to disturbances caused by climate.

c. Sustainable resource management: Conservation strategies advocate for the responsible and efficient utilization of natural resources to guarantee their continued availability and prevent their exhaustion or deterioration. Sustainable methods, such as agroforestry, organic farming, and fisheries management, aim to achieve a harmonious balance between human needs and the objectives of conserving ecosystems. These practices entail implementing methods that reduce the negative effects on the environment, promote biodiversity, and preserve ecosystem services. Facilitating sustainable resource management necessitates the establishment of governmental frameworks, market incentives, and community engagement to promote the adoption of environmentally conscious practices.

d. Ecotourism and recreation: Protected regions and natural landscapes entice visitors and those who enjoy outdoor activities, resulting in financial gains from ecotourism ventures such as observing wildlife, trekking, and engaging in nature-oriented recreation. Ecotourism may generate economic advantages for local communities while also fostering environmental care and raising awareness about conservation. Developing sustainable ecotourism entails fostering local community engagement, economic benefits, enforcing environmental, social regulations for tourism operations, and investing in conservation efforts and infrastructure to facilitate sustainable tourism expansion. Ecotourism functions as a means of supporting conservation efforts, providing education, and promoting community development. It emphasizes the interdependence between economic interests and environmental conservation.

To strike the proper equilibrium between economic development and environmental preservation, people need complete comprehension about how human prosperity relies on ecosystem health. Economic branches that focus on agriculture, industry and infrastructure expansion require natural resources together with proper ecosystem services, including clean air, water, good soils, climate controlling system and biodiversity. Economic growth without control results in environmental harm such as deforestation, together with pollution as well as biodiversity destruction that damages both sustainable development and human quality of life. A participatory decision-making system needs to be established because it will enable marginalized communities, along with indigenous peoples and future generations, to take part in forming policies concerning both development initiatives and environmental issues [16, 17]. The designed mechanisms need to promote complete transparency, equity and maintain sustainable resilience for the long term. Policy creation and planning processes that unite economic, social and environmental factors maintain ecological stability by preventing development from destroying nature.

9.4 What is the Resistance to Change in Sustainable Land Use Patterns?

A complete strategy should be applied to tackle the resistance toward sustainable land use patterns. The solution requires government policy initiatives along with institutional reforms and social engagement campaigns and funding programs for personnel development. The actions need to direct intervention at existing inertia sources in order to transform land management activities toward more inclusive and equitable methods that respect the environment. The organizations must encourage teamwork for different parties and form coalitions to create creative and flexible educational and governance systems that embrace positive transformations in communities.

a. Economic interests: Members of the economic sector such as land developers, agricultural producers and extractive industries resist change because they can only foresee lower profits or big expenses of new technology or practice. The stakeholders who deal with land usually consider sustainable rules to be costly alongside complex requirements because they themselves benefit from natural wealth. Urban developers will confront zoning restraints, which regulate expansion in the urban areas, and farmers will be reluctant in utilizing the conservation tillage due to the fact that they fear higher costs and production effects on their crops. The government needs to develop fiscal tools that will support sustainable operations, financial programs and market protocols that will motivate conservation practices in order to counteract stakeholder resistance to sustainable practices.
b. Cultural and social norms: Several correlates of culture and tradition support practices of land use, which lead to unsatisfactory environmental outcomes. Communities normally oppose changes to traditional cultural practices, like the removal of trees for farming and land clearing for building, as such practices are central to their beliefs, because they have their basic lifeline through them and because they doubt the motives of outside actors. Sustainable promotion of land use involves consideration of local traditions along with education dissemination and showing the long-term value of those methods. The mixture of community involvement for participatory planning activities, along with traditional knowledge integration with modern conservation solutions, effectively guides towards their realization.
c. Insufficient awareness or education: Sustainable land use acceptance, along with systems of implementation, becomes difficult as stakeholders do not know what they will gain by doing so. Individuals with a stake in particular businesses do not understand the complete impact of previous reckless practices on the environment and society, as opposed to the other type of practices where one might receive actual

advantages. Small farmers usually do not know how sustainable farming mechanisms are good for both soil quality and harvest productivity in the long run. The success of support-building activities together with behaviour change promotion depends on effective education and outreach programs. The programs should be scoped to sustainable practice dissemination through delivery of educational content through practical training solutions and identification of successful sustainable land use cases.

d. Institutional constraints: The potential of developing effective land use planning and management is limited due to poor institutional governance in combination with bureaucratic inefficiencies and institutional fragmentation. Many barriers prevent the formulation of cohesive solutions for sustainable land utilizations because government agencies fail to build teamwork, and they also have differing orders while ownership of land is vague. Collaborating institutions have challenges because there are areas of dual jurisdiction control, where environmental departments and agricultural departments operate, which results in clashing rules and enforcement problems. It becomes necessary to address issues on coordination among agencies and proper streamlining of both bureaucratic process and land ownership definitions, and also improve local institutions.

e. Risk aversion: Important people in the community demonstrate resistance to change because they are concerned about the new area and the uncertain effect after different forms of land use have been carried out. Certain risk-averse types of people will pass over or take a moment before trying out new advancements, regardless of the fact that these advancements have positive aspects related to efficiency of resources, ecological sustainability and potential for adaptability. Farmers do not want to make organic farming transitions because of the uncertainty of future market requirements and price levels. Monitoring pilot projects with demonstrations, development of insurance plans, and reward distribution mechanisms will help individuals address the issues of being risk-averse as they begin to understand the long-term sustaining benefits of these practices.

f. Short-term thinking: Organisations and key officers usually opt for immediate economic gains such as industrial activities earnings, agricultural growth and resource harvesting as against enduring sustainability targets. This involves controversially focusing on quick financial gains by which certain industries are given the authority to engage in mass logging and experience few assessments on its long-term environmental effects. This includes deforestation, availing of a habitat, temporary interruption of the water cycle and adverse effects on indigenous communities that rely on forest systems for their survival. The approach yields poor decisions because the interconnectedness of environmental and human systems is overlooked, hence environmental breakdowns and lifestyle deterioration. The elimination of this pattern requires the development of forward-looking sustainable thinking, which must be reflected throughout the perspectives of planners, policymakers and investors. Over time, once educators emphasize systems thinking principles besides environmental awareness programmes, stakeholders will learn about environmental stewardship benefits.

g. Technological barriers: The sustainability of land use practices is hampered considerably because of limitations of technology, which impacts areas that have limited access to proper technology, infrastructure and trained expertise. The financial resources and support systems which would allow implementation of advanced sustainable solutions are out of reach for the smallest farmers as well as rural areas and nations that are still under development. The cost barriers of initial investments and insufficient access to power grid and internet structures along with staffing issues in maintaining such technology compounds extensive problems. Smallholding farmers lack access to solar-powered irrigation systems because the high initial expenses and insufficient technical support prevent their acquisition. The establishment of sustainable technology infrastructure will enable everyone to access environmental tools that enhance sustainability, which leads to fairer and more successful land planning processes.

h. Path dependence: Because of past choices in land management, certain methods are now strong and supported by many existing infrastructures and institutions. Such outdated systems make it very difficult to implement more environmentally friendly systems. Because well-established programs, development infrastructure and related financial systems are in place, many people are opposed to changing things.

i. Lobbying and opposition: When policies are seen as endangering their profit and market position, firms involve themselves in lobbying to fight them. Leading-edge lobbying organizations share questionable arguments and file suits to prevent the introduction of rules that aid sustainable land management. It is commonly seen that fossil fuel companies take steps to delay or reject rules that help to use more renewable energy. Handling these influences calls for establishing influential networks for sustainable policy and growing broader public understanding. Also, including everyone in the policymaking process makes land use governance more open, accountable and ready for the future.

j. Political factors: Political matters are very important in deciding land use and they often make it harder for sustainable laws to be implemented. Many strong interest groups and influential lobbyists attempt to block efforts to change how land is used in a way that helps the environment. It is often difficult to implement new rules and enforce existing ones, since many organizations prefer not to move away from their usual practices. In particular, developers and big agriculture regularly ask politicians to allow more ecological harm in wetland and forest areas. Removing these challenges involves the grouping of different members of society to support sustainable changes, openness and clarity in making policies and ensuring equal and fair enforcement of environmental laws.

k. Power dynamics: Unfair relations of power shape how land is used, which often leads to disagreements between groups and ongoing opposition to change. Strong leaders in society—mainly corporations, landowners and powerful groups—often cause other opinions that support good land management to be ignored. Because of this, issues in agriculture are mostly discussed in favor of large companies, and the voices of people who depend on agriculture are usually suppressed. Because of this misbalance, fair policies are hard to design and sustainable and inclusive land use plans are hard to put in place.

l. Reliance on subsidies: Using subsidies and stimulus measures can encourage the government-driven increase in activities that negatively affect the environment by distracting people from these effects and making them easier and cheaper. If governments compensate for the environmental damage caused by fossil fuel, industrial farming or changing land use, they indirectly attract more people willing to invest by making it more economical for them. Using fossil fuel subsidies, farmers are able to mechanize land clearance for less cost, which ends up overusing the land and causing damage to the environment. Just as before, choosing monoculture for crops and farming without considering the environment causes the land to get less fertile and depleted with time. Sustainability-driven fiscal actions that promote clear government moves, proper environmental cost monitoring and involve all concerned parties can deal with these challenges. Choosing environmentally favourable ways to use money can establish balanced use of land that preserves nature and the economy, creating a development plan for the future.

m. External pressures and global trends: Local land use decisions have problems in the achievement of sustainability due to global economic changes, market legislation and geopolitical changes. Other organizations outside the local community such as multinational companies, international financial organizations and foreign governments facilitate monetary growth through policies that undermine ecological conservation and social equity. The sustainable trade practices require policy harmonization between international and regional entities with the aim of ensuring that the economic activities serve the environmental and social sustainability objectives.

9.5 Strategies for Overcoming Resistance

To mitigate negative responses to change in regard to sustainable land use patterns, stakeholders have access to a bundle of tactics that center on overcoming deep-rooted barriers, achieving unanimity, and creating a favourable environment for new measures [18, 19]. A few essential tactics are described below.

a. Stakeholder engagement and participation: Land use decisions should involve a range of interests, thoughts and experiences from different stakeholders. When everyone is included, people, organizations and the government feel more united in handling problems and trust one another. Successful programs that organize and manage how land is used are possible only when local communities, indigenous groups, civil groups, businesses and government colleagues all take part. Engagement can be helped by methods including public consultations, stakeholder groups and planning workshops. With these mechanisms, policymakers can ensure that their strategies for land use support fairness, fit the local situation and help achieve sustainability in development results.
b. Incentives and rewards: Establishing reward systems backed by financial compensation becomes an effective method for stakeholders to adopt sustainable land use practices. The encouragement of sustainable practices programs is characterized by funds for sustainable farming, tax breaks and funding for restoring ecosystems and sustainability certifications. More interest would be generated in the methods of soil conservation when the farmers are given financial support for the implementation of the same. Also, government-driven economic efforts that support sustainability can lead to much faster progress in green strategies. Because of this, sustainable practices are adopted across different sectors and support long-term healthy growth.
c. Pilot initiatives: Starting pilot projects is important because it gives land users and community members direct experience with new ways of managing the land. Such projects give participants a chance to learn about the real effects of sustainable practices on the environment and the economy. Results and key learnings gained through pilot projects need to be reported and publicized as much as possible through events, the internet and printed sources to have a bigger effect. Also, to repeat and grow existing successful models, it is necessary to provide integrated support including technical help, training programs and the necessary resources, including seeds, equipment and money. Diversity in geography and society calls for these robust components to make sustainable land management accessible to more land users.

d. Empowerment and capacity building: Empowerment of local communities; indigenous peoples, women, youth and marginal groups to participate in the decision-making process is critical for building social and environmental fairness. Capacity-building programs could help these organizations to put together resources and lobbying for their rights, as well as bring to implementation projects conducted by the community. For instance, when providing instruction and helping these groups (i.e., indigenous) in sustainable forest management, their ability to protect and benefit from their territories can be enhanced. Initiatives towards empowering people should also strive to enhance the leadership capacities and institutions of inclusive governance.
e. Political leadership and commitment: It is important for leaders at the highest levels to lead the development of sustainable policies for using land. Getting politicians on board is essential for driving advocacy which helps design and implement new rules for managing land and other resources. To strengthen political will, leaders and workers within government who support change and work on creating supportive conditions for ongoing sustainability and durability should be promoted. One key strategy for major progress is for political leaders to pledge to introduce and maintain thorough reforms in land use planning. They let others know what the organization values, making it easier for national development to align with environmental and sustainability priorities.
f. Adaptive management: Adaptive management and learning need to be included in order to have effective land use planning. This includes preference for techniques that will enhance continuous improvement, experimentation, and learning. Through the promotion of active involvement of stakeholders in monitoring and assessment of the results of their work, adaptation of tactics according to feedback received, and exchange of information and experience, it is possible to get a more powerful and effective process. An example of adaptive management in a watershed restoration project is periodic observation of water quality, changing necessary restoration strategies, and distributing these findings to other watershed managers.

g. Education and awareness building: Dissemination of knowledge, provision of teaching and cultivation of skill development are cardinal activities to increase awareness and understanding of the benefits of sustainable land use. Educational activities include workshops, seminars, field demonstrations and online courses. These are initiatives meant to equip stakeholders with the required information and skills towards embracing and practising sustainable skills. Training programs on agroecology or conservation agriculture can improve farmers production while increasing protection of resources for nature.
h. Regulatory reforms: Improving sustainability goals and correcting market failures that lead to unsustainable behaviours are fundamental reasons why stakeholders should pursue more stringent land use rules, zoning ordinances, and environmental legislation. The effective regulatory framework needs to include the strict enforcement of existing regulations and the mix of emerging policies and standards that push for adherence to sustainability practices. Regulatory reforms should also address growing concern about land tenure security as well as access to resources equity, such that regulations seek to advance sustainable land management.
i. Collaborative partnerships and networks: Launching cooperation, networking and knowledge sharing between stakeholders is important for the support of learning, innovation and collective action. Strategic collaborations, alliances, or multi-stakeholder platforms can bring together diverse people or entities with complementary knowledge and assets. Such partnerships have the ability to address common challenges and achieve common goals more effectively. An important case of working together is the cooperation of environmental NGOs, government agencies and the private sector in land restoration projects. They prove that when stakeholders join forces, significant outcomes can be achieved for the environment, since they offer expertise, regulations and funding of a comprehensive land restoration project.
j. Communication and advocacy: Good communication tools and specific advocacy efforts can raise knowledge, bring people together and keep the issue of sustainable land use at the forefront. Reaching people through different channels

such as media outreach, public events and educational materials, allows for involvement by a large audience which helps make real changes happen. An example of this is a social media campaign raising awareness about sustainability in farming which can reach a wide bunch of people and encourage them to act collectively. Similar to other sustainability actions, policy support should target players in government with the goal of enacting laws that favour responsible land management.

9.6 Dilemma of Immediate Economic Gains vs. Long-Term Sustainability

The reason real-situ and future-wellbeing conflicts exist is that immediate profit gain usually clashes with preserving limited resources and useful services for the future. Below are topics with both approaches in detail.

9.6.1 Immediate Economic Gains

a. Profit maximization: Frequently, companies and individuals put their near-term gains ahead of long-term security, which may have negative results for the future. Sometimes businesses that do forestry, mining and intensive agriculture claim rapid gains, with the advantages often going to impoverished communities by using up natural resources. Nevertheless, these approaches usually fail to notice the ongoing and combined harm done by depleting resources and harming nature, which give rise to social issues. An example is that cutting down trees to deforest the land might help in the short term but leads to damage to nature, the environment and the climate, which causes long-term harm to the land.
b. Job creation and economic growth: Industries such as mining, building, and manufacturing offer employment opportunities for workers and make significant contributions to the local economies, especially in locations abundant in natural resources or emerging nations. Nevertheless, the immediate economic advantages can be achieved at the cost

of long-term viability, as they frequently entail methods that deteriorate land and natural resources.

c. Investor expectations: Investors and shareholders often prioritize immediate returns on investment, placing a high value on organizations that can achieve rapid profit growth. Corporate decision-making typically centres around quarterly earnings, stock market performance, and shareholder value, which encourages companies to prioritize rapid financial advantages rather than sustainable land use practices. The prioritization of short-term goals can result in choices that optimize immediate financial gains but have the potential to compromise long-term viability, such as the excessive extraction of resources without adequate environmental protection measures.
d. Consumer demand: It is influenced by consumer preferences and market dynamics. This demand is focused on products and services that value convenience, price, and rapid pleasure, rather than considering sustainability factors. Poor practices such as accelerated product turnover, an intentional approach that permits products manufactured in greater volume than those that are disposed of only after repeated use, and patterns of throwaway consumption all help to deplete resources, create trash, and destroy environments. For example, the fast-fashion industry produces cheap, fashionable clothes in a hurry, encouraging the purchase and then disposal of clothes. This tendency has serious environmental implications and unsustainable land use.
e. Infrastructure expansion: Expanding roads, highways, seaports and cities is part of the necessary growth of any country's infrastructure. Generally such projects boost the immediate economy with quicker transportation, increased foreign trade and new job options. Even so, these changes often greatly alter the natural environment, which reduces habitats, increases pollution and causes ongoing harm to nature. Therefore, when infrastructure is increased, keeping the ecosystem balanced and using land sustainably becomes very difficult.
f. Agricultural intensification: Due to pressing demands for more food and increased profits, farmers usually choose to raise more crops by planting single types and by applying lots of fertilizers and pesticides. Although these innovations

help crops in the beginning, in the long run they can lead to erosion of the soil, contamination of water supplies and decreases in biodiversity. These results harm the long-term sustainable development of farming and run counter to the sustainability principles needed in using land.

g. Urban development: It is accelerated by economic expansion, resulting in immediate prospects for real estate development, construction, and associated sectors. Urban expansion facilitates the provision of homes, commercial spaces, and infrastructure, hence bolstering local economies. Nevertheless, this expansion of cities frequently invades natural habitats, diminishes areas of vegetation, and adds to environmental problems such as urban heat islands, air pollution, and heightened garbage production, upsetting sustainable land utilization patterns.

9.6.2 Long-Term Sustainability

a. Resource conservation: The objective of sustainable land use patterns is to preserve natural resources, safeguard ecosystems, and sustain ecological equilibrium over an extended period of time. Implementing conservation measures such as reforestation, soil conservation, and biodiversity protection is crucial in guaranteeing the sustained accessibility of vital resources such as clean water, fertile soil, and biodiversity for future generations. These behaviours contribute to the maintenance of ecosystem services that are essential for human well-being, including water purification, pollination, and climate regulation.

b. Climate resistance: Sustainability over the long term can be attained by actively tackling climate change and its effects by lowering pollution, collecting carbon and making systems less vulnerable to climate risks and disasters. Agroforestry, sustainable agriculture and ecological restoration benefit climate resilience by boosting the services of nature and preventing problems from harsh weather. Agroforestry systems are important because they catch and lock carbon into the soil, improve its fertility and assist in growing many different plants. Using these techniques helps the economy and the environment, providing support for short-term needs and ensuring ecological balance in the future.

c. Social equity and justice: Long-term sustainability requires a solution to the problem of social inequalities, promotion of all-round development, and safety of the rights and welfare of the marginalized population and future generations. In planning land use, it is important to emphasise fairness, participation and social justice. This may be done by enfranchising local people, protecting indigenous rights and benefiting from an equitable distribution of benefits and costs. Community-led conservation programmes can provide local employment, build community resilience and save cultural heritage so that the rewards of conservation are equitably distributed.
d. Long-term economic stability: Going forward, sustainability in terms of land use helps to maintain long-term economic stability by promoting vigorous and diverse economies. Investments in green infrastructure, renewables, and sustainable agriculture have the potential to produce steady revenue, reduce dependency on the volatility of the market and create jobs. Economic stability promotes community well-being, reduction of poverty and sustainable development.
e. Conservation of biodiversity: Systematic conservation of biodiversity is essential to ensure the services of ecosystems, genetic resources, and ecology as a whole. Long-term sustainability initiatives prioritize the preservation of various species and ecosystems to guarantee ecological stability and resilience. Biodiversity conservation facilitates essential processes, including pollination, pest regulation, and genetic variation, which are crucial for agriculture, medicine, and the well-being of ecosystems.
f. Water resource management: Sustainable land use patterns will incorporate activities to ensure efficient protection and administration of water resources. Water quality, water flow regulation, and aquatic ecology are maintained using techniques of watershed management, wetland restoration and irrigation. Effective practices of water management are required for ensuring accurate supplies of water for agriculture, potable water and industrial purposes, protecting the ecosystems and avoiding water-conflict.
g. Soil health and fertility: Soil health and fertility need to be maintained in order to help farms remain productive for

years and to encourage a sustainable ecosystem. Land use methods like crop rotation, cover cropping, tilling the soil less and organic farming help soil improvement by bringing out minerals, making more nutrients available and raising the activity of soil biology. As a result of these methods, the soil becomes better structured, less erosion happens and the land becomes able to support high yields from agriculture. Healthy soils are needed for trapping carbon, so they act as climate change protectors and help keep food secure. Because of this, soil-conserving methods should be included in the planning of farmland.

h. Decreased long-term expenses: Even though paying for sustainable land practices is initially costly, such spending can save significant amounts in the future by reducing environmental damage, loss of resources and social issues. Doing things like contour farming and putting in terraces while adding organic manure to the soil greatly lessens soil erosion. If preventative efforts are put in place, expensive cleanups are less likely and farming can continue to be effective over time. Hence, proper stewardship of land makes the land healthier and helps with long-term savings and value for the economy.
i. Intergenerational responsibility: Preserving the environment for generations to come is the main idea behind intergenerational responsibility which guides land use policies. The idea supports sustainable land management that is focused on taking care of the environment, conservation and development in the long run. Applying these principles to land planning and decisions helps societies secure benefits of natural ecosystems, including a range of species, healthy soil and dependable climates, for now and for generations to follow. It promotes lasting, strong and fair ways to manage land.
j. Economic diversification: Economies can not only diversify their industrial setup but can also offer new avenues of economics and get a hold on the capacity of competing in the long run by investing in environment-friendly land use. There are opportunities for brilliant solutions, job creation and growth in the economy through the exploitation of green technologies, renewable energy sources, eco-tourism and sustainable agriculture, while reducing dependence on

finite resources and the degradation aspects that accompany them.

k. Improved ecosystem services: Adoption of sustainable land uses supports the improvement and long-term delivery of ecosystem services such as cleanup of air and water, regulation of climate, and maintenance of soil fertility. These patterns ensure a permanent supply of basic ecosystem services that are vital for human health, agriculture and general welfare, provided that natural habitats are conserved and the ecosystems are sustainably managed.

9.7 Conclusion

The course to sustainable land use patterns is doubtlessly complicated and complex, replete with interweaving barriers that demand holistic, integrative approaches to achieve successful outcomes. Environmental desolation, economical constraints, socioeconomic imbalances, and governing concerns are all interwoven to form a complex pattern of interrelatedness that impacts terrestrial ecosystems and societies across the world. This requires a deep understanding of how environmental, social, economic, and cultural elements interact with the land use patterns. It also highlights the global connectivity of land patterns use and necessity for multinational cooperation to rebalance sustainability issues. Communities' essential role in decision-making and inclusion of governance systems that empower underrepresented groups of people and ensure equitable access to resources and accommodation are key towards overcoming these obstacles. Technology should be harnessed to correct inequalities, enhance access to data as well as enhance innovations in sustainable land management. This includes ethics, cultural sensitivity and the esteem of indigenous knowledge systems as the basis of sustainable land use patterns that acknowledge that land has inherent value and its life is interconnected. Sustainable land use challenges can only be addressed with collaborative, participative, adaptable approaches to pursue resilience, equity, and prosperity for people and planet. These are overwhelming issues that provide an opportunity for a community to act, innovate and build resilience of the coming generations to live a more sustainable future. One can build a world in which land is well managed, and ecosystems and communities prosper in harmony with nature, by embracing a complex reality, proactively developing cooperation, and advancing holistic, context-appropriate solutions.

References

1. Akıncı, H., Özalp, A.Y., Turgut, B., Agricultural land use suitability analysis using GIS and AHP technique. *Comput. Electron. Agric.*, 97, 71–82, 2013, https://doi.org/10.1016/j.compag.2013.07.006.
2. Bandyopadhyay, S., Jaiswal, R.K., Hegde, V.S., Jayaraman, V., Assessment of land suitability potentials for agriculture using a remote sensing and GIS based approach. *Int. J. Remote Sens.*, 30, 4, 879–895, 2009, https://doi.org/10.1080/01431160802395235.
3. Liang, Y., Du, M., Wang, X., Xu, X., Planning for urban life: A new approach of sustainable land use plan based on transit-oriented development. *Eval. Program Plann.*, 80, 101811, 2020, https://doi.org/10.1016/j.evalprogplan.2020.101811.
4. Mohamed, A., Worku, H., Lika, T., Urban and regional planning approaches for sustainable governance: The case of Addis Ababa and the surrounding area changing landscape. *City Environ. Interact.*, 8, 100050, 2020, https://doi.org/10.1016/j.cacint.2020.100050.
5. Cobbinah, P.B., Asibey, M.O., Gyedu-Pensang, Y.A., Urban land use planning in Ghana: Navigating complex coalescence of land ownership and administration. *Land Use Policy*, 99, 105054, 2020, https://doi.org/10.1016/j.landusepol.2020.105054.
6. Guo, Z., Hu, Y., Zheng, X., Evaluating the effectiveness of land use master plans in built-up land management: A case study of the Jinan Municipality, eastern China. *Land Use Policy*, 91, 104369, 2020, https://doi.org/10.1016/j.landusepol.2019.104369.
7. Campo, D., González, A., Gazzola, P., Onyango, V., The mutualism of strategic environmental assessment and sustainable development goals. *Environ. Impact Assess. Rev.*, 82, 106383, 2020, https://doi.org/10.1016/j.eiar.2020.106383.
8. Anthony, T., Shohan, A.A.A., Oludare, A., Alsulamy, S., Kafy, A.-A., Khedher, K.M., Spatial analysis of land cover changes for detecting environmental degradation and promoting sustainability. *Kuwait J. Sci.*, 51, 2, 100197, 2024, DOI: 10.1016/j.kjs.2024.100197.
9. Liu, J., Jin, X., Xu, W., Yang, F., Wang, S., Zhou, Y., Assessing trade-offs and synergies among multiple land use functional efficiencies: Integrating ideal reference and key indicators for sustainable landscape management. *Appl. Geogr.*, 158, 103037, 2023, DOI: 10.1016/j.apgeog.2023.103037.
10. Han, B., Jin, X., Sun, R., Li, H., Liang, X., Zhou, Y., Understanding land-use sustainability with a systematical framework: An evaluation case of China. *Land Use Policy*, 132, 106767, 2023, https://doi.org/10.1016/j.landusepol.2023.106767.
11. Al Arjani, A., Modibbo, U.M., Ali, I., Sarkar, B., A new framework for the sustainable development goals of Saudi Arabia. *J. King Saud Univ. Sci.*, 33, 6, 2021, https://doi.org/10.1016/j.jksus.2021.101477.

12. Cao, M., *et al.*, Multi-Scenario Simulation of Land Use for Sustainable Development Goals. *IEEE J. Sel. Top. Appl. Earth Obs. Remote Sens.*, 15, 2119–2127, 2022, DOI: 10.1109/JSTARS.2022.3152904.
13. Riekki, J. and Mämmelä, A., Research and Education towards Smart and Sustainable World. *IEEE Access*, 9, 53156–53177, 2021, DOI: 10.1109/ACCESS.2021.3069902.
14. de Jong, L., *et al.*, Understanding land-use change conflict: a systematic review of case studies. *J. Land Use Sci.*, 16, 223–239, 2021, https://doi.org/10.1080/1747423X.2021.1933226.
15. Riveira, S. and Maseda, R.C., A Review of Rural Land-Use Planning Models. *Environ. Plann. B: Plann. Des.*, 33, 165–183, 2006, https://doi.org/10.1068/b31073.
16. Mugari, E., Mamabolo, E., Mathebula, N., Mogale, T.E., Mashala, M.J., Mabitsela, K., Ayisi, K.K., Barriers and enablers to implementing on-farm sustainable land management (SLM) practices among smallholder farmers in Mphanama, Limpopo Province, South Africa. *Sci. Afr.*, 28, Art. no. 02750, 2025, https://doi.org/10.1016/j.sciaf.2025.e02750.
17. Park, K. and Choi, S.H., Analysis of urban flood risk for the implementation of sustainable land use measures. *Clim. Serv.*, 38, Art. no. 100568, 2025, https://doi.org/10.1016/j.cliser.2025.100568.
18. Mekuria, W. and Girma, R., Assessment of land degradation neutrality to guide Sustainable Land Management practices in Ethiopia. *Environ. Challenges*, 19, Art. no. 101137, 2025, https://doi.org/10.1016/j.envc.2025.101137.
19. Pratzer, M., Maillard, O., Baldi, G., Baumann, M., Burton, J., Fernandez, P., Levers, C., Meyfroidt, P., Tasquer, M., Vallejos, M., Kuemmerle, T., Considering land use complexity and overlap is critical for sustainability planning. *One Earth*, 8, 5, Art. no. 101247, 2025, https://doi.org/10.1016/j.oneear.2025.101247.

10
Case Studies of Successful Sustainable Land Use

Abstract
This study investigates successful sustainable land use practices by conducting detailed case studies from a variety of regions, such as the Sahel region of Africa, the Great Plains in the United States, the Sahel region of Costa Rica, the Loess Plateau in China, Ralegan Siddhi in Maharashtra, India, and Sikkim, India. These cases share common themes that are essential for sustainable land management, despite the fact that their geographical, climatic, and socioeconomic contexts diverge. The integration of comprehensive strategies that address environmental, economic, social dimensions, robust policy, institutional support, active community engagement, capacity building, economic incentives, livelihood diversification, environmental restoration, conservation, enhancement of climate resilience through adaptive practices and renewable energy use are among the key factors. Collectively, these case studies illustrate the transformative potential of sustainable land use practices in the restoration of degraded lands, improvement of agricultural productivity, and enhancement of local livelihoods. They provide a blueprint for the global scaling up of sustainable land use practices, valuable lessons, underscoring the necessity of comprehensive, inclusive, adaptive approaches to the cultivation of resilient, sustainable ecosystems and communities.

Keywords: Adaptive practices, environmental restoration, community engagement, land use practices, land management, sustainable

10.1 Introduction

It is very important to encourage sustainable land use in order to save the natural environment, stimulate economic development, and develop social well-being. All over the world, people have developed ingenious ways of governing their land effectively so as to attain ecological balance, increased productivity and preservation of cultural heritage. This chapter examines a series of case studies from different regions, including, Costa Rica, China,

Kundan Meshram and Chatrabhuj. A Sustainable Approach to Land Use, (223–268) © 2026 Scrivener Publishing LLC

Kenya, the Sahel region in Africa, the Great Plains of the United States, and India. These case studies serve as examples of good sustainable land use methods. Such examples provide distinct views on how a variety of values, including cultural relevance, local knowledge, empowerment, equity, and long-term sustainability, can be effectively integrated into land management strategies.

The case studies in Costa Rica's Osa Peninsula and in Ralegan Siddhi in the state of Maharashtra, India, focus on combining cultural relevance and sensitivity into the planning of land use. By the participation of local actors and by adherence to traditional practices, these measures were able to attract a community response and participation, thus ensuring sustainable tourism measures, biodiversity conservation and effective watershed management. Likewise, the trend toward sustainable agriculture in Sikkim, India, creates a case for the adoption of traditional agricultural skills, which have helped increase credibility and effectiveness of organic farming techniques.

There is an example set for China's Loess Plateau and Sikkim, where the emphasis is on empowerment and capacity building. It demonstrates how training programs/initiatives can work on people in the locale to accept and sustain new practices. These efforts have led to improved land productivity, ecological restoration and general success of sustainable land-use initiatives [1–4]. Community involvement is an identified issue that can be observed in Murang'a County, Kenya, and the region of Sahel in Africa. The involvement of the local farmers in agroforestry practices in Murang'a County has produced a sense of ownership and accountability, positively encouraging the viability of the projects in the long term. In the Sahel region there is an effective response to multiple land use needs, conflict and community cohesion through participatory approaches in land management. The case study from the Great Plains in the United States teaches the relevance of accountability and social cohesiveness in sustainable land management. The initiatives have created avenues for trust and mutual support as ranchers and local communities have been incorporated in developing and implementing the sustainable grazing practices, which motivate successful as well as compliant land use strategies.

These case studies prove the necessity of a profound and diverse solution to provide sustainable land use. They demonstrate that culturally sensitive, indigenous knowledge, community empowerment, fairness, long-term involvement and social harmony can lead to successful enduring management strategies for the land. It is hoped that the case studies contribute major insights to policymakers, practitioners and academics in the field of sustainable land use.

10.2 Case Study I: Successful Sustainable Land Use in Costa Rica's Osa Peninsula

10.2.1 Background

The Osa Peninsula in the southwestern part of Costa Rica is famous for its outstanding biodiversity within its 700 square miles, containing 2.5% of the world's species. The most biologically intense spot on Earth is right here, Corcovado National Park, which *National Geographic* has noted. The region supports a wide variety of living species, including jaguars, scarlet, macaws and a vast number of plants unique to the region. Although of great ecological value, the Osa Peninsula has faced immense environmental challenges due to deforestation, agricultural expansion and unchecked urbanization. These obstacles threaten the exceptional biodiversity of the region and the welfare of the local residents who depend on natural resources and ecotourism [5–7].

10.2.2 Challenges

In the early 2000s, a lot of endangerments of the Osa Peninsula resulted from human activities:

- Deforestation: Large tracts of woodland were being deliberately cleared to create room for cattle ranching, palm-oil plantations and other forms of agricultural activity. The process of deforestation caused the loss and fragmentation of habitats which to species constituted a serious threat and a real danger of extinction.
- Agricultural expansion: The spread of monoculture crops in the form of palm oil crop led to soil eroding away, loss of biodiversity, and increased carbon emissions. Unplanned development, disrupting the ecosystems, and bringing about pollution, uncontrolled construction and infrastructural projects led to the undesired consequences.
- Economic pressures: The local communities often had to resort to activities that had negative impacts on the environment. As these formed a means of livelihood, they had no other sustainable and lucrative means of earning of livelihood. The task entailed coming up with a land-use plan that would help preserve the biodiversity of the peninsula at the same time providing sustainable economic offerings to the local community.

10.2.3 Solution

To overcome these challenges, the government of Costa Rica, in partnership with local communities, Non-Governmental Organizations (NGOs) and foreign partners, developed a broad-based sustainable land use strategy. The solution comprised the following essential elements:

- Increased protected areas and buffer zones creation; Additional rungs were introduced in Corcovado National Park boundaries, and some regions were declared protected. Buffer zones were developed around protected areas to facilitate regulated and sustainable use of resources for the central conservation zones. Hence, the need to establish a transitional area to ease strain on the central conservation zones.
- Promotion of agroforestry and sustainable agriculture: Farmers were motivated to adopt agroforestry strategies, where planting of trees is integrated with their crops. This not only improved the quality of soil but also created favourable conditions for wildlife. Diversified cropping systems promotion was designed to minimize dependence on one cash crop. Farmers grew a wide variety of crops such as cocoa, coffee and tree fruits and timber species. As a result of receiving organic farming training, the method of using toxic chemicals was minimized; better environmental and more profitable agricultural outputs followed.
- Development of ecotourism: Many training programs are offered to local communities on many fronts such as hospitality, guiding, and conservation education. Investments were made towards the provision of eco-friendly establishments like lodges and trails. The move was with the intention of reducing the negative impact on the environment. The Osa Peninsula was promoted as a high-end ecotourism location, attracting tourists from all over the globe who sought to enjoy its pristine natural locations.
- Implementation of Payment for Ecosystem Services (PES). The PES program awarded financial incentives to landowners whose land had forest, and those who participated in conservation activities. They had their property compensated in terms of the ecological services from the property, such as the provision of carbon sequestration as well as water purification. The Costa Rican government, international donors

and private sources supported the initiative, which provided a sustainable financing system.

- Community engagement and education: A wide array of community outreach and education initiatives was adopted in order to improve public awareness, deep knowledge of conservation and sustainable practices. Members of the local population were able to improve skills in sustainable land management and alternative livelihoods through workshops and training sessions. The community members took part physically in the decision-making process, such that the solutions were culturally appropriate and regionally popularised.

10.2.4 Results

The integrated approach of sustainable land-use strategy demonstrated dramatic results:

- Reduction in deforestation rates: The Osa Peninsula has experienced a high level of reduction in rates of deforestation over the last years, which indicates a great environmental success. Most of this positive trend is explained by the strategic growth in legally defined protected areas and the use of ecologically sympathetic buffer zones around the core conservation areas. These have been crucial in the conservation of important ecosystems, limiting illegal logging and an overall improvement of forest governance. Innovative descriptions that rely on intensive use of high-resolution satellite imagery, in combination with systematic ground-level ecological monitoring, have convincingly revealed a persistent downward trend in forest loss. This decline not only assists in the conservation of the biodiversity hotspots but also fulfils the global changes of the climate mitigation by improving the ability of carbon sequestration capacities and lowering the risks of the greenhouse gas emissions related to land-use change.
- Preservation of Biodiversity: The Osa Peninsula has seen a strong rebound of wildlife numbers, an indication of the thriving comprehensive conservation efforts. Animals previously on the brink of local extinction have shown signs of recovery in population numbers and a slight reoccupation of habitat. This resurgence can be attributed to specific

biodiversity conservation measures, including large-scale reforestation activities, habitat corridor development and pressure reduction from anthropogenic pressures such as poaching and habitat fragmentation. Ecological restoration projects have played a critical role in the return of native vegetation, stabilization of ecosystems and creation of an environment that will promote breeding and survival of species. Furthermore, the strengthening of legal protection and community conservation programs has improved the long-term viability of these attempts while guaranteeing the protection of flora and fauna in one of the most biologically diverse areas in the world.

- Advantages of ecotourism in terms of the economy: Ecotourism thrived, offering a sustainable and consistent source of revenue for local people. The area's immaculate natural beauty and abundant wildlife attracted tourists. The expansion of ecotourism has resulted in the emergence of employment opportunities in guiding, hospitality, and conservation, thereby diminishing the dependence on harmful agricultural methods.
- Enhanced farming techniques: Farmers who implemented sustainable techniques saw enhanced soil health, elevated crop yields, and diverse sources of income. Agroforestry systems show robustness in the face of climate fluctuations and pest infestations.The adoption of organic farming practices decreased environmental pollution and improved the quality of agricultural products, hence creating opportunities for the expansion of organic markets.
- Improved social welfare: The PES program offered landowners immediate financial advantages, improving their means of living and motivating them to engage in conservation endeavours. The promotion of education and active involvement in the community nurtured a culture centred around the principles of sustainability and the responsible management of the environment. The local communities came to understand the ecological and economic importance of conservation more deeply.

10.2.5 Summary

The effective lasting land-use policy in the Osa Peninsula in Costa Rica is an example of the compatibility with reciprocal enhancement of both environmental conservation and economic advancement. Serving as the pivot between ecological preservation and human growth, the region managed to create an equal balance, preserving biodiversity, promoting sustainable agriculture and investing in ecotourism. This case study is a blueprint of what can be adopted in other areas with similar challenges, revealing the importance of all-encompassing community-driven strategies for sustainable land use. From the experience of the Osa Peninsula it is apparent that it is possible to save the natural legacy of the Earth and at the same time improve the welfare of local citizens, provided that one has implemented effective techniques and teamwork.

10.3 Case Study II: Restoring and Promoting Sustainable Land Use in the Loess Plateau, China

10.3.1 Background

The Loess Plateau, which lies in the part of northern China, has an approximate surface area of 640,000 square kilometers. It stands out for its soil that is apt to be eroded, as well as silt and dust carried by the wind. Historically, this region had fertile soils and supported big farming. However, many years of deforestation, over-grazing, and unsustainable agriculture methods have, unfortunately, led to high degrees and extensive levels of soil erosion, degradation and decline in agricultural production. During the mid-20th century, the plateau was experiencing chronic poverty, frequent droughts and devastating dust storms led to huge environmental and economic challenges in the plateau [8, 9]. The effects of the decline in the Loess Plateau had local and regional ramifications because it created problems in the local area and played a part in more general problems such as the sedimentation of the Yellow River. This, in its turn, worsened flooding downstream.

10.3.2 Challenges

For a long while, the Loess plateau has been facing a variety of environmental and social issues and the main problem has been serious soil erosion. Erosion is mainly the result of deforestation and other poor farming habits which has caused the topsoil to spread out across the land through different kinds of runoff. With soil depletion continuing, the lands have become less fertile, human farming has suffered, people find it harder to eat well and rural communities remain very poor.

The soil quality degradation negatively affected the yields of the crops, degrading the farming state to being unsustainable. It was harder and harder for agricultural workers to support themselves as the land became steadily less productive. Such environmental degradation was reflected in the reduction of vegetative cover and biodiversity and in the acceleration of the processes of land degradation and desertification. The feedback loop of poor land management, limited plant cover and continued erosion even aggravated the resilience of the landscape.

These pressures have brought huge social consequences. The continuous drop in agricultural production has made it difficult for many residents to stay where they are. This has led to a huge wave of migration to try and follow better economic opportunities, thus leaving the affected areas depopulated and the remaining community structures under stress. Social cohesion erosion has complicated coordinated conservation and development endeavours.

Further, the abysmal state of the plateau has left it more and more vulnerable to climate extremes. Droughts and bouts of intense rainfall, which are becoming increasingly common due to climate change, have also increased erosion and water scarcity. This increased environmental volatility calls for the simultaneous implementation of sustainable land management practices, ecological as well as socioeconomic restoration plans and community-based resilience-building strategies to interrupt the cycle of degradation and socioeconomic vulnerability.

10.3.3 Solution

The Chinese government only started the Loess Plateau watershed rehabilitation project during the 1990s, with support from international institutions such as the World Bank. The project's aim was to improve the ecological welfare of the plateau and improve the standard of living for its residents through effective use of land on a sustainable basis. The solution comprised several essential elements:

- Reforestation and afforestation: Intensive afforestation projects were carried out in an effort to increase soil stabilization and restore vegetation. Those tree species with high growth rates and native plants were selected based on their ability to survive in the local ecosystem. In all about 100,000+ hectares have been forested with a mix of native and exotic species, and this was done to boost diversity. Terracing was utilized on steep inclines to mitigate soil erosion and establish cultivable areas for the cultivation of trees and crops. This technique converted inclines into terraced flat areas, resulting in a decrease in flow of water and increase absorption of water into the ground.

- Conservation of soil and water: The implementation of terraces, check dams, and sediment control structures effectively mitigated water runoff, minimized erosion, and enhanced soil moisture retention. A total of 10,000 kilometers of terraces were built. Contour farming techniques were advocated to synchronize agricultural practices with the natural topography, hence reducing soil disruption. These methods involved strategically growing vegetation along the existing contours of the ground in order to minimize the amount of water flow and prevent soil erosion.
- Ecologically responsible farming practices: Agricultural practitioners were urged to embrace sustainable farming techniques, such as crop rotation, intercropping, and organic farming, in order to improve soil fertility and decrease reliance on chemical inputs. The implementation of crop diversification strategies has resulted in enhanced food security and its associated risks. The management of livestock grazing involved the implementation of measures to restrict and confine it to certain regions, with the aim of avoiding excessive grazing and facilitating the regeneration of vegetation. Rotational grazing systems were implemented as a means of sustainable pasture management.
- Economic diversification: One of the foundations of the Loess Plateau restoration initiative was the encouragement of economic diversification processes intended to minimize the region's exposure to conventional practices of agriculture, which are an environmental menace. The project actively propagated the search for alternative livelihoods, such as beekeeping, mushroom cultivation and the cultivation of medicinal herbs, which were a good fit to the ecological conditions of the region. These initiatives generated additional income for the households, which led to locals becoming more economically resilient and alleviating pressure on degraded farmlands. Apart from small-scale livelihood diversification, the project also facilitated the growth of rural-based industries such as fruit processing, industries

based on local artisanal handicrafts and industries based on eco-tourism. These efforts not only provided new employment opportunities for the people but also helped to retain the local population by providing the economy with new prospects in the region.

- Involvement of the community and education: Planning and implementation of restoration efforts were done with active involvement from local communities. Training programs and workshops were conducted in order to train farmers in sustainable land management practices. It was important for community-based groups to facilitate participation and sustain the momentum of the project. The initiative provided money incentives and technical support to ensure that the community is involved and the project is sustainable over the years. The incentives included subsidies for the adoption of sustainable practices and microloans for small-scale businesses.
- Investigation and surveillance: The project was dependent on the ongoing study and monitoring to determine the success of the restoration strategies and the implementation of the adaptive management scheme. An effort was made by scientists and local experts to monitor the status of soil, vegetation cover and water quality.

By collecting the data after the monitoring efforts, gaps in approaches to be improved were identified, and best practices were then disseminated to other sites that faced similar difficulties.

10.3.4 Results

The Loess Plateau watershed rehabilitation project has achieved amazing success in effectively reconfiguring the landscape and improving the quality of life for millions of dwellers.

- Soil erosion reduction: Due to the introduction of massive reforestation efforts and conservation of the soil technology, the rates of soil erosion have come down. The renewed vegetation acted as an erosion controller, which reduced the effects of wind and water erosion. Research demonstrated a decrease in sediment flow of more than 70% in certain regions. The implementation of terraces and sediment control structures successfully mitigated the occurrence of runoff and sediment erosion. These methods also contributed to a reduction in the amount of silt entering the Yellow River, thereby mitigating the occurrence of flooding downstream.

- Rehabilitation of flora and fauna: The initiatives to replant forests and establish new forests resulted in a significant augmentation of vegetation coverage. The previously desolate plateau transformed into a verdant and fertile landscape. The vegetation cover had a significant increase, rising from approximately 17% to more than 34%. The reintroduction of native plant species and restoration of ecosystems led to an enhancement in biodiversity, which in turn supported a revival of local wildlife populations. The region experienced the reappearance of species that had previously vanished, such as a variety of avian, mammalian, and insect species.
- Enhanced agricultural productivity: Implementing sustainable agricultural techniques improved soil health, leading to higher crop yields and a more varied agricultural output. Farmers achieved enhanced food security and increased incomes. Enhancements in soil fertility and water retention led to a notable upsurge of up to 50% in crop yields in certain regions. The implementation of contour farming and other conservation techniques expanded the amount of cultivable area and mitigated the likelihood of crop failure caused by soil degradation. Improved irrigation techniques and enhanced water management procedures have provided benefits to farmers.
- Improved means of living and decreased levels of poverty: The project generated employment possibilities and offered alternative income streams for local populations through initiatives like as afforestation, terracing, and sustainable agriculture. The introduction of employment opportunities in emerging industries played a significant role in diminishing migration and establishing stability within communities. The region experienced economic improvement as a result of increased agricultural output and restored ecosystems, leading to a reduction in poverty levels and a decrease in migration. The average income in project regions experienced a 60% boost.

- Advantages for the environment and climate: The reestablished plant life aided in the process of carbon sequestration, thereby alleviating the impacts of climate change. The Loess Plateau has emerged as a notable carbon sink, effectively mitigating emissions. According to estimates, the areas that were restored were able to capture and store millions of tons of carbon dioxide every year. The mitigation of dust storms and soil erosion resulted in enhanced air quality and a decrease in the occurrence of natural disasters. Restored lands experienced enhanced microclimates, which subsequently facilitated improved agricultural conditions.
- Enhanced community resilience: The emphasis on community engagement and education empowered the local inhabitants to effectively oversee their resources in a manner that ensures long-term viability and enables them to effectively respond to environmental fluctuations. Communities developed more resistance to the effects of climate change and economic instability. The project's accomplishments were sustained, and farmers were able to share expertise and provide support to each other through the establishment of local cooperatives and community groups.

10.3.5 Summary

The Loess Plateau in China is an example of how community-based, integrated restoration measures and sustainable land use have a deep capacity to rehabilitate environments. This project successfully reduced soil erosion, promoted sustainable agriculture, and engaged the local community actively, leading to successful reclamation of degraded land, conservation of biodiversity and a major change in the living standards of millions of people. The Loess Plateau case study has become an ideal model for other areas that have similar challenges and proves the power of ecological restoration to drive sustainable development and support the community's resilience to climate change. The project stresses that through effective tactics, dedicated partnerships and participation of the community, it is possible to rehabilitate a degraded landscape and obtain lasting environmental and social benefits.

10.4 Case Study III: Sustainable Agroforestry in Murang'a County, Kenya

10.4.1 Background

The Murang'a County, in central Kenya, is in the highlands of the country. People are amazed by the quality of its soil as well as its favourable temperature, which suits agriculture. Despite all these benefits, however, the area has faced numerous challenges, including soil degradation, loss of forest, and deplorable conditions among small-scale farmers. Traditionally, the use of single cropping and the clearing of land through the application of fire has led to wastage of soil, its nutrients and poor productivity of crops [10, 11]. Overpopulation has worsened these challenges, such that the rate of using natural resources has increased. The region that takes its main source of economy from agriculture suffered a slump, as farmers realized a decline in the harvest and revenues.

10.4.2 Challenges

Murang'a County faced several key barriers, as described below.

There is soil degradation due to long-standing monocropping and unsustainable practices, high rates of soil erosion, poor fertility of the soil, and increasingly poor yields of crops. Insufficient soil management procedures inflicted nutrient depletion and reduced the land's capacity to produce crops. The need to use arable land and fuelwood has led to extensive cutting down of trees, causing permanent soil erosion and adversely affecting indigenous ecosystems. Deforestation has also enhanced the rate of biodiversity loss and disturbed the local water rhythm. Smallholder farmers, who formed the majority of the population, were not only poor but also experienced food insecurity due to poor agricultural yields, hence low incomes. A large number of houses were living below the poverty line without income to meet their basic needs.

Climate change has caused a very unpredictable nature with droughts and too much rain, reducing the effectiveness of traditional farming methods while making local people more susceptible to damage. Climate change aggravated preexisting environmental and socioeconomic concerns. Farmers, from time to time, were to be denied access to modern agricultural technology, finances, and knowledge regarding sustainable practices. These limitations alike precluded their ability to increase output and adapt to changed situations.

10.4.3 Solution

To address these difficulties, multiple projects were implemented to advocate for agroforestry as a sustainable land-use method in Murang'a County. Agroforestry is a practice that combines the cultivation of trees and shrubs with traditional agriculture, resulting in various advantages for the environment, economy, and society. The solution comprised several essential elements.

- Overview of agroforestry practices: Agriculturalists were urged to cultivate trees in conjunction with their crops and along the perimeters of their fields. This encompasses versatile tree species that offer fruits, firewood, animal feed, and timber. Commonly found species comprised Grevillea, Calliandra, as well as fruit-bearing trees such as mango and avocado. Alley cropping is a technique that involves planting rows of trees or shrubs between rows of crops. This method helps to prevent soil erosion and enhance soil fertility by promoting nitrogen fixation and adding organic matter. This strategy also generated supplementary revenue through the sale of tree goods.
- Conservation of soil and water: The implementation of terraces and contour bunds effectively mitigated soil erosion and enhanced soil water retention. These structures impeded the speed of water flow, facilitating increased water infiltration into the ground, which was essential for sustaining soil moisture levels during arid periods. Rainfall harvesting systems were implemented to collect and retain rainfall for the purpose of irrigating crops during periods of drought, hence enhancing water accessibility for agricultural purposes. These systems encompassed the establishment of lakes, reservoirs, basins, and cisterns to guarantee a dependable water supply.
- Training programs and workshops: Seminars were organized to enlighten farmers about the advantages of agroforestry and sustainable farming practices. Agricultural practitioners acquired knowledge on tree nursery administration, methods for preserving soil quality, and strategies for integrated pest control. Extension officers offered ongoing assistance and expert guidance. Demonstration plots were created with the purpose of exhibiting effective agroforestry techniques and functioning as educational locations for farmers.

The plots demonstrated the tangible advantages of incorporating trees into agricultural operations.

- Advancement of varied means of making a living: Agroforestry generates supplementary revenue streams by selling tree products such as fruits, timber, honey, and medicinal plants. The act of diversifying lowered the financial risk that was connected to crop failures and enhanced the earnings of households. Farmers were incentivized to participate in value-added endeavours, such as the processing and sale of agroforestry products, in order to augment their economic gains. This encompassed instruction in post-harvest management, packing, and the creation of market connections.
- Participation of the community and provision of support from institutions: Community-based organizations and cooperatives were created to enable collaborative efforts, exchange of information, and pooling of resources. Along with that, they provided forums for reciprocal mutual aid and sharing of problem solutions. Working with both governmental and non-governmental organisations provided a financial and technical input, ensuring the sustainability of agroforestry projects in the long term. These collaborations guaranteed the search for financial resources for the initiatives, access to research and technology, and encouraged policy advocacy promotion.
- Investigation and surveillance: The project had to be undertaken with continuous study and monitoring in order to guarantee the effectiveness of the restoration approaches as well as the implementation of adaptive management strategies. A collaborative exercise was undertaken by scientists and local specialists to monitor the state of soil, vegetation cover, and the quality of water. The information derived from monitoring operations was used to enhance approaches and share best practices with other areas experiencing similar problems. This information needed to be shared so that knowledge practices could be expanded.

10.4.4 Results

The application of agroforestry practices to Murang'a County led to some remarkable positive results:

- Project increases soil health and strengthens agricultural yield: Strategies in agroforestry are proven to control soil erosion, bolster the supply of nutrients and increase soil organic matter and recycling of nutrients. Those who practice agroforestry in agriculture report boosts in crop yields and changes for the better in their soil composition. Planting trees that fix nitrogen has helped tool soil fertility using biological activities. Planting nitrogen-fixing plants and shrubs allowed farmers to supply their land with nutrients which led to a decrease in fertilizer use and savings. It made it possible to implement stronger and environmentally friendly farming approaches and also lessened the amount of pollution.
- Increased biodiversity and ecosystem services: Integrating trees into farm landscapes has increased biodiversity in the region by adding habitats that will support many species of plants and animals. This strengthened the resilience and recovery of the ecosystem while also supporting the delivery of pollination and pest control services. Vegetation including trees and bushes played an important part in regulation of microclimate through minimizing the rate of temperature fluctuations and wind velocities.
- Enhanced household incomes and food security: The introduction of various income streams based on agroforestry products facilitated the improvement of household incomes and reduced dependence on a single crop. The economic resiliency enabled families to allocate resources towards education, healthcare, and other fundamental necessities. Enhanced agricultural production and diverse food sources have bolstered food security, thereby mitigating the danger of hunger and malnutrition. Households enjoyed an expanded range of nourishing food options, which positively impacted their general well-being.
- Strategies for addressing and responding to climate change: Agroforestry has made a significant contribution to mitigating climate change by capturing and storing carbon in trees and soil. The augmented forest cover additionally facilitated the regulation of local climate conditions, enhancing the area's resilience to climatic extremes. Enhanced soil moisture retention and diversified agricultural systems bolstered farmer's capacity to adapt to climate variability, hence diminishing their susceptibility to droughts and high rainfall.

Trees served as a protective barrier against severe weather occurrences, offering a degree of safeguarding for crops.

- Enhanced community organizations and empowerment: Community-based groups and cooperatives strengthened farmers through the establishment of platforms that facilitated collective action, resource sharing, and advocacy. These institutes promote the sharing of knowledge and assist in expanding agroforestry practices. Participation of farmers in training programs and extension services strengthened their capacity and assurance to adopt and sustainability agroforestry systems, promoting resilience and self-sustainability. Communities that were granted power and authority were more effective in their advocacy of their needs and interests; thus they could operate upon local laws and practices.
- Enhanced ecological consciousness and responsible management of the environment: The primary contribution of the project was the development of an increased level of ecological consciousness among local communities. The initiative used substantial outreach, training and participatory education activities to develop a local engagement and traditional knowledge-based culture of environmental stewardship. Farmers and land users came to appreciate the long-term ecological and economic value of long-term practices including agroforestry, contour farming, and soil conservation practices. This heightened awareness of the ecological issue was transformed into a wider commitment to the responsible utilization of land and natural resources. Communities started intensively engaging in reforestation efforts, water conservation and biodiversity protection activ ities. The project highlighted the synergy between healthy ecosystems and human well-being, causing land management approaches that reviewed the win-win conditions of ecological balance, soil regeneration and climate resilience to be slowly adopted.
- Public education: Public education made a critical difference in shaping attitudes towards conservation and sustainable land use within the region. Several forms of awareness-raising activities, such as community workshops, school-based programmes, village meetings and media campaigns, were undertaken to give residents information on the need to conserve natural resources and protect the

local environment. These educational efforts highlighted the consequences of land degradation and the merits of following a system of sustainability, which could make ecological issues relevant and decisive to implementation at the grassroots level. Thus, there was an evident change in public perception as community members presented an increasing sense of environmental responsibility. This turned out to be an invaluable contribution to the general spread of more sustainable approaches to land management and increased local involvement in conservation initiatives, such as tree planting, erosion control and water resources protection. As the initiatives integrated environmental consciousness into routine choices, the efforts also gave the people power and encouraged a people-wide commitment to the future maintenance of the land.

- Improved social unity and cooperation: The project has made a considerable contribution to the consolidation of social cohesion and the development of cooperative interaction between local populations through the creation of community-based organizations and agricultural cooperatives. Such groups were key interfacing platforms for collaboration where farmers could share knowledge and pool resources, along with having a collective way of dealing with issues of land management and livelihoods. The growth of these social networks strengthened the mutual support system and encouraged the emulation of best practices, which increased environmental and socioeconomic resilience. One of the critical components of this progress featured a project-oriented, inclusive, and participatory perspective in which women, youth, and marginalized community members were engaged in decision-making processes and in project implementation. Not only did the initiative contribute significantly to social equity, but it also developed a significant range of perspectives within community planning and environmental stewardship through equitable representation and participation.
- Policy influence and advocacy: The successful outcomes of agroforestry projects in Murang'a County greatly improved the policy towards sustainable agriculture and land use at the local and national levels. Policymakers recognized the

benefits of the agroforestry system and began incorporating provisions for encouraging it into the agriculture and environmental legislative framework. Through advocacy, community-based organizations and their partners successfully managed to get finance and technical assistance for developing agroforestry projects. Moreover, these activities promoted the adoption of ideal approaches in other areas, promoting broader environmental and economic benefits.

10.4.5 Summary

The way in which the agroforestry method has been implemented in the Murang'a County of Kenya demonstrates the potential of sustainable land use strategies for resisting environmental degradation, helping to increase agricultural output, and supporting livelihoods. Farmers received a wide range of benefits from integrating trees, shrubs, etc., in their farming systems, including better soil health, biodiversity, food security, and a better ability to cope with the impact of climate change. The relevance of community involvement, skills, knowledge enhancement, and institutional backing in promoting sustainable utilization of land is underlined by the success of agroforestry in Murang'a County, Kenya. This case study represents a blueprint for other areas with similar problems and shows how agroforestry could be valuable for the promotion of sustainable development and protection of the environment.

10.5 Case Study IV: Sustainable Land Management Practices in the Sahel Region of Africa

10.5.1 Background

The semi-arid Sahel region, which extends from Senegal to Sudan, has experienced severe land degradation with the effects of climate change, population growth and poor agricultural methods. The ecological sensitivity is reinforced by the significant diversity in rainfall over the area, as well as in space and time [12, 13]. Over the last several decades, repeated droughts in the Sahel have intensified the problem of food shortage and poverty among locals. Land degradation has diminished agricultural production, accelerated desertification and increases sensitivity of the area to fluctuations in climate.

10.5.2 Challenges

The Sahel region faces a cascade of interrelated crisis issues that threaten to disrupt both environmental health and the well-being of its citizens. The most urgent problem is the significant amount of land degradation, which is largely due to activities as overgrazing, deforestation, and overagriculture. Repetitive overuse of land by livestock negatively affects the protection of soil from degradation, while deforestation eliminates important root networks that prevent the soil from eroding. That is also worth mentioning as dependency on monoculture, lack of attention to crop rotation, and poorly designed irrigation systems threaten the integrity of soil and the accumulated advantage of essential nutrients. Such integrative effects manifest themselves as accelerated degradation of soil, decline in agricultural production and large-scale alteration of land that makes many areas unfit for agricultural and grazing activities.

Such degradation means that the capacity to produced food is affected. Weaker soils produce lower agricultural output and poorer pastures, which compromise regional access to food and livelihoods dependent on livestock, major food providers and sources of employment. Consequently, massive populations of communities are left in a state whereby they suffer constant food scares and mounting poverty, with smallholder farmers and nomadic pastoralists being particularly seriously affected by their dependency on the land.

Increased sensitivity of the region to climate variability aggravates these existing challenges. Unpredictable occurrences, few and short periods of rains, and droughts often mess up the agricultural cycle and restrict access to water in the Sahel. Thus, erratic agricultural and grazing availability surfaces, marring and deepening food systems and worsening famine in drier seasons.

One major worry is the rapid growth of population and therefore the intensity of natural resource demand, which is already stressed. With increased population, there is increased consumption of food, firewood, housing and arable land. Consequently, unsustainable land management practices are also commonly developed, which causes land degradation to occur at an increased rate. Conversion of forests for farming, draining of wetlands and exploitation of marginal lands increase environmental erosion.

10.5.3 Solution

As a response to these problems, several projects were undertaken to stimulate the wider use of sustainable land management (SLM) practices in the Sahel region. The primary aim, however, was to connect old wisdom with modern methodologies to rejuvenate depleted land and enhance the sustainability of livelihood. The solution comprised essential elements as described below.

- Agroforestry and Farmer-Managed Natural Regeneration (FMNR): These two techniques are used in the management of land resources. Agriculturalists were incentivized to adopt FMNR, a practical and green method for tree and shrub restoration based on native roots and seeds. By preserving and controlling natural regeneration of trees and bushes in agricultural territory, this approach promotes the rational use of resources. Agroforestry system promulgations relate to a combination of trees and crops which increase soil fertility and prevent soil erosion while achieving more food, fuel and benefits. Typical agroforestry species comprise nitrogen-fixing trees like Faidherbia albida and Acacia.
- Conservation of soil and water: Methods such as contour bunds, stone lines, and zaï pits were utilized to enhance water infiltration, diminish runoff, and augment soil moisture. These methods aided in the restoration of deteriorated lands and improvement of agricultural output. Infrastructure such as rainwater harvesting systems, which consist of tiny dams and water catchments, was built with the purpose of collecting and retaining rainwater for irrigation during periods of drought. The methods for cultivating crops and raising livestock were environmentally friendly and could be maintained over the long term. The promotion of conservation agriculture methods, such as limited tillage, crop rotation, and the utilization of cover crops, aimed to preserve soil health and enhance crop productivity. These strategies supported the protection of soil structure and the balance of nature in the soil. Pest and disease management strategies of eco-friendly Integrated pest management (IPM) were instituted and use of chemical pesticides was reduced significantly.
- Enhancing capabilities and providing outreach services: Educators sessions and training workshops were organized to train farmers on sustainable land management practices and how they benefit from agroforestry and FMNR. A constant supply of experts was facilitated by extension services to provide advice and guidance. Farmer field schools and

field demonstrations were established to demonstrate exemplary practices in SLM and promote information exchange among participants.

- Community engagement: Formal groups and cooperatives were established with the objective of promoting community wide partnership, ensuring resource allocation and lobbying for conducive legislation. These collectives were instrumental in the promotion and the provision of guarantees that there would be continued use of SLM strategies for the future. The partnership of public and private interests provided critical financial and technical benefits that helped carry out and grow sustainable land management practices in the region.

10.5.4 Results

The incorporated use of sustainable land management by the Sahel region led to positive results.

- Land rehabilitation: Extensive adoption of FMNR and agroforestry has been successful in rehabilitating a wide range of land environments previously denuded. Farmers observed more trees growing, improved soil and reduced soil erosion. The application of soil and water conservation methods has increased the soil water holding capacity. This in turn enhances large agricultural output and drought resistance. Previously overlooked pieces of land are now returning to very productive agricultural fields.
- Enhanced food security and livelihoods: Adoption of agroforestry practices has improved rural households food security, decreased poverty, increased agricultural productivity and increase in income generation. With trees, crops and livestock thriving side by side on the same parcel of land, agroforestry has encouraged enhanced soil fertility, improved microclimatic conditions and improved water retention, all leading to increased agricultural output and improved livestock health. With the introduction of agroforestry, farmers gained the ability to sustain their output during climatic instability because of the ecological benefits the cover of tree covers brings. Other than improving farm productivity, the market development of tree-based agricultural products like edible fruits, nuts, medicinal herbs and

fuel wood opened various and sustainable sources of income for households. By spreading their sources of income among tree products, households obtained more diverse sources of income, which reduced their dependence on one crop, and therefore influenced the households' financial stability positively. Revenues from commercialization of forest resources helped finance critical areas including children's education, access to health care, household amenities, and more variety of food types. As a result of their history in harvesting and selling tree products, women in particular benefited from this diversification, with more money control and power within their communities. Agroforestry accelerated more than ecosystem renewal and economic efficiency, which was a core component of rural prosperity, poverty reduction, and community-well-being.

- Increased biodiversity and ecosystem services: By planting trees and shrubs within agricultural lands, farmers provide varied habitats to a variety of plants and animals life. Such practices improved the stability of an ecosystem and also promoted benefits such as pollination and efficient pest control. Trees and shrubs had an effect on the local microclimate and helped to avoid the extreme fluctuations of temperature and low speeds of winds. This enhancement favoured improved crop growth, but at a reduced crop water loss by evaporation.
- Strategies to address and respond to climate change: Both agroforestry and Farmer-Managed Natural Regeneration (FMNR) have enhanced the storage of carbon in soils and trees contributing a crucial step in climate change mitigation. Vegetation increase was very useful in controlling the local climate and enhancing the microclimate at the surrounding area. The farming community has been empowered to better adapt to changing climatic conditions owing to greater soil moisture conservation and various farming practices adaptation.
- Community engagement: Community-based groups and cooperatives gave those farmers an opportunity to informally organize, share resources and lobby for their interests. These organizations promoted information flow and also promoted the development of sustainable cultivation techniques. The adoption of training programs and extension

services enhanced farmers' knowledge and confidence in being able to sustain SLM practices, which indeed encouraged and promoted sustainable resilience and self-sufficiency.

- Policy influence and advocacy: The effectiveness of SLM programs in the Sahel region is central in the development of policy in sustainable agriculture and land use at both national and localized level. Armed with the appreciation of the merits of Sustainable Land Management (SLM) practices, authorities started introducing support measures into laws for agriculture and environment. Advocacy efforts on the part of community-based organizations and partners brought SLM initiatives the needed financial and technical resources.

10.5.5 Summary

Implementing sustainable land management practices in the Sahel has shown how integrated solutions can fight against soil degradation and boost food production as well as improve human well-being. Through a combination of proven time-tested methods and contemporary scientific approaches, the farmers in the region have restored their soil, increased nutritional security and made their populations more resilient to climate situations. The effectiveness of SLM projects in the Sahel demonstrates that a local component, greater capacity building, and strong institutional frames are fundamental for sustaining land use practices. This experience provides a model to other societies that are dealing with the same engagements to show how SLM approaches can support sustainable growth and environmental conservation. The lessons from the Sahel experience are that any effort at making lasting and important change is based on the integration of environmental, economic and social concerns into comprehensive strategies.

10.6 Case Study V: Sustainable Grazing Management in the Great Plains, USA

10.6.1 Background

This region of the United States, which is home to its famous fertile grasslands and productive agricultural yields, is renowned as the "breadbasket." However, too many grazing animals, soil loss and declining soil quality are posing major threats to its stability in both farming and nature. Things have become even worse as more farms use modern methods and climate change disturbs the environment, resulting in lower productivity and loss of various native animals and plants [14, 15]. Concerns about the environment have made ranchers and land managers in the Great Plains switch to sustainable grazing techniques to restore faulty ecosystems and safeguard the grasslands.

10.6.2 Challenges

The main obstacles encountered by the Great Plains region are described below.

The Great Plains faces a wide range of environmental and socioeconomic threat scenarios that create difficulties in the maintenance of the sustainability of its ecosystems and in providing a basis for rural livelihoods. One of the biggest problems is the excessive soil erosion resulting from the unending, poorly managed grazing habits. As a result of inadequate rotational grazing practices, the native grasses have been exhausted, soil has become more compacted and the potential for erosion has increased, thus reducing the land capability of having both healthy vegetation and livestock. The adverse effects on the land undermine its potential for production and sustain successive declines in the fertility of soil and water retention.

One of the major problems is the decreased biodiversity, mostly due to the substitution of natural grasslands by widespread monoculture farming and ongoing inadequately regulated grazing activities. Monocultural farming practice and poor grazing management have systematically reduced the variability of species and have significantly undermined the ecological stability and health of the region. The decline of native species has impaired critical ecosystem services such as pollination, pest control, and nutrient cycling, which underpin sustainable lands in the long term.

Climate change has exacerbated existing challenges of the Great Plains, including an increase in average temperature, lengthier periods of drought with changes in precipitation. Such transformations have helped in the establishment of more regular and extreme weather, degradation of grass cover and water resources therein, consequently reducing the strength and adaptability of grassland ecosystems.

Land use patterns are greatly affected by financial issues. Especially with unclear market conditions and limited funds, ranchers are mostly concerned with quick results and maximizing their earnings. Concentrating on quick outcomes generally makes it harder to focus on long-term measures such as rotating grazing, planting native vegetation or restoring riparian areas, as these often need much time and money before offering obvious results. Thus, the high cost to set up sustainable systems and the slow delivery of benefits prevent many people from using them.

Finally, lack of information, access to technical guidance, and limited financial opportunities have limited the wider spread of sustainable grazing and land management practices. Even getting training, understanding new conservation models, or seeking funds for improvements like fencing, water system, or restoration of soils has proven difficult for many ranchers. Without institutional support and community extension, the rate of advancement towards sustainable land management practices ranges significantly in the area.

10.6.3 Solution

To address this, pastoralists and land stewards of the Great Plains are adopting a number of sustainable grazing practices that seek to restore grassland vitality and provide perennial yield. The solution comprised the following essential elements.

- Rotational grazing: To solve the problem of overgrazing, ranchers changed their way of grazing animals to rotational grazing, where animals are moved in a planned manner from one pasture to the other. Through this method, it is possible to restore grasslands, stimulate the growth of active plant communities and decrease soil erosion. Through carefully planning different grazing approaches, it was seen that meeting the life cycles of local grass species and the nature of each environment could help to share grazing pressure more evenly.
- Adaptive Multi-Paddock (AMP) Grazing: The AMP grazing technique involves subdivision of large grazing areas into smaller ones and rotating livestock to follow whatever nature would have dictated when it comes to grazing. This approach promotes increased grassland productivity and enhanced resilience through the development of deeper roots and improved soil states. By using AMP grazing systems, ranchers were able to customize grazing to deal with the existing weather conditions, availability of fodder and changing ecological scenario.
- Native prairie restoration: Native grasslands were to be restored through plant varieties reintroduction and suppression of invasive species. This included replanting native grasses and the co-management of non-native species using

some mindful grazing control measures with mowing. The restoration activities were carried out by ranchers with support of the government organizations and non-profit institutions, including technical expertise and monetary aid.

- Conservation of soil and water: Effective conservation of soil and water has been realized through the use of riparian buffers, establishment of water reservoirs, and enforcement of erosion control. Through the above strategies, the risks of lost water within the land were decreased, and water quality was improved, which translates to the improvement of the environment for wildlife. Organizers wanted to ensure soil integrity and increase its quality by implementing contour ploughing and no-till farming techniques.
- Surveillance and data gathering: Ranchers used measuring tools as instruments for estimating soil health, growth of forage, and general cattle health. The use of data as a lens on their strategies allowed them to make reasoned decisions and change their management toward reasoned actions. Joint work with research organizations and extension programs allowed ranchers to implement state-of-the-art discoveries and methods and continue the improvement of their practices.
- Community involvement and instruction: Ranchers created these cooperative groups and maintained educational projects to allow the exchange of information and knowledge. Through these networks, cooperation was enhanced, and there was promotion of the adoption of sustainable grazing systems. Notable improvements came when more ranchers accepted efficient land management thanks to the field days, workshops and demonstration projects.

10.6.4 Results

The adoption of sustainable grazing management strategies in the Great Plains has resulted in notable favourable consequences.

- Enhanced vigor of grassland ecosystem: The adoption of rotational grazing and adaptive multi-paddock (AMP) grazing practices has significantly enhanced the ecological quality and functional resilience of grassland ecosystems in the Great Plains. This method involves deliberate rotation of livestock in separate grazing paddocks so that the vegetation gets time to recover, roots get stronger, and the soil becomes more stable. Rotational grazing patterns prevent overgrazing in the same areas continuously, all the while allowing native grasses and forbs to replenish efficiently. On the whole, improvement in soil structure (porosity, reduction in compaction) has favoured the development of roots, organic matter, and improved water retention capacity. Improved availability and quality of forage have facilitated higher livestock yields that promote more environment-friendly grazing practices. Results from land surveys and scientific assessments have shown that attempts to restore native prairies have produced enhanced biodiversity and ecosystem functioning. Besides, returning these habitats has renewed valuable ecological services such as increasing water intake, reducing surface flow, soil wash, increasing carbon storage in plant tissue and soil organic matter. By attaining these results, grazing lands are becoming more robust against adverse impacts, including drought and land degradation, thus helping maintain their long-term ecological soundness.

- Increased biodiversity: The introduction of sustainable grazing methods has significantly enhanced the variety of plants and animals on the Great Plains ecosystem. In transitions away from persistent overgrazing to approaches, such as rotational and adaptive grazing, land managers have provided native grasses, wildflowers, and perennial plants the opportunity to thrive. These plants play a fundamental role in the structure of prairie ecosystems, offering needed shelter, food, and home to a wide range of wildlife. As plants became more varied and restored, so did the complexity of the ecosystem, thus allowing for more pollinators such as bees and butterflies and ground-nesting birds, small mammals and invertebrates. Land managers and conservationists have noticed the increasing number of native wildlife species, and many species on the decline are recovering, showing ecological recovery. Greater species diversity also enhances functional biodiversity by improving soil composition and nutrient recycling, as well as pollinator communities important for natural ecosystems and agricultural yield. Greater biological diversity in the region is resilient to climate fluctuations and the pest risks, and also good for future ecological and economic health.

- Enhanced soil health and carbon sequestration: The Great Plains of North America have experienced notable restoration of soil health since the adoption of advanced grazing practices such as the rotational and adaptive multi-paddock systems. The integration of these management practices, which reduce over-grazing and provide continuous ground cover, has been the catalyst for the build-up of organic matter on soils, which is key to the improvement of soil health and biological activity. The addition of plant residues, in combination with rapid root development, has caused an increase in soil organic carbon, a fundamental component of soil quality. In addition, implementation of these practices results in improved soil structure with increased porosity and reduced soil compaction. The improved stability of the structure of soil provides an advantage in terms of increased microbial populations that are crucial for decomposing organic material, release of nutrients and protection of the ecological balance of soils. Due to these practices, ranchers are reporting more water infiltration, better utilization of nutrients and reduced frequency of surface water and sediment runoff. Remarkably, well-managed grasslands have proven themselves also to be effective carbon sinks. Sustainable grazing systems increase biomass production and root absorption volumes, which will facilitate the downward and upward absorption of carbon dioxide from the soil and aboveground vegetation. Through carbon storage these ecosystems significantly cut greenhouse gases in agricultural settings. This long-term effect supports environmental flourishing and the flexibility of rangeland ecosystems to suit the changes in climatic conditions.
- Advantages in terms of economics: The adoption of sustainable grazing practices has translated to huge environmental returns and significant economic benefits to Great Plains ranchers. The increased availability and quality of forage, and corresponding healthier vegetation and soil, has lowered the need for expensive supplemental feed, a major cost factor in raising livestock in general. High levels of nutritious and high-yielding forage availability throughout the grazing period have led to healthier cattle, faster weight gains, and better reproductive responses, thereby increasing ranch profitability. Furthermore, improved pasture

management has allowed ranchers to increase animals per acre, without harming the natural ecosystem negatively. This translates to increased yield on every acre of land and improved sustained profitability. Diversification of revenue has greatly enhanced economic stability. Several ranchers who practice sustainable land management have been able to successfully earn income through payout modalities for ecosystem services, including reimbursement for carbon sequestration, conservation of habitats, participation in government or NGO-based sustainability projects. Through its extra income opportunities these programs help to stabilize ranchers' budgets, reduce their exposure to market risks and encourage the continuity of sustainable agricultural practices. In summary, the healthy performance of sustainable grazing confirms that sustaining the economy and conservation of natural resources are inseparable goals.

- Adaptability to climate fluctuations: The introduction of sustainable grazing techniques improves the ability of grasslands to cope with changing climates because deeper roots are fostered as well as an enhanced recycling of soil moisture. In times of drought and harsh weather, ranchers reported improved quality in their pastures. Using adaptive management as the cornerstone of their approach, ranchers could easily adapt to changing atmospherics and make their farms more resistant to climate-related issues.
- Enhanced community and knowledge exchange: The movement towards sustainable grazing in the Great Plains has been greatly aided by the rise of community-based programs and cooperative networks of ranchers. Such joint frameworks have played a great role in the creation of social unity, so producers can share insights, pool resources, and solve common impediments together. Continuing discussions at forums and workshops, peer learning activities, and grazing cooperatives have encouraged cooperative support and stimulated the emergence of new grazing practices. In these platforms, there was a culture of knowledge exchange, and ranchers had the opportunity to benefit from the experiences and mistakes of other ranchers in the community. These platforms allowed ranchers with minimal technical resources to obtain and apply state-of-the-art approaches and adaptive practices. At the same time, the ranchers enjoyed the

power of a multitude of education programs, governmental extension agency support, and strategic partnerships with universities and NGOs working toward enhancing technical capability. These resources inform ranchers on the maintenance of soil condition, rotation grazing, monitoring their properties, recording carbon emissions and enhancing their drought resilience. This allowed ranchers to adopt sustainable grazing with an imagination of adapting these given a new situation. These synchronized activities have improved land preservation as well as facilitation of sustainability for rural communities with extended impacts towards the sustainability of the environment and society.

10.6.5 Summary

Sustainable grazing regimes in the Great Plains represent an example of how strategies can be successfully applied to the problems of land degradation, promotion of productive grasslands, and improvement of the well-being of the local community. The introduction of rotational and adaptive multi-paddock grazing, soil and water conservation enabled ranchers to capture a variety of benefits. Such gains are evident in the form of more robust grasslands. Incorporating sustainable grazing methods in the Great Plains has contributed to the growth of ecosystems which are able to withstand climate change, and many species are showing greater resistance to it. This underlines how involvement of stakeholders, communities, organizations and good governance supports the goal of sustainable land use. The Great Plains project shows that appropriate grazing benefits the environment, local communities, and helps control climate change. The example teaches us that one must consider ecology, economy and society when making land management decisions.

10.7 Case Study VI: Sustainable Land Use and Watershed Management in Ralegan Siddhi, Maharashtra, India

10.7.1 Background

The small village of Ralegan Siddhi is located in the Ahmednagar region of India, Maharashtra state. It has gained recognition for its remarkable transformation from a drought-prone and environmentally degradable region to a successful model of sustainable development [16, 17]. With the help of Anna Hazare, a social activist, Ralegan Siddhi residents pursued an integrated approach toward sustainable management of land and watersheds areas. With this strategy, environmental restoration was accomplished, and there was a significant improvement in the standard of living and economic prosperity of the local people. Working together and employing certain plans allowed the community to achieve major revitalization in Ralegan Siddhi.

10.7.2 Challenges

Recurrent droughts, occurring with little water, resulted in further social crisis. Effects of such issues loom far into the agriculture, daily regimes and general economic order. Water shortage has stifled the growth of crops, reduced agricultural yield, and made water livelihood activities of communities vulnerable. Simultaneously, soil and land conditions have deteriorated due to poor strategies on land use and rampant deforestation. Intensive destruction of forests that occurs to a large extent because of increased fuelwood consumption and development of agricultural lands has significantly reduced forested areas. The reduction of forest coverage has intensified soil erosion, thus reducing efficiency in soil quality and agricultural potential. Deterioration of fertile soil with the subsequent negative effects on crop growth has led to a continuous decline in agricultural output and increasing food insecurity. The negative impacts on the environment caused by those activities have reduced agricultural harvests and, therefore, have also limited economic growth. The combination of environmental pressures and economic growth has generated significant social worries, and one of these is quite apparent in the growing immigration. Due to economic hardships, a lot of rural people have expressed desire to search for improved opportunities by migrating from their villages to urban areas. As a result, these migration patterns have undermined the local economy and broken social ties, thereby increasing poverty in rural communities. Having to rely on third-party support for a long time is now a major concern. Though government initiatives and humanitarian help reduce some immediate threats, they do not address the reasons why people stay vulnerable for years. Because of this long-term reliance on help from abroad, the system is unstable and does not have funds or strong institutions to solve the main reasons behind its persistent difficulties.

10.7.3 Solution

The development of Ralegan Siddhi encompassed a comprehensive strategy centred around sustainable land utilization, water preservation, afforestation, active community involvement, and the expansion of livelihood options. The solution consisted of several essential components, as described below.

- Management of a watershed: Rainwater harvesting involves the construction of contour bunds, check dams, and percolation tanks to collect rainwater and replenish underground water sources. These structures were deliberately positioned to collect and retain rainwater, minimizing runoff and facilitating more water absorption into the soil. Gully plugging involves the installation of barriers, known as gully plugs, to prevent the erosion of soil and help keep moisture in the soil. This process also contributes to the replenishment of groundwater. Construction of reservoirs and water reservoirs that gather and retain rainwater serves to irrigate crops and provide water for livestock during seasons of drought.
- Conservation of soil and water: The adoption of contour ploughing and terracing techniques aims to mitigate soil erosion and enhance water retention. These techniques aided in preserving soil structure and mitigating nutrient depletion. Utilizing organic mulching and compost to enhance soil fertility and moisture retention in the context of organic farming. Adopting organic farming techniques decreased reliance on chemical fertilizers and pesticides, while enhancing the quality of soil and crops. Vermicomposting is the process of using worms to transform organic waste into compost that is rich in nutrients. This helps improve the health of the soil and reduces trash.
- Reforestation and agroforestry: Afforestation refers to the deliberate act of planting native tree species with the aim of restoring forest cover, stabilizing soil, and enhancing the local microclimate. The selection of neem, babul, and banyan trees was based on their ecological advantages and ability to withstand adversity. Agroforestry refers to the practice of combining trees with agricultural crops in order to establish farming systems that are both more robust and productive. Agroforestry generated supplementary revenue by selling lumber, fruits, and other tree-derived commodities.
- Ecologically sound farming practices: Crop diversification involves the introduction of a range of crops, especially those that are resistant to drought, to increase the ability to withstand climate changes and improve availability of food.

The promotion of crop rotation and intercropping strategies aimed to preserve soil fertility and micro-organisms. Water-efficient irrigation involves the adoption of techniques such as drip irrigation and other approaches that optimize water usage and minimize loss.

- ➢ Renewable energy and sustainable livelihoods: Biogas facilities are installed to transform biological waste into clean energy for cooking, hence reducing the need for firewood and increasing interior air quality.
- ➢ Solar energy: Utilization of solar energy systems for illumination and water extraction diminishes reliance on non-renewable resources and diminishes energy expenditures. Livelihood diversification refers to the process of creating alternate sources of income, such as engaging in dairy farming, horticulture, and handicrafts. This strategy aims to minimize economic reliance on agriculture and broaden the range of revenue sources available.
- ➢ Engagement of the community and development of their skills and abilities: Community mobilization refers to the proactive engagement of the local community in the process of designing, executing, and upholding sustainable practices. Residents were informed and trained through village meetings and workshops to enhance their awareness and develop their skills. Self-Help Groups (SHGs) are established with the purpose of promoting collective action, mobilizing resources, and providing mutual support. These groups also aim to foster the formation of cooperative societies and were crucial in maintaining the initiatives and advancing community well-being.
- ➢ Training and education: Organized education and training activities aimed at equipping communities to deliver in sustainable agriculture, water conservation and use of renewable sources of energy. Advocating for policies means proposing certain policies and pushing them to be implemented. Institutional support involves piloting organizations or institutions with the help and materials required to achieve their goals.

10.7.4 Results

As the result of the introduction of sustainable practices of land use and watershed management in Ralegan Siddhi several significant and positive effects have been achieved.

- Ensuring the availability and sustainable management of water resources: The application of measures such as check dams, percolation tanks, and trenches has sufficiently revived groundwater replenishment. Implementation of these projects has provided the community with a secure and sustained water supply for daily consumption, farming irrigation and watering of livestock. During years of scarce rainfall, the village has been able to sustain significant resilience against drought-affected areas. The resilient strategy for water management has guaranteed access to key resources and upheld the community against discouraging climatic threats.
- Soil health and agricultural productivity: Adoption of practices like terracing, contour bunding and organic fertilizer application has greatly assisted in reducing soil erosion. These steps have played an essential role in increasing farm yields and quality of the product. As a result, food security has been achieved and dependency on food from outside has reduced, while local farmers have found the economic situation more stable. The increase in soil health has enabled farmers to grow several crops, which improves not only their resilience to climatic shocks, but also improves their overall earnings.
- Restoration of forests and the preservation of biodiversity: Reforestation and afforestation projects are the primary goals of ecological restoration in Ralegan Siddhi. Villagers, by introducing indigenous trees and preserving natural plant life, have halted ongoing deforestation and recovered local species richness. The result of these efforts is that several plant and animal species have come back, most of which were lost due to habitat destruction. Restoration of forests has enhanced important ecosystem functions, namely capturing of carbon, regulating water cycles and conserving soils, fundamentally enhancing all-around improvement

of ecological welfare. Increased biodiversity has been good for the local environment; at the same time, it has ensured the well-being of the community's culture and economy because now sustainable harvesting provides for livelihood and conservation objectives. With these accomplishments, it becomes evident that well-managed land and watersheds are very critical in facilitating environmental health, farm production and the livelihood of communities. The lessons of Ralegan Siddhi describe how communities can become sustainable and resilient, based on the good management of land usage and watersheds.

- Enhancing economic growth and reducing poverty: Household incomes were boosted, and poverty was decreased as a result of diversified livelihoods and enhanced agricultural output. The community attained economic autonomy, as its inhabitants participated in many revenue-generating endeavours. The migration of people to urban regions lessened as inhabitants discovered viable employment options within the village, thereby enhancing the local economy and fostering communal cohesion.
- Societal empowerment and the well-being of the community: Residents' sense of ownership and responsibility were promoted through community involvement in decision-making processes. Self-help groups and cooperative societies have provided empowerment to women and marginalized groups, thereby promoting social fairness and inclusion. Enhanced availability of renewable energy, clean water, and sustainable sources of income improved the overall standard of living, promoting better health, education, and well-being. This ensured the long-term viability of the environment and the ability to withstand and recover from climate-related challenges. The implementation of an integrated strategy to land and water management enhanced the village's ability to withstand the impacts of climate variability and extreme weather events. Sustainable practices fostered environmental sustainability and diminished the ecological impact of the community.
- Impact: The successes of Ralegan Siddhi could serve as inspiration and give a practical vision to other communities at home and abroad. The approaches and tactics employed at Ralegan Siddhi have been replicated with success in many

other communities confronted with similar challenges. The village has become an important example of sustainable rural development for policymakers and development specialists, which has affected wider policy directions and programs.

10.7.5 Summary

The successful transformation of Ralegan Siddhi from a poor, drought-ridden village to one of sustainable development shows that it is possible to combine land use methods and watershed management to resolve environmental and economic challenges. Through old ways and new ideas, the village made significant achievements in water conservation, soil, farm production, and the quality of life of its people. Ralegan Siddhi's change reflects the need for teamwork, skills training, and the embrace of all-around practices as a necessary strategy for sustainable agriculture and prosperity. The experience of Ralegan Siddhi teaches other areas that suffer from analogous problems that the application of sustainable processes can lead to environmental restoration, economic development and increased social welfare.

10.8 Case Study VII: Sustainable Agriculture and Land Use in Sikkim, India

10.8.1 Background

One of the greenest states in the world can be found in northeastern India, where Sikkim has received international renown for its revolutionary ways of sustainable agriculture and land use. Sikkim became India's first fully organic state in 2016. This milestone was achieved through the full switch to new farming practices, all aimed at promoting environmental welfare, showing long-term economic stability, and contributing towards social welfare [18, 19]. The change was facilitated by governmental activities, community participation, and the employment of the new forms with the aim of creating a long-term ecological farming strategy [20, 21].

10.8.2 Challenges

Sikkim endured many challenges as it transitioned into land use that is more sustainable.

- Chemical dependence: Constant use of chemical fertilizers and pesticides resulted in soil damage, poor quality of water and health hazards to people both in rural and urban settings. The long-term abuse of these chemicals, which were aimed at the good insect and microorganism numbers, reduced the fertility of the soil.

- Soil erosion and degradation: High-yield agriculture and the rugged topography led to serious soil fertility degradation and soil degradation. As practised agriculture standards were difficult to implement in the hilly region, the monsoons resulted in serious soil erosion and loss of the important topsoil for crop output.
- Biodiversity loss: Ordinary farmer activities threaten the local biodiversity, negatively affecting plant and animal populations. Habitat simplification resulted from monoculture and the use of chemical inputs, and ultimately, crop diversity decreased, as well as the number of wildlife species.
- Climate vulnerability: Sikkim's topography rendered it susceptible to the effects of climate change, including irregular rainfall, landslides and changes in weather behaviours. There was an increase in the occurrence of parasites and diseases, soil erosion intensified, and there were also negative effects on crop yields as a result of climatic changes.
- Economic sustainability: Farmers looked for evidence that they could expect economic sustainability and viability from the switch to organic production for several years. Farmers faced huge financial challenges when transitioning from conventional practice to organic methods, including the associated costs of organic certification and an immediate fall in crop production during conversion.
- Capacity building: There was a well-known problem associated with the farmers' need for extensive education and training in order to obtain the necessary skills for organic farming. Many farmers struggled with adopting organic techniques because they lacked skills, and they had to be helped to learn about sustainable practices, disease management and soil care.
- Market access and infrastructure: It was essential to have a strong and reliable system for the sales of organic products. For organic produce to flow through markets easily and at a higher price, farmers required functioning supply chains, storage, and a viable transport network.

- Consumer awareness: One of the effective measures to stimulate demand was to present to potential consumers the unique benefits of organic farming. It will not be easy for farmers to woo customers who are willing to pay the high prices needed for organic agriculture to be viable in the absence of sufficient market interest.

10.8.3 Solution

Different connected methods were employed to transform Sikkim completely organic.

- Government Support and Policy Initiatives:
  - Sikkim organic mission: The mission began in 2003 to address the promotion of the system of organic farming methods and minimizing the dependency factors towards the use of chemical fertilizers and pesticides. The state government of Sikkim helped the organic farmers to be actively supported with strong policy backing in the form of financial support, certification services, and access to the market.
  - Legal framework: The Sikkim government passed the Organic Agricultural Act in 2014, which established a legal framework to govern organic farming and control the manufacture, sale and distribution of organic commodities.
  - Farmer training and capacity building: Comprehensive training programmes were established to teach farmers pest control measures, soil management, application of organic farming methods and composting. These projects were very important in establishing both trust and expertise in organic farming practices among farmers.
  - Demonstration farms: Through establishment of model organic farms and demonstration plots, farmers were able to see and practice successful organic methods in real-life situations.
- Water and Soil Conservation:
  - Terracing and contour farming: The use of terracing and contour farming practices is to improve the retention of water and also to control soil erosion. The fact that Sikkim was hilly made these practices very worthwhile.

- Organic soil amendments: Soil fertility and structure were improved by using organic compost, vermicompost and green manure. Using these amendments, soil health restoration was achieved, and agricultural productivity increased substantially.
- Enhancement of biodiversity: Agroforestry entails the integration of trees and vegetation in agricultural-type landscapes, within varied dimensions of improving soil quality, providing shade and enhancing biodiversity of the environment. Agroforestry supplemented households incomes because it provided products for sale such as fruits, berries and other tree-based products. Methods that encourage mixed crops and strategies for controlling pests, enhancing soil condition, and increasing adaptive capacity to changing climate patterns are known as crop diversification. Preservation of the region's cultural heritage and biodiversity was achieved by way of producing traditional and indigenous products.

➢ Pest and Disease Management:
  - Biological control agents: The use of good insects and other parasites, such as microbes, combated pest and disease implications. By so doing, this strategy promoted a healthy ecosystem that saved the use of synthetic pesticides.
  - Integrated Pest Management (IPM): Sustainable pest control management is achieved through a synergy of cultural, mechanical, biological and organic means of varmint control.
  - Organic certification: Producers get access to support for securing organic certification so that their goods meet local and international organic standards. The certification created a way into the premium markets and led to an increase in the prices for organic produce.
  - Market development: This involves the development of distinct organic markets and brand propagation of organic products in line with advertising ideas. The government facilitated direct contact between producers and consumers, making a stable market for organic products.
  - Subsidies and incentives: Financial grants, subsidies to buy organic supplies, and help with building the required

infrastructure are offered to make the economic struggles of a change to organic farming easier.

- Sustainable Livelihoods and Renewable Energy:
  - Diversification of livelihoods: This offers support for the diversification of income through growing opportunities in horticulture, dairy farming, and handicrafts, thus weaning farmers off economic dependency in agriculture.
  - Solar energy: Through the incorporation of solar technologies in water inclusion, among other functions, dependence on traditional energy sources is reduced and cost as well.
  - Biogas plant: Biogas plant installations to minimize firewood usage and maintain better indoor air quality in the conversion of organic refuse into clean cooking energy are in plan.
- Community Engagement and Capacity Development:
  - Community mobilization: The local community proactively participated in planning, implementing and maintaining sustainable projects. Village meetings and seminars were conducted to enhance awareness and build capacity of residents.
  - Self-Help Groups (SHGs): SHGs and cooperative societies were established to promote collective involvement and joint funding with joint responsibilities within the community. They played a vital role in the implementation process of the initiatives and contributing to continued welfare in communities.
  - Training and education: The citizens are always educated and trained on sustainable agriculture, water management and renewable energy by virtue of ongoing initiatives and programs.
- Institutional Support and Policy Advocacy:
  - Government collaboration: This involves the establishment of partnerships with government agencies to ensure that sustainable development initiatives receive technical assistance, funding, and policy support.
  - Non-Governmental Organizations (NGOs): With synergization with NGOs, organizations are able to access important advocacy resources and expertise that can be used to support sustainable management of lands and community development.

10.8.4 Results

Sustainable agriculture and land use practices contributed to substantive improvements in Sikkim, which are shown below.

- Advantages for the environment: Sikkim, with the adoption of organic farming practices, has experienced restoration of fertility and structure of soil and microbial activity that has resulted in high agricultural yield and soil health. An improvement in the quality of water in rivers and streams occurred following a reduction in chemical runoff, thus providing a gain for people and wildlife. Biodiversity is maintained upon the adoption of agroforestry and crop diversification methods as the practices threaten conservation of various forms of species and promote ecological balance.
- Economic benefits: The willingness of the market to pay a higher price for products from the organic sphere promoted the financial interests of farmers and increased the economic security of participants in the crop sales process. The state's organic reputation rose, and market interest and demand in its produce also grew. Sikkim's emphasis on sustainable agriculture and organic farming made the state an ideal environment for ecotourism, which stimulated the economy through the new inflow of money.
- Health benefits: By resorting to organic farming, there has been a dramatic reduction of the use of dangerous agrochemicals, which decreases environmental injury and risk exposure to toxic material for the people engaged in the cultivation of food and its consumption. Farmers and consumers have achieved health benefits due to reduced exposure to synthetic pesticides and fertilizers. Decrease in the cases of chemical-associated illnesses such as respiratory disorders, dermatitis and chronic diseases has contributed to population healthcare and reduced needs for healthcare resources.
- Community empowerment: Sikkim's strategy was highly community-oriented because it entailed strong training, educational and capacity building among its communities. These initiatives enabled farmers to acquire the skills and pieces of information required in embracing as well practising organic farming sustainably. The feelings of empowerment and accomplishment have risen drastically among

community members, which in turn encourages both collective benefits and sustainable land management. Enrolment and leadership positions in the agriculture program have increased for women and other marginalized groups to promote inclusive agriculture.

- Climate resilience: Organic farming and agroforestry are being used to address how agricultural systems are resilient to both climate change and extreme weather events. Improving the quality of soil by means of organic composting as well as mulching improved the ability of the land to retain moisture and withstand drought spells. Furthermore, agroforestry's coverage helps regulate microclimates and guard against too much rain and soil washout. Consequently, rural communities are now more prepared to manage climate change uncertainties.
- Policy advocacy: Sikkim's full-scale transition towards 100% organic farming has provided developers and policymakers with a good model for sustainable growth. Sikkim's reality demonstrates how grassroots engagement, governmental momentum, institutional support, and carefully designed policies are key to meaningful, sustainable development. Sikkim's success can now be analysed as an example and inspiration for other regions that want provincially friendly and economically profitable agriculture.

10.8.5 Summary

The practice of sustainable agriculture and land use in Sikkim, India, demonstrates a critical success concerning environmental management and agriculture policy. Sikkim succeeded in becoming the first state in the entire country to adopt 100% organic farming through the use of traditional farming techniques and continuous efforts to capacitate and train local farmers to practice organic farming methods. The transformation led to great environmental benefits, including reduced chemical use, improved soil fertility, and higher biodiversity. Because of organic standards, the state can now offer economic perks to farmers by allowing them to earn more from special markets, meet the rising demand from customers and take part in wider trading activities. Community member's relationships strengthened because the farmers gained valuable new skills which led them to feel more connected, supportive and proud of the group. Government policies and support mantle of backing were a cornerstone in facilitating the transformation so that it could be widely accepted. The initiative successfully combined with the old ways of doing things and affirmed their appropriateness to the culture and acceptability by the people. This demonstration provides a pattern for other districts looking to embrace long-term farming strategies. The implication of the Sikkim model is that wider acceptance of the kind of strategies designed may prove rewarding for conserving the environment while advocating for sustainable practices globally.

10.9 Conclusion

The global scope of case studies reveals the considerable effect of the initiatives for sustainable land management. Initiatives such as these have shown great success in restoring degraded land, supporting the local economy, and improving agricultural yields through the use of integrated, communal, and ecological approaches. The case studies make it clear that sustainability supports economic growth, helps build regional policies, brings together community and sustainability and requires strategies to be flexible and work together. The combined functioning of these elements underwrites the development of sustainable, productive, and resilient land-based systems that serve as a reference for other areas that hope to achieve comparable results. The implications gained from these case studies can be of positive significance to the global community for the development of sustainable land use, as it deals with land degradation, climate change and food security. By adopting these principles, one can promote the ecosystem vitality and community prosperity, supporting a more sustainable and inclusive future.

References

1. Abera, W., Tamene, L., Tibebe, D., Adimassu, Z., Kassa, H., Hailu, H., Mekonnen, K., Desta, G., Sommer, R., Verchot, L., Characterizing and evaluating the impacts of national land restoration initiatives on ecosystem services in Ethiopia. *Land Degrad. Dev.*, 31, 37–52, 2020.
2. Johnstone, N. and Bishop, J., Private sector participation in natural resource management: what relevance in developing countries? *Int. Rev. Environ. Resour. Econ.*, 1, 1, 67–109, 2007, https://doi.org/10.1561/101.00000002.
3. Kassam, A., Friedrich, T., Derpsch, R., Global spread of conservation agriculture. *Int. J. Environ. Stud.*, 29–51, 2018.
4. Cao, S., Chen, L., Yu, X., Impact of China's grain for green project on the landscape of vulnerable arid and semi-arid agricultural regions: a case study

in northern Shaanxi Province. *J. Appl. Ecol.*, 46, 536–543, 2009, https://doi.org/10.1111/j.1365-2664.2008.01605.x.

5. Shrestha, S., Bouffard, M., Langstaff, J., Lee, C., Pauline, E., Assessing Threats to River Water Quality and Mangrove Health Based on Watershed Land Use on the Osa Peninsula, Costa Rica, NASA Applied Sciences, 2020, https://disasters.nasa.gov.
6. Shannon, A., Connecting the Dots of Conservation: Mapping Wildlife Corridors in Costa Rica with NASA Earth Observations, NASA Applied Sciences, 2020, https://appliedsciences.nasa.gov.
7. Hernando, A., Tejera, R., Velázquez, J., Sustainable Tourism around Ecosystem Services: Application to a Case in Costa Rica Using Multi-Criteria Methods. *Land*, 12, 3, 628, 2023, https://doi.org/10.3390/land12030628.
8. Zhang, X., Wang, Y., Yuan, X., Shao, Y., Bai, Y., Identifying ecosystem service supply-demand imbalance for sustainable land management in China's Loess Plateau. *Land Use Policy*, 123, 27–37, 2016, https://doi.org/10.1016/j.landusepol.2022.106423.
9. Shan, Y., Xie, J., Lei, N., Dong, Q., Soil Moisture Characteristics of a Typical Slope in the Watershed of the Loess Plateau for Gully Land Consolidation Project. *E3S Web Conf.*, 199, 00006, 2020. EDP Sciences https://doi.org/10.1051/e3sconf/202019900006.
10. Kiboi, M.N., Ngetich, K.F., Fliessbach, A., Muriuki, A., Mugendi, D.N., Soil fertility inputs and tillage influence on maize crop performance and soil water content in the Central Highlands of Kenya. *Agric. Water Manage.*, 217, 316–331, 2019, https://doi.org/10.1016/j.agwat.2019.03.014.
11. Oduor, N., Kiboi, M.N., Muriuki, A., Adamtey, N., Musafiri, C.M., Ngetich, F.K., Soil management strategies enhanced crop yield, soil moisture, and water productivity in Nitisols of the Upper Eastern Kenya. *Environ. Challenges*, 5, 100375, 2021, https://doi.org/10.1016/j.envc.2021.100375.
12. Reynolds, T.W., Waddington, S.R., Anderson, C.L., Chew, A., True, Z., Cullen, A., Environmental impacts and constraints associated with the production of major food crops in Sub-Saharan Africa and South Asia. *Food Secur.*, 7, 795–822, 2015, https://doi.org/10.1007/s12571-015-0478-1.
13. Mortimore, M., Adapting to drought in the Sahel: lessons for climate change. *Wiley Interdiscip. Rev. Clim. Change*, 1, 1, 134–143, 2010, https://doi.org/10.1002/wcc.25.
14. Epstein, H.E., Lauenroth, W.K., Burke, I.C., Coffin, D.P., Ecological responses of dominant grasses along two climatic gradients in the Great Plains of the United States. *J. Veg. Sci.*, 7, 6, 777–788, 1996, https://doi.org/10.2307/3236456.
15. DeAngelis, A., Dominguez, F., Fan, Y., Robock, A., Kustu, M.D., Robinson, D., Evidence of enhanced precipitation due to irrigation over the Great Plains of the United States. *J. Geophys. Res.: Atmos.*, 115, D15, 2010, https://doi.org/10.1029/2010JD013892.

16. Deshmukh, R., Managing water for sustainable Agriculture: The case of Ralegan Siddhi in India, 2016.
17. A.K. Mehta and T. Satpathy, Escaping poverty: the Ralegan Siddhi case. SSRN 1538891, 2008.
18. Mishra, P.K., Rai, A., Abdelrahman, K., Rai, S.C., Tiwari, A., Analysing challenges and strategies in land productivity in Sikkim Himalaya, India. *Sustainability*, 13, 19, 11112, 2021, https://doi.org/10.3390/su131911112.
19. Mishra, P.K., Rai, A., Rai, S.C., Land use and land cover change detection using geospatial techniques in the Sikkim Himalaya, India. *Egypt. J. Remote Sens. Space. Sci.*, 23, 2, 133–143, 2020, https://doi.org/10.1016/j.ejrs.2019.02.001.
20. Kumar, P. and Rai, S.C., An overview of existing contours for promoting different strategies for livelihood security in Sikkim Himalaya. *Curr. Res. Environ. Sustainability*, 3, 100034, 2021, https://doi.org/10.1016/j.crsust.2021.100034.
21. Chaudhary, M. and Lama, R., Community based tourism development in Sikkim of India—A study of Darap and Pastanga villages. *Transnatl. Corp. Rev.*, 6, 3, 228–237, 2014, https://doi.org/10.5148/tncr.2014.6302.

11
Community Engagement in Land Use Decision-Making

Abstract
This chapter showcases how community participation and planning can affect decisions about land use. It demonstrates why involving the community leads to more sustainable and locally relevant land use decisions by taking local ideas and views into account. The report looks into approaches that can help eliminate obstacles to community involvement, particularly the adoption of digital resources and the analysis of social media's role in informing the public about land use decisions. Additionally, it examines the importance of tracking the effectiveness of community feedback in decision-making, considering both numerical data and people's opinions. It underlines the importance of having open decision-making systems, involving everyone, and helps build trust, collaboration, and accountability. Even though community engagement has many benefits, there are still hurdles when trying to incorporate people's feedback, including difficulties in communication, differing viewpoints, lack of resources, and unequal power. Finally, it stresses the importance of planning to solve these difficulties and making use of community knowledge to foster environmentally sound land use.

***Keywords*:** Community, digital platforms, sustainable land use, sustainability

11.1 Introduction

Sustainable land use choices require that communities be involved from the start. The process joins together policymakers, practitioners, and all those who will experience the consequences of land use changes. In the past few years, people have realized that it is crucial to align land use planning with what local communities want. Often, the effects of land use decisions made without local citizens' input are short-lived and not very sustainable. Community engagement is founded on the assumption that regional communities can contribute special knowledge and viewpoints to land use decisions. Members of the community should be included in

Kundan Meshram and Chatrabhuj. A Sustainable Approach to Land Use, (269–300) © 2026 Scrivener Publishing LLC

planning and decisions to ensure that the land use restrictions are developed with knowledge of local customs and needs [1–3]. As result, including groups from the community in decision-making helps ensure those decisions are seen as valid and are more likely to be followed.

These values are needed to guide the process so that everyone joins in and all participants are treated equally. Inclusivity is just one of the ways that community engagement is managed well. When everyone, and especially those from disadvantaged or vulnerable groups, can be involved in decision-making and voice their ideas, inclusivity ensures that all significant groups of people are present. Clarifying important information on land use plans, rules, and choices in plain language and making them easy to access helps secure transparency [4–6]. In addition, being responsive involves paying attention to the opinions shared by the community and using them to shape decisions. Valuing different perspectives, understanding of knowledge, and assorted cultures among members allows everyone to trust one another and work together more effectively. Many different strategies and approaches are used to bring communities into the land use decision-making process. Such methods consist of regular face-to-face meetings as well as modern online engagement options. To seek the community's viewpoint, foster dialogue, and join forces on problem-solving, workshops, town hall meetings, focus groups, and participatory mapping projects are typically organized. In addition to social media, citizen advisory groups, public forums, and stakeholder engagement sessions are formal mechanisms that offer continual chances for people to work together. On the other side, social media and internet forums support wider participation and enable more people to join in [7].

Even though community engagement in land use decision-making brings a variety of benefits, it also brings many obstacles and challenges. Challenges such as unequal distribution of power, cultural gaps, language barriers, and different interests may make it difficult for people to work together, which may end in mistrust and dissatisfaction among those participating. Without the necessary support in terms of resources and time, community participation can be unsuccessful in generating major improvements. Addressing power inequalities, fostering cultural understanding, and building trust and collaboration allow practitioners the experience and assets of a wide stakeholder group for community engagement [8–10]. These advancements in technology and methods of communication have also improved opportunities for communities to participate in shared decision-making. Thus, there has been a rise in the overall inclusiveness, performance, and long-term viability of land use initiatives.

11.2 Participatory Planning

Participatory planning as a decision-making method ensures that a diverse group of people, such as local communities, can join in making land use laws, plans, and actions through collective and open participation. Participatory planning aligns with communities affected by land use decisions by offering the chance for them to guide, shape, and develop planning solutions together with decision-makers. It is based on the understanding that local communities possess significant knowledge, opinions, and goals that may help create more suitable and environmentally sustainable land use patterns [11]. The main advantage of participatory planning in land use planning is that it supports more legitimate, practical, and durable policy outcomes. When communities are involved in planning, land use aims and activities are more likely to match the distinctive needs, values, and ambitions of those living in the area. Besides, taking part in participatory planning contributes to people feeling more ownership, responsibility, and confidence, which results in greater commitment and support for the plan. Using this method augments communities' ability to respond and adapt to complex land use issues such as urbanization, climate change, and management of natural resources. Using this method helps create just, ecologically mindful, and sustainable ways of managing land that continue to meet the changing needs of the present and future populations.

There is growing importance in understanding the role of citizen involvement as a means to strengthen participative processes for policy-making and community advancement [12–16]. The chapter presents an in-depth analysis of widely used forms of public participation.

11.3 Theoretical Underpinnings of Participatory Methods

The purpose of participatory methods, influenced by deliberative democracy, is to help people speak up on issues that impact them and for public officials to understand each other better. They may still have issues due to poor moderation, not all people taking part, and difficulties related to access. Through citizen forums, people can share views and opinions with other participants to recommend solutions for complicated situations. In order for citizen forums to be successful, dedicated resources are needed along with well-trained facilitators who can support them all the time. No matter which of these is used, all play a part in developing a more open and responsive government as long as they are organized the right way.

11.3.1 Workshops

Workshops help in many ways, but they can sometimes be limited by certain difficulties. Organizing workshops appropriately generally requires a great deal of planning and management skills. The skills needed to organize a workshop are planning, coordination, and excellent facilitation. If proper planning and facilitation are lacking, a workshop may run off track and not meet its intended outcomes. Another issue is that those organizing the workshop may not know how to properly support or engage the group, especially when there are attempts to make sure everyone gets a chance to speak. Having groups interested and engaged is essential for a workshop. Engagement levels can be an issue in a workshop, and this might depend on how relevant the topic is, where and when it happens, and how it is set up. Keeping participants engaged is important, as time constraints may cause ideas to remain on a surface level, and this can be a concern if potential barriers come from some participants dominating the meeting. Trying to finish a workshop in a limited time often means the discussion can only cover easy topics, with no attention given to more complex matters. Problems such as scheduling issues, technical problems when using remote formats, and accessible space or equipment may make running workshops more difficult. It is often not easy to turn workshop ideas into policies or action due to the lack of ongoing support and difficulties associated with institutional boundaries.

11.3.2 Town Hall Meetings

People living in the community have an open chance to interact directly and face-to-face with leaders or public officials during town hall meetings. Town hall meetings allow all residents of a locality to take part in an open exchange about important community-related issues. First, government authorities give updates and information at a town hall meeting, and then attendees have the opportunity to interact with them on a one-on-one basis.

11.3.3 Citizen Forums

It is not easy to organize and carry out citizen forums. Arranging and holding citizen forums often requires a lot of time, money, and expert help. Broad and equitable representation is particularly difficult, as it often means finding creative and careful ways to get a group of people from all walks of life involved. As the topics in such forums have several aspects and are difficult, participants must be knowledgeable or develop such skills during the process. Thus, to be successful, the topics discussed in forums should be supported with ongoing learning and support. Discussions in citizen forums can easily focus on issues that seem abstract without proper guidance and help. Therefore, for citizen forums to work well, they should be carefully arranged, managed by experts, and have structures in place to produce policies from what was discussed.

Every form of participation, including seminars, town hall meetings, and citizen forums, provides particular advantages but also faces certain barriers. Workshops are highly effective at encouraging creative problem-solving in a group setting, provided that they are facilitated by someone skilled. One benefit of town hall meetings is that they allow easy interaction between citizens and officials, but it is often tough to address the needs of a diverse and large group. Citizen forums support wide-ranging and inclusive interactions, but preparing and supporting them is a resource-intensive endeavor.

11.4 Involvement of Communities in Decision-Making

Participating communities in the planning of land use increases the trustworthiness of decisions by making people feel they have ownership, trust, and responsibility. Having communities participate substantially in decision-making processes is required to maintain the legitimacy of those decisions and avoid disputes within the community. In the following, a detailed explanation is provided on how the engagement of communities increases the credibility of decisions.

a) Representation and voice: The guarantee of representation and expression is one of the most important elements of community participation. Taking part in the decision-making process, communities help ensure that a wide array of views and expectations are taken into consideration for land use plans. Ensuring that all inhabitants are included means that important land use decisions are based on the community instead of being made by outsiders or the elite. If communities participate in decision-making, vital land use decisions are more likely to reflect the actual needs and perspectives of the community, not just those of ruling groups.

b) Transparency and openness: It is also necessary to make decisions in a clear way so people can always learn what is happening and can easily ask for their opinions or see why decisions were made. Community engagement also helps authorities share information about land use planning with people, so stakeholders can better understand the reasons for decisions and explain their own feedback in a sensible way. Stakeholders can give more helpful feedback and understand why decisions were made if they get all the needed information and have a clear explanation. Transparency in decision-making helps build trust, makes things simpler to understand, and helps people work together on planning. Transparent decision-making processes help people be able to share their concerns right away, and this makes it likely that everyone can agree on how land is used, and there will not be problems later on.

c) Accountability and responsibility: Greater chances for citizens to help with planning encourage planners to be more responsible and accountable because everyone is watching them. Citizens are included in the process of planning and supervising decisions on how to use land. By making the decision-making process more visible, people become aware of when decision-makers fail to act well, which encourages them to act in the community's best interest. So, public participation becomes a regular step in the process, which helps decision-makers keep the interests of the community in mind. In addition, when the community shares responsibility, it increases the chance that policies will be sustainable because people are more likely to support decisions they have influenced. Because of this shared responsibility, land use controls are supported by the community and are also objectively based.

d) Social cohesion and trust: Community engagement in decision-making promotes social cohesion, solidarity, and trust among stakeholders. When communities engage in decision-making processes, they are provided with an opportunity to participate in discourse, collaboration, and consensus-building alongside other stakeholders. This connection fosters the development of relationships, networks, and social capital, hence strengthening trust, cooperation, and mutual understanding among stakeholders. Decision-making processes that involve collaboration and partnership are seen as more genuine because they demonstrate group efforts to tackle common difficulties and attain shared objectives.

e) Adaptive management and flexibility: Involving the community in determining land use decisions is beneficial in ensuring adaptability and flexibility. Engaging local communities enables decision-makers to respond to new information and unplanned challenges throughout the whole process. Continuous engagement and monitoring by residents help officials receive quick updates on what changes to land use are being achieved in the community. As a result, planners have the opportunity to fine-tune their decisions, making sure that the policies continue to be appropriate, effective, and attentive to the needs of the community. Adaptive management is intended to draw from real-world interaction and adjust approaches iteratively, rather than insisting on fixed, previous decisions. When land use plans integrate feedback from the community, they are seen as more credible, on account of showing that concerns from the public are taken seriously. This ability to adapt policies enhances not just their results but also reassures the community that their presence is important for determining environmental change.

f) Cultural relevance and sensitivity: Indicative of inclusive land use planning is sensitivity to cultural backgrounds. This means acknowledging and weaving cultural values, traditions, and practices that help define a community's relationship to the land. For a number of cultural groups, land has deeper significance and includes religious, historical, and cultural dimensions besides merely being a physical space. When communities are included in the decision-making process, planners can prevent plans that could damage or disregard aspects of the land that are important to the people there. Making sure cultural heritage and identity are kept safe results in less division and helps people in the community feel more united and prouder. If the people and their culture are included in the planning process, the decisions are likely to be trusted and respected, and they can help unite the community with those deciding on its future.

g) Local knowledge and expertise: Local people's experience and abilities help a lot in decision-making processes that require the community's input. Residents usually have informed opinions about the land, weather, local resources, and old ways of using that land. In many cases, communities' personal knowledge is far greater and clearer than the external data available, so it plays an important role in land use policies. If planners use the insight from local communities, they are more likely to make improved land use decisions. Utilizing community inputs for decision-making enriches the quality of choices while also making them seem more credible and relevant to the general public.

h) Empowerment: By involving communities in decision-making, they are able to take control of their future and become more involved in government activities. By taking part in workshops, training sessions, and participatory mapping activities, communities learn the necessary abilities, confidence, and resources for decision-making. Increased empowerment increases the acceptance of decisions because communities become actively involved in determining their future and in moving their communities ahead.

i) Equity and social justice: Promoting equity and social justice depends on involving the people who are most vulnerable in all stages of decision-making. Making land use decisions may lead to impacts that are not the same for everyone, particularly affecting vulnerable groups like the poor, women, indigenous communities, and other disadvantaged communities. If planners reach out intentionally to such groups, they can realize and resolve the embedded forms of discrimination and inequality in land management. Comprehensive plans are more just, fair, and equitable when multiple groups and perspectives play a part in the planning process. When equity is prioritized, decisions become more credible since they indicate assurance of impartiality, dignity, and respect for human rights.

j) Conflict resolution: Involving communities greatly contributes to both conflict resolution and consensus-building. Conflicts often arise when land use is changed, and these include disputes between different groups, for example developers, conservationists, farmers, and indigenous people. These processes ensure that people in the community are able to communicate their issues, air their grievances, and work collectively to find appropriate solutions. By involving communities in decision-making, conflicts may be prevented, and disputes are changed into opportunities for collaborative problem-solving.
k) Long-term sustainability: Achieving sustainable land use plans over time requires the full participation of the community. Community participation in decision-making usually results in greater responsibility and ongoing support for the implementation and ongoing maintenance of policies. Besides, choosing to involve the community in decision-making results in options that are better equipped to withstand changes and adapt to new circumstances, as they are based on local challenges and aspirations. Emphasizing sustainability increases how well decisions are viewed and supports public trust that the planning contributes both to current needs and to those of posterity. The emphasis on sustainability improves the perceived credibility of judgments, as they are regarded as being made with the utmost consideration for the well-being of future generations.

Incorporating communities into land use decision-making enhances the legitimacy of decisions by giving priority to cultural appropriateness and sensitivity, leveraging local knowledge and expertise, empowering communities, fostering fairness and social equality, ensuring long-term sustainability, and promoting adaptable management and flexibility. By actively including the community, planners and decision-makers can enhance the perceived credibility and approval of decisions among stakeholders, leading to more effective and environmentally friendly land use outcomes.

Figure 11.1 exhaustively delineates the numerous advantages of involving local communities in decision-making processes. Local knowledge benefit underscores the importance of utilizing the distinctive insights and profound comprehension that community members possess regarding

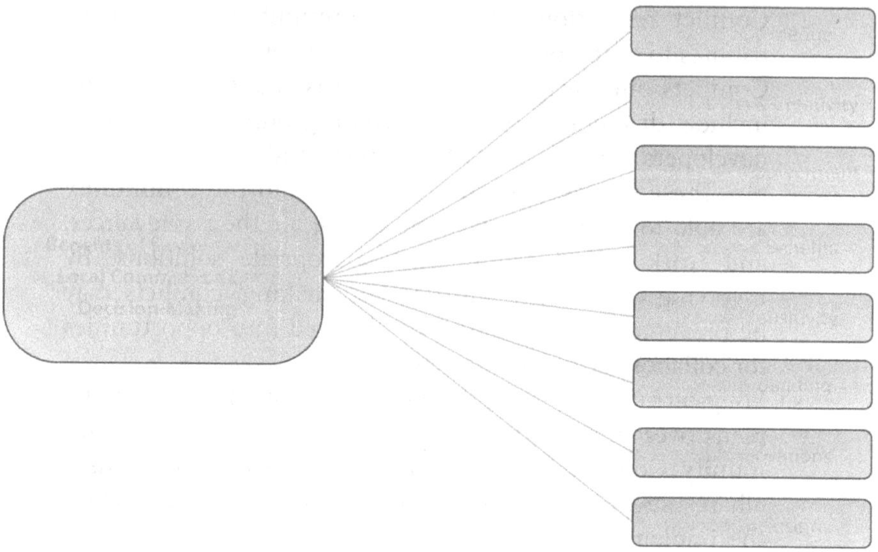

their environment, traditions, and obstacles. This local expertise has the potential to result in more effective and contextually appropriate decisions. By empowering local communities, innovative approaches and creative problem-solving can be unleashed, as community members are frequently in the best position to devise novel solutions that are tailored to their unique requirements and circumstances. When communities are involved in decision-making, individuals are given the opportunity to participate more fully in the management and making of policies. This engagement enables citizens to shape decisions related to their lives and to ensure that authorities answer for the decisions taken. When people are fully involved in decision-making, they tend to build greater civic responsibility and political mindfulness. Therefore, the outcomes are generally described as more democratic, more representative, and responsive to the range of needs and expectations in the community. Community involvement in making decisions affects the economy of local areas. Local populations' direct participation in economic policy-making and land use planning brings about decisions that are better suited to their real needs and available opportunities. It can therefore result in programs that encourage starting local businesses, create work opportunities, and protect the environment. Consequently, these economic strategies that involve all members lead to enduring growth, autonomy, and an improvement in residents' well-being. An additional major advantage of involving communities is that policies and responses are suitable and respectful of local cultures. When local populations are involved, their cultural practices, values, and heritage are given due regard and implementation in decision-making. As a result, the outcomes in land use and development are effective on practical levels and also respect the local cultural and social norms. In addition, decision-making participatory processes enable communities to build their knowledge, acquire useful skills, and increase their confidence for future roles in initiatives. Communities increase their knowledge and ability to take part in both planning and launching projects through workshops, training, and collective planning. Therefore, this empowerment helps create ongoing commitment and leadership among community members. In addition, community involvement makes it possible for new local innovations to emerge. By participating in identifying challenges and creating solutions, residents commonly come up with innovations that are well suited to their context and socio-economic situation. When a diverse group of community members participate, it ensures that the opinions of all communities, especially women, youth, indigenous peoples, and those who are usually overlooked are acknowledged and taken into account. Such an inclusive approach increases the fairness and credibility of decisions, promotes belonging among citizens, and builds up trust with institutions.

11.5 Importance of Locals in Decision-Making

For development that is fair, protects the environment, and involves the community, local people must be part of making land use decisions. For development to be successful, communities should be let in to help decide how land is managed because they understand the local area and issues. It is often the case that local communities cannot get involved because of various difficulties, despite the fact that this would be very helpful for development. The following sections give a detailed description of the problems standing in the way of community participation.

a. Insufficient awareness and information:
Those without access to necessary information find it hard to be meaningfully involved in decisions about land use. A large number of individuals living in the community do not know about current or planned planning processes due to poor dissemination of relevant information. Information that is circulated often uses technical language and refers to complex documents in planning, so it can be hard for people without training in planning to make sense of it. It follows that people become baffled and less

motivated to engage. Residents are sometimes confused about when and how to participate, as this information is not always easy to find. Official announcements and government websites may not be sufficient for engaging all sections of the community, especially people who live in rural or disadvantaged settings. As a result, a lot of people do not learn about their rights or how decisions impact them and end up making less effort and decisions that the public does not agree with.

b. Technical complexity:
Land use planning can be very complicated for members of the community who have never trained or worked in that area. It is common for the average resident not to understand zoning, building codes, evaluating the environment, different ways land is owned, and legal organizations. It is not easy for most residents to see how a zoning change might change things such as traffic, noise, or a lack of green space, unless the issues are explained carefully. It is difficult for most people to see how things in the neighbourhood could change because of zoning, unless they have the right knowledge and experience to understand these matters. Many people from marginalized groups find it harder to interact with planners because the field is too technical for them.

c. Lack of resources:
A lack of resources means that local communities and grassroots organizations frequently miss the chance to join in land use planning. These organizations often have little money, small teams, and lack the necessary skills to join in on land use planning processes. It is very hard for them to schedule such activities as outreach, briefing people, or training members of the community, due to limited funds. A shortage of people and volunteer members stops local organizations from being actively involved in making land use decisions. Not having such access limits their ability to argue against proposals and take part properly in discussions. Therefore, underserved communities see their voices excluded, and those with greater money and organizational backing shape the land use decision-making, contributing to ongoing inequality and a lack of empowerment.

d. Socioeconomic barriers:
Socioeconomic status is an important factor that determines a person's ability to take part in decision-making related to land use. Low-income or marginalized individuals usually encounter multiple obstacles that make it hard for them to engage. Many people cannot afford to go to public

meetings, particularly when they take place during the day or require people to travel. Time pressures are a major problem for people, given that many have to balance several jobs or important duties, leaving little time for public participation. Expectations of participation are shaped by the social exclusion and a lack of access to influence, so people in these communities may find it hard for their voices to be acknowledged. Widespread distrust between these communities and planning bodies is common, traced to earlier experiences of being excluded or having the promises of inclusion broken. Consequently, this lack of trust prevents people from engaging and plays a role in decisions that do not sufficiently address the needs of these communities.

e. Power imbalance:
Disparities in power between developers, government officials, and local communities are an important barrier to fair land use decision-making. Developers and officials usually have greater financial backing, technical experience, political power, and connection to decision-makers than groups of local residents. The result is that members of local communities often struggle to express their reservations about plans that may influence where they live. Frequently, the main reasons behind land use planning are mostly related to making money, getting votes, or helping with big developments, which often means that community members feel like their opinions do not matter a lot. When members of a community talk to authorities, they often feel their opinions are overlooked if what they suggest is not important to the planning process. As a result, the community's role in decision-making is minimal, and decisions often pay little heed to the environment.

f. Lack of adequate feedback mechanisms:
It is difficult to promote community involvement in planning land use because communities struggle with providing and accepting feedback. Many communities do not have adequate methods for people to participate in planning for future developments. Even so, when they can give feedback, individuals feel that their words do not have an impact on what happens. This view is supported further if people believe the concerns they raised were not significant and just part of a routine consultation. There are situations when individuals feel their opinions are unimportant, and as a result, they do not participate in land use planning in the future. This usually makes people uninterested, pessimistic, or unwilling to accept decisions made by the government or planners.

g. Language and cultural barriers:
Having language and cultural barriers can stop people from different groups from being active in decisions over land and city planning. Usually, all the planning documents, meetings, and discussions focus on the dominant language, leaving behind those who do not understand or cannot speak it well. Some people in the community struggle to either take part in or understand official land use planning talks because they do not have sufficient access to information in their own language. Most people from various cultures do not have the same idea of their role in community matters. This is especially true for community members who are part of some cultural groups, as they might feel uncomfortable speaking up during public meetings with authority or mixed groups. Because of this team, women, elders, and minority groups usually do not participate enough.

h. Restricted participation in decision-making procedures:
Land use decisions are made within formal institutions like public hearings or planning commission meetings that have fixed dates and places. People experiencing work commitments, transportation difficulties, or physical limitations may find that attending these meetings is not easy. In particular, when meetings are arranged for standard working times, people who must work full-time may be unable to join. Further, the actual meeting venues can be difficult for people with disabilities or public transport users, which makes it harder for them to take part. The way that public consultations are set up may indirectly exclude particular groups if the information is difficult to understand or if there are no alternative ways to participate, like online options. Hence, such obstacles often cause underrepresentation in land use decisions, potentially threatening the equity and accessibility of the planning stage.

11.6 Strategies to Overcome

For land use choices to involve the community well, an approach is required that concentrates on inclusiveness, accessibility, transparency, and teamwork. Below are a few techniques to effectively address these challenges.

a. Enhancing information accessibility:
Disseminate transparent, and easily understandable details regarding land use proposals, planning procedures, and timetables for decision-making across various platforms, such as websites, social media, community newsletters, and public notices. Utilize clear and concise language along with

visual aids to successfully convey intricate planning concepts and recommendations, ensuring that information is comprehensible to a wide range of individuals. Provide translation services, offer interpretation assistance, and create multilingual materials to effectively communicate with residents who have poor English ability or are non-native speakers.

b. Advocating for the implementation of inclusive strategies for engagement:
Take initiative to engage with marginalized communities, grassroots organizations, and community leaders to ensure their perspectives are acknowledged and respected in the decision-making procedures. Implement culturally sensitive approaches to involvement that honour a wide range of viewpoints, beliefs, and ways of communicating, so promoting inclusivity and a shared feeling of responsibility among all community members. Organize and oversee small-group talks, focus groups, and community forums to establish secure environments for open and respectful conversations, as well as to encourage constructive input among people with diverse perspectives.

c. Utilizing technology to promote inclusive participation:
Employ digital platforms, online questionnaires, interactive maps, and social media tools to expand participation, collect feedback, and enhance transparency in land use decision-making processes. Ensure that digital engagement tools are designed to be intuitive, easily available, and inclusive, catering to people with diverse levels of digital literacy, internet connectivity, and technological resources. Deliver training and technical assistance to residents and stakeholders, instructing them on how to successfully navigate and utilize online engagement platforms.

d. Promoting cooperative governance and partnerships:
Help groups such as government agencies, community organizations, developers, and residents work together to address problems, exchange what they have, and solve issues in land use as a group. Develop advisory panels, task teams, or community planning boards with people from lots of different groups so that everyone can talk more easily and come up with better solutions together. Incentivize groups, industries, and various sectors to join forces and make the most of each other's resources to achieve even better outcomes.

e. Incorporating fairness and principles of social justice:
It is important to make sure fairness, diversity, and inclusivity are part of land use planning to ensure communities are fair. This means that it is

important to ensure the views and needs of these groups are addressed before making decisions about land use. Helping communities get involved helps history repeat and adds new viewpoints to improve land use planning. Planners use different surveys to determine how certain land use changes will affect different groups of people. These tools help point out where resources, environment, and public services are not the same, and this helps us find any hidden inequities. This means that early in the planning process, planners can use these tools to make zoning rules simpler for affordable housing and include more green space in places where it is lacking. Ensuring these principles are used in planning allows land use to become fairer and helps create strong and united communities.

f. Improving the accessibility of decision-making processes:
It is important to schedule meetings and hearings at different times and in places that are easy for everyone in the community to reach. It means having meetings at different times, for example in the evenings, on weekends, and at places where people can use public transportation. Such a calendar makes it easier for people with different work schedules or personal obligations to join in. It is also very important to have remote options so that people who have trouble getting to meetings because of mobility, childcare, or health worries can participate as well. Offering virtual meetings, livestreams, and call-in opportunities gives everyone a chance to take part in making decisions, regardless of where they are. Decision-makers make it possible for all residents, especially those who are usually left out, to voice their views on land use decisions by providing multiple engagement methods.

g. Promoting transparency and accountability:
Supporting transparent and responsible decision-making in planning helps people trust the process and be involved in their surroundings. When residents can get meeting agendas, key documents, and decision criteria in a timely way, they are better prepared to join planning conversations. By giving residents this information, they are able to understand what is being decided and how it will affect them. Also, establishing a way to observe, record, and follow up on community feedback and outcomes makes land use planning more transparent. The establishment of such systems allows residents to see the effects of decisions and determine whether commitments made earlier are honoured. Frequently soliciting opinions from residents and others about the engagement process helps find places where

things can get better. If decision-makers welcome a variety of opinions and keep making their processes better, they will become more transparent and responsive to the public.

h. Enhancing the knowledge and skills of the community:
Building up community knowledge and skills is vital for securing robust participation in land use planning and development activities. This can be made possible through workshops, webinars, and training programs that concentrate on educating the community about land use planning, zoning rules, and how communities develop. With these programs, residents will better understand the fundamental ideas, which allows them to take part in land use conversations and decision-making with confidence. To aid participation, making easy-to-understand materials, including toolkits, guides, and informational pamphlets, is very important. It is important that these documents introduce simple definitions, cover citizens' rights, and walk users through activities like submitting feedback, joining in hearings, and advocating for the community.

i. Enhancing community networks and social capital:
Organize networking events, community meetings, and social activities to encourage connections, cultivate trust, and enhance social ties among inhabitants, thereby establishing a shared sense of identity and unity. Promote the establishment of committees, associations, and watch groups led by citizens, which can act as forums for residents to communicate, express concerns, and cooperate on local projects linked to land use and community development.

j. Facilitating and promoting grassroots advocacy and community organizing:
Offer resources, training, and technical support to grassroots organizations, advocacy groups, and resident-led campaigns that mobilize community members, amplify local voices, and lobby for legislative reforms or community-driven solutions to land use concerns. Establish networks of solidarity, mutual assistance, and collective effort that enable citizens to mobilize, plan, and campaign for changes that promote social, economic, and environmental justice in land use decision-making.

k. Fostering involvement and interaction among different generations:
Facilitate intergenerational discourse, collaboration, and knowledge-sharing to bridge generational gaps and ensure that the opinions and

insights of young people, older adults, and individuals of all ages are respected and incorporated into land use decision-making processes. Create youth councils, intergenerational planning committees, and mentorship programs that enable young leaders and elders to engage in planning conversations, exchange perspectives, and actively contribute to creating the future of their communities.

l. Implementing participatory planning tools and techniques:
Utilize participatory planning approaches, such as visioning workshops and design charettes, to actively involve communities in interactive activities that allow them to envision and collaboratively develop future land use scenarios. This approach promotes a sense of ownership and commitment to the planning process. Employ community-based mapping, storytelling, and photovoice tools to gather local knowledge, document personal stories, and empower marginalized groups in influencing land use decisions that align with their values, goals, and actual experiences.

m. Enabling substantial discourse and resolution of conflicts:
Implement moderated dialogue sessions, mediation processes, and conflict resolution mechanisms to effectively handle conflicts, bridge gaps, and foster consensus among stakeholders with divergent perspectives or conflicting interests. Facilitate occasions for organized discussion, attentive hearing, and courteous sharing of thoughts that value comprehension, compassion, and cooperative resolution of problems, cultivating a culture of positive involvement and joint decision-making.

n. Enhancing the ability and enabling the empowerment of communities:
Deliver education, training, and technical support to citizens, community organizations, and neighbourhood associations to improve their comprehension of land use planning procedures, entitlements, and chances for involvement. Promote the advancement of community-driven planning projects, participatory budgeting programs, and grassroots advocacy campaigns that enable residents to choose priorities, suggest solutions, and push for change in their local areas. Facilitate the establishment of alliances and cooperative relationships among municipal authorities, educational institutions, non-governmental organizations, and community-based groups to maximize the utilization of resources, knowledge, and connections for joint initiatives and the enhancement of capabilities.

11.7 Incorporating Local Knowledge in Decision-Making

The inclusion of local knowledge in decision-making processes improves planning methods, increases community involvement, and promotes sustainable development results. Decision-makers can enhance the development of land use strategies by incorporating the valuable insights, experiences, and perspectives of residents. This collaborative approach fosters inclusivity, relevance to the local context, and effectiveness. It empowers communities, safeguards cultural heritage, and advances the overall well-being of society in terms of social, economic, and environmental aspects. Integrating indigenous knowledge into decision-making processes is essential for a variety of reasons.

a. Advancing cultural sensitivity and inclusivity:
Using and honouring indigenous knowledge in the planning and decision-making process strongly advances sensitivity and inclusion. It makes clear that the planners are honouring the traditions, culture, and values important to local communities. An inclusive and fair platform is built when decision-makers welcome cultures and honour different ways of thinking and acting. This method brings people closer together and inspires them to look after shared sites. If people notice that their cultural identity is included in plans and policies, they tend to participate more and feel responsible for the final goals. Supporting collaboration with cultural sensitivity keeps conflicts at bay and supports projects that match the identities and needs of those in the community. When cultural sensitivity and inclusion is fostered, results are both fairer and more lasting, and they strengthen the bonds between people in the community.

b. Enhancing social capital and resilience:
When community knowledge is included in decision-making, it boosts both the unity and resilience of the community. Participating in the exchange of traditional knowledge, local experiences, and place-specific understandings makes communities more connected and trusting. By exchanging knowledge, people are strengthened individually and also learn to cooperate with others. This kind of active participation helps create social capital that is needed for a community's long-term well-being. Besides, when people are appreciated for their knowledge and involved in decisions that matter to them, it increases their sense of belonging and inclusion.

Engaging everyone in this way helps unite people and strengthens mutual help, creating the conditions for stronger communities. In these communities, people are more ready to face changes and overcome problems related to the environment and society. By including local wisdom, communities are better able to take the lead in managing and improving their future.

c. Increasing credibility and confidence:
Trust, credibility, and accountability in public administration can only be built if leaders engage with local communities. When leaders consult with community members and hear what they say, it makes it obvious that everyone is expected to participate, rather than only be told how things happen. These methods make it possible for people and government leaders to communicate, reach a stronger agreement, and talk openly about their responsibilities. Using the knowledge and feedback from people in the community in decision-making signifies that leaders support true democracy. They make it easier for everyone to join in and help create better, fairer, and longer-lasting results in a community. In addition, involving people in decision-making makes governing more authentic, since their needs are better reflected, which helps build a society that is both stable and harmonious.

d. Enhancing efficacy and longevity:
Local knowledge is necessary for making land use interventions effective and sustainable because it helps to match them to the setting and culture. Local and indigenous knowledge is based on many years of experience and can tell us a lot about ecosystems, resource use, and culture. Drawing on local knowledge, policymakers are able to create and carry out interventions that suit the specific needs of local people and shift as needed over time. Land use planning that includes adaptable practices and traditional knowledge reflects what communities care about and seek for the future. When initiatives include everyone and listen to their views, local communities are more likely to take ownership, and the projects are more likely to succeed and last. By making use of local knowledge, one can develop better and more lasting policies and programs that protect the land and improve life for current and future generations.

e. Promoting innovation and creativity:
Local insight plays a strong role in encouraging innovative and creative approaches to decision-making where land use is complicated. By making use of residents creativity and problem-solving ability, decision-makers can find solutions that might not be noticed regularly. Applying local,

grassroots methods allows communities to overcome issues related to the environment and changes in their economy. Sustainable development plans can better fit each place, leading to success and acceptance within the community. Community resilience increases and social changes are encouraged when government agencies consider and value local creativity in their main planning. This way of planning leads to better decisions and helps communities adapt to new circumstances.

f. Contextual awareness:
Understanding what people in a place live through and value provides decision-makers with new insights into land use choices. It consists of learning about people's day-to-day activities, history with the land, their cultural traditions, and the ideas and dreams they have kept for a long while. Knowing about what people and their environment go through and value helps decision-makers see the link between the conditions around them and how people live. As a result, decisions related to development are better suited to the community's needs and traditions.

g. Local knowledge:
Drawing on people's local knowledge helps find opportunities and problems that technical sources may miss. People in the community have their own daily experiences and insights that offer fresh ways to see the local environment, picking up on details that experts may not notice. Community-based insight can bring to light unused natural assets, important landmarks, or local networks of help, and also bring attention to challenges such as seasonal hardships, social conflicts, or environmental dangers that might be missed. Decision-makers benefit from tacit knowledge in the planning stage and can better comprehend what affects land use decisions. With this better insight, it is possible to build plans that suit the real conditions and support adaptation over time. Also, it gives local communities a voice by respecting their knowledge and working together to find solutions, resulting in improved and sustainable land use.

11.8 Digital Platforms for Engaging Communities in Land Use Decisions

Due to improvements in technology and the need for clear and open decision-making, people are using digital platforms more often to involve the community in land use.

a. Communication and information sharing:
If people can access important documents and decisions as soon as they are shared, they will be prepared for meetings and discussions in their community. Therefore, locals are informed about the topic and its outcomes, so they can join the discussions. Having good communication helps ensure communities take part in how land is used. Thanks to webpages, social media sites, and mobile programs, people can view updates and information about current land use proposals. Therefore, people can keep up-to-date with important decisions and discuss them with their neighbors. Residents can keep asking questions, share their ideas, and discuss problems thanks to web apps and social media platforms. This encourages people to join in, and their input is made sure to impact the decision-making process. Using chatbots and online forums, people can update each other, ask questions, get responses, and discuss different land-use matters as they arise. People can use these tools to speak up, receive answers, and trust the community more.

b. Participation from a distance:
Online platforms let people join in on decisions, even if they cannot get to meetings because of where they live or if they have trouble getting there. Residents can join in and take part in discussions or decisions by using things like online meetings or video calls from wherever they are, as long as they have a connection. Ensuring this makes it possible for people who live far away or have trouble getting to meetings because of disabilities or anything else, to still get involved from where they are. Allowing people to join remotely widens engagement and makes sure that people with different opinions can be heard. It matters most for those in communities that are not usually well-represented or who have been left out of usual decision-making. Digital participation allows decision-makers to give all residents, wherever they are, an opportunity to help shape the policies and land use plans that concern them.

c. Data acquisition and analysis:
Digital platforms make it possible to gather and examine information that helps with land use decisions. Online tools, including surveys, visualization platforms, and mapping applications help quickly gather data on the needs of the community, environmental problems, and people's land use choices. These tools make it easier for many people from the community to share their views with decision-makers quickly. Data collected through

the internet allows decision-makers to more easily understand what is most important to people and guide planning. In this way, GIS tools allow decision-makers to look at and map what the community input has to say. After data is collected and GIS tools are used, decision-makers can understand and see all kinds of information, patterns, and environmental issues easily. A better and more flexible planning outcome can be reached when decision-makers use information collected online about community priorities and environmental issues to decide on land use.

d. Crowd-sourcing and community mapping:
These platforms and tools enable communities to participate in planning by creating maps of their resources, land, and sites of meaningful places. With such information, decision-makers can be sure to include local knowledge in their planning activities. Nowadays, communities can share their concerns and views by mapping the pattern of natural resources, hazards, and places they care about. The use of these sites allows communities to highlight the risks, natural resources, and changes in land use where they live. Communities can make use of crowd-sourcing to alert planners to potential issues that may go unnoticed. As a result, decisions about land use plans and actions are more likely to consider the concerns of community members.

e. Education and capacity development:
Thanks to digital platforms, it is now possible to train communities and develop their skills on land use, saving the environment, and important policies. Thanks to online training, webinars, and educational videos, residents of communities are taught about land use and why it is important to look after the environment. This means people can learn from home or wherever they are, at any time. If you use games and other interactive resources, people will find it easier and more enjoyable to learn about land use issues. People are drawn in more and gain knowledge by using quizzes, simulation tools, and interactive case studies. Therefore, communities can learn about ways to help the environment and also pick up the skills they need to talk about the issue and help make good decisions.

f. Transparency and feedback mechanisms:
Tools on the internet make it easier for officials to get opinions from residents and use them in their decisions. Thanks to digital tools, sharing opinions and comments on possible land use plans is now much easier for everyone. By using these, people can contribute their views to the planning, which ensures they are listened to. Being honest about community input

leads to better and more effective decision-making. Sharing how community feedback shaped decisions helps decision-makers focus on what matters to the community. When decision-makers communicate their choices based on feedback, they help gain the trust of locals and encourage more people to participate.

g. Civic engagement and advocacy:
Using digital tools, residents can come together and press for change in land use issues on the internet. By using digital storytelling, petitions online, and campaigning on social platforms, residents can share their concerns and stories with a broad group of people. They make land use problems more noticeable to others and help spread information about them to many who could be unaware. People in the community can directly contact leaders by using digital advocacy and ask for changes that support the environment. Online tools allow people living in a community to unite, ask more people to get involved, and make sure their efforts have a lasting impact on important environmental, social, and economic decisions. Consequently, digital platforms help spread the message of local campaigns and make them more effective in achieving positive land use outcomes for the good of everyone and the environment.

Digital platforms offer several options for communities to be engaged in the process of making decisions about how land is used. These platforms give communities the ability to participate more in shaping their land and environment by backing transparency, advocacy, data use, remote engagement, and information exchange. Still, one should make certain that digital engagement initiatives are available and easy for all members of the community to use.

11.9 Role of Social Media in Shaping Public Opinion on Land Use

Through social media, perceptions are shaped, dialogue is encouraged, and many voices are heard on issues related to land use. Facebook, Twitter (now X), Instagram, and LinkedIn are important tools for getting information out, keeping people updated, and spreading awareness about different land use matters, policies, and projects. Distributing articles, reports, images, and videos at a rapid pace allows users to interact with many people, helping raise awareness of environmental concerns, urban planning, and conservation. One of the main advantages of social media is that it brings together local people, government officials, NGOs, environmentalists, and experts so they can interact immediately and directly. These platforms make it easy for everyone to talk about and work on solutions to problems involving land use. By using hashtags and mobilizing people through online petitions and viral messages, support for urgent actions and awareness of development challenges can quickly grow, putting pressure on decision-makers to reconsider land use proposals. Additionally, organizations and governments use these tools to observe public opinions, notice what is popular, and observe ongoing online discussions in real time. Data from real-time social media use helps officials to make informed decisions, communicate better, and understand what the public thinks. At the same time, while social media helps people be involved and promote change, it can also spread untrue statements, misinformation and wrong explanations. When used carefully, social media can help drive greater transparency and accountability in land use processes and involve more people in decision-making.

11.10 Effectiveness of Community Input in Decision-Making

It is very important to assess how much community input improves inclusivity, transparency, and responsiveness to diverse stakeholders in decision-making. Good engagement with the community justifies decisions because it also produces more sustainable and widely supported results. Community participation should be analyzed by looking at both numbers and personal experiences from the individuals involved.

Quantitative metrics:
One can look at measurable indicators to understand how wide and effective community engagement is. Some examples are counting participants, checking the spread of backgrounds among the participants, and tracking how regularly and consistently people engage. Other indicators can track how many consultations have happened, how well people turn up for public forums, how people take part in workshops or surveys, and what percentage of community suggestions are accepted in the final results. These figures show whether a lot of people are participating and if they keep participating as time goes on.

Regular surveys and feedback mechanisms:
Community members can be given regular surveys and feedback forms to find out how they feel and what they think about being involved in decision-making. They give community members a chance to describe how successful the strategies are, whether information is easy to access and understand, and how well their ideas have been considered. Reviewing survey responses can show what is keeping people from joining, so engagement methods can be improved on an ongoing basis. Repeated use of surveys can keep track of how views in the community are changing and what impact the engagement work has over time.

Qualitative analysis:
One must carry out qualitative analysis to grasp the influence and richness of the input given in community meetings. It means looking at notes or writings from meetings, comments made by people, and the things they said at different events before putting all that information into groups to see which topics or points came up most often. Through the use of qualitative analysis, it is more likely that decision-makers can determine the subjective feelings and biggest concerns of the community. Opinion analysis allows decision-makers to see if opinions matter much and helps to highlight any special or personal experiences brought up by participants. Unlike statistical methods, this kind of study helps find subtle thoughts and experiences of the participants.

Stakeholder interviews:
Interviews and focus groups allow for a better understanding of community life. By conducting interviews with such people, it becomes possible to gather information from community members regarding their practices of engagement. They help to reveal the community's opinion about transparency, the level of impact community members feel, and what main issues or expectations remain unmet. Such talk helps decision-makers to review relationships with stakeholders and update their ways of working together.

Case studies and success stories:
Case studies and success stories help to show the real impact of community input. Community engagement's positive impact can be clearly seen in situations where community input led to a policy change, infrastructure work, or improvements in delivered services. Case studies of community engagement highlight what succeeds, help solve problems, and provide models others can use in other situations. Case studies and success stories

build trust among involved people and prove why participatory governance is necessary.

Outcome evaluation:
Another important step in measuring success is outcome evaluation. This means measuring how community engagement helped change the plans, resources, rules, and services. This is done by checking if community contributions brought about changes in the original plans, reallocation of resources, new regulation rules, or better services. Outcome evaluation needs to look at long-term gains, such as a stronger community, higher knowledge of civic matters, or better environmental or social results.

Feedback loops:
Implement feedback mechanisms to actively seek continuous input and evaluation from community members at every stage of the decision-making process. Urge stakeholders to consistently provide feedback regarding the efficacy of engagement activities, the promptness of decision-makers, and the perceived impact on outcomes.

Comparison with benchmarks:
Evaluating against recognized benchmarks, standards, or best practices, assess the community input methods and outcomes in relation to public involvement and stakeholder engagement. Assess the degree to which decision-making procedures conform to ideals of inclusion, transparency, accountability, and responsiveness.

Participatory assessment:
Engage community people directly in the assessment process through participatory evaluation activities, workshops, or co-design sessions. Motivate stakeholders to introspect on their experiences, exchange feedback, and collectively recognize strengths, flaws, and prospects for enhancement.

Continuous improvement:
Utilize evaluation findings to guide ongoing improvement initiatives focused on strengthening the efficacy of community input in decision-making processes. To get better at engaging with the community, it is important to put new suggestions into action, improve the ways in which the community is reached, and spend resources on abilities the organization needs.

By mixing these approaches, those involved can check how well the community is involved in decision-making, spot where improvements are needed, and make sure that community participation actually matters.

11.11 Quantitative and Qualitative Metrics for Evaluation

Both quantitative and qualitative metrics help to understand how the community shapes decision-making. Each type of metric provides something different; quantitative ones give us clear, number-based data, while qualitative ones help us understand how and why people feel and think about things in the community. Table 11.1 gives a closer look at both types of metrics.

Table 11.1 Metrics for evaluation.

Quantitative indicators	Qualitative indicators
Participation Rates: Track the number of people from the community who attend meetings, take surveys or participate in public forums.	Engagement Depth: Consider the degree to which each person actively and smartly takes part in discussions.
Decision Implementation: Evaluate how many community recommendations are implemented.	Quality of Dialogue: Observe respect, openness, and politeness in conversations between participants.
Response Rates: See how often the authorities consider and reply to concerns raised by community members.	Perceived Influence: Make sure those affected believe they have power and are valued in making decisions.
Input Volume: Determine the overall number of comments, suggestions or proposals submitted.	Thematic Analysis: Find out what concerns or topics arise often in what the community says.
Engagement Frequency: Maintain a record of how many times engagement events are organized.	Impact Assessment: Look at how much the community is involved in shaping the town's decisions or policies.
Demographic Representation: Check that participants are chosen to represent people from various parts of the community (such as age, gender and ethnicity).	Stakeholder Satisfaction: Find out participants' reactions to the process and the results.

11.12 Challenges in Incorporating Feedback

All sides should be committed, work together, and make good use of feedback. Decision-makers can improve decision-making by identifying and handling obstacles and using opportunities at hand. As a result, there will be benefits for the local area and the wider society.

Diverse stakeholder perspectives:
Understanding and connecting the different interests of several stakeholders is tricky because people from different areas and backgrounds often want different things when it comes to making land decisions. Bringing all these people together to agree takes a lot of discussions and give-and-take, and it takes a lot of time and resources. It also does not help that people from different cultures and backgrounds often see the land in different ways. People from different sectors and backgrounds usually want different things when it comes to making land use choices.

Power dynamics:
It can be hard to take public feedback into account when leaders and those in power hold most of the decision-making authority. Power differences often lead to situations where the opinions of people who are not well represented are not taken into account, mainly because governments, corporations, or large groups hold a lot of sway in decisions. It is important to have processes that include all feedback, on an equal level, regardless of a person's income or background.

Legal and regulatory restrictions:
Giving stakeholders an actual voice in land use decisions can be curtailed by the rules or laws in place. When deciding what to do about different stakeholder needs, decision-makers must go by strict laws and regulations. Since decision-makers are often bound by rules and regulations, it can become difficult for them to give stakeholder opinions the attention they deserve.

Resistance to change:
Often, trying to include stakeholder input in land use decisions takes longer because people are often slow to change old methods, decisions, or processes, because they are worried about what might happen. People involved in land use decisions and the people who have a say in them are often slow to make changes because they are not sure everything will work out well. Trust must be created, conversations need to be promoted, and

people should be made aware of the positive outcomes of using new land use planning methods. Effective collaboration among stakeholders combined with the management of changes, can help make sustainable land use practices a success.

Communication barriers:
Adequate communication and information exchanges play a key role in making sure stakeholder input is incorporated into land use plans. Communication barriers caused by language, culture, or technology make it difficult for stakeholders to exchange ideas and information. Ensuring trust, collaboration, and that all stakeholders are treated equally, with access to information and decision-making, depends on having clear and open ways of communication.

Timing and timeliness:
Integrating stakeholder input into land use decisions depends a lot on how soon the feedback is collected and used. Dealing with feedback from stakeholders slowly will weaken how much it impacts decisions in environments where decisions must be made quickly. To make sure feedback reaches decision-makers in time and is used in decisions, decision-makers must work to engage stakeholders early on. Reducing delays and making feedback integration work better can be achieved by being proactive in communication and stakeholder engagement.

Feedback fragmentation:
The presence of conflicting information from a range of constituents makes it much harder to organize and sort out feedback. It is challenging for decision-makers to pick out common themes and key priorities when many stakeholders present different views and suggestions. Feedback synthesis is much easier when there are strong analysis systems, engagement techniques, and tools that can help people reach agreement.

Communication overload:
Decision-makers often face a lot of feedback from many sources when they try to organize and synthesize it. With so much information coming in, it is hard for decision-makers to pick out what is most important, and making decisions becomes a real challenge. Decision-makers can sort through a lot of information and focus on what matters most for sustainable land use planning by using analytical tools and decision-support systems that display, gather, and rank data.

11.13 Conclusion

Sustainable and fair land use management needs decisions that are made inclusively. Participatory planning, which engages the community, makes use of local experiences to find suitable solutions. Listening to what people in the community know gives valuable information about the culture, society, and nature. Community participation and accountability improve when people use digital platforms such as social media. Even so, planners must use both digital and traditional tools to deal with biases. Assessing progress involves measuring different aspects of how deeply stakeholders are involved and the outcomes. Effective community engagement harnesses collective wisdom for landscapes that benefit present and future generations.

References

1. Gerlak, A.K., Guido, Z., Owen, G., Rodriguez McGoffin, M.S., Louder, E., Davies, J., Smith, K.J., Zimmer, A., Murveit, A.M., Meadow, A., Shrestha, P., Joshi, N., Stakeholder engagement in the co-production of knowledge for environmental decision-making. *World Dev.*, 170, 106336, 2023, doi: 10.1016/j.worlddev.2023.106336.
2. Adelle, C., Pereira, L., Görgens, T., Losch, B., Making sense together: The role of scientists in the coproduction of knowledge for policy making. *Sci. Public Policy*, 47, 1, 56–66, 2020, doi: 10.1093/scipol/scz046.
3. Akpo, E., Crane, T.A., Vissoh, P.V., Tossou, R.C., Co-production of knowledge in multi-stakeholder processes: analyzing joint experimentation as social learning. *J. Agric. Educ. Ext.*, 21, 4, 369–388, 2015, doi: 10.1080/1389224X.2014.939201.
4. Cash, D.W., Clark, W.C., Alcock, F., Dickson, N.M., Eckley, N., Guston, D.H., Jäger, J., Mitchell, R.B., Knowledge systems for sustainable development. *Proc. Natl. Acad. Sci.*, 100, 14, 8086–8091, 2003, doi: 10.1073/pnas.1231332100.
5. Geekiyanage, D., Fernando, T., Keraminiyage, K., Assessing the state of the art in community engagement for participatory decision-making in disaster risk-sensitive urban development. *Int. J. Disaster Risk Reduct.*, 51, 101847, 2020, doi: 10.1016/j.ijdrr.2020.101847.
6. Varol, C., Ercoskun, O.Y., Gurer, N., Local participatory mechanisms and collective actions for sustainable urban development in Turkey. *Habitat Int.*, 35, 1, 9–16, 2011, doi: 10.1016/j.habitatint.2010.02.002.
7. Haaland, C. and Konijnendijk van Den Bosch, C., Challenges and strategies for urban green-space planning in cities undergoing densification:

A review. *Urban For. Urban Greening*, 14, 4, 760–771, 2015, doi: 10.1016/j.ufug.2015.07.009.
8. Gaillard, J.-C., Cadag, J.R., Gampell, A., Hore, K., Le Dé, L., McSherry, A., Participatory numbers for integrating knowledge and actions in development. *Dev. Pract.*, 26, 8, 998–1012, 2016, doi: 10.1080/09614524.2016.1226263.
9. Chen, W., Cheshmehzangi, A., Yu, J., Mangi, E., Heath, T., Zhang, Q., An analysis of patterns of public engagement in China's community micro-rehabilitation projects: A case study of Guangzhou. *World Dev. Sustainability*, 3, 100108, 2023, doi: 10.1016/j.wds.2023.100108.
10. Li, X., Zhang, F., Hui, E.C., Lang, W., Collaborative workshop and community participation: A new approach to urban regeneration in China. *Cities*, 102, 102743, 2020, doi: 10.1016/j.cities.2020.102743.
11. Arciniegas, G. and Janssen, R., Spatial decision support for collaborative land use planning workshops. *Landscape Urban Plann.*, 107, 3, 332–342, 2012, doi: 10.1016/j.landurbplan.2012.06.004.
12. Arciniegas, G., Janssen, R., Rietveld, P., Effectiveness of collaborative map-based decision support tools: Results of an experiment. *Environ. Modell. Software*, 39, 159–175, 2013, doi: 10.1016/j.envsoft.2012.02.021.
13. Boroushaki, S. and Malczewski, J., Measuring consensus for collaborative decision-making: A GIS-based approach. *Comput. Environ. Urban Syst.*, 34, 4, 322–332, 2010, doi: 10.1016/j.compenvurbsys.2010.02.006.
14. Jankowski, P., Towards participatory geographic information systems for community-based environmental decision making. *J. Environ. Manage.*, 90, 6, 1966–1971, 2009, doi: 10.1016/j.jenvman.2007.08.028.
15. Armstrong, M.P., Perspectives on the development of group decision support systems for locational problem-solving. *Geogr. Syst.*, 1, 1, 69–81, 1993.
16. Balram, S., Dragicevic, S., Meredith, T., Achieving effectiveness in stakeholder participation using the GIS-based collaborative spatial Delphi methodology. *J. Environ. Assess. Policy Manage.*, 5, 3, 365–394, 2003, https://doi.org/10.1142/S1464333203001413.

12

Remote Sensing Techniques for Land Use Pattern Extraction

Abstract
The use of satellite images from remote sensing is necessary for the identification of changes in land use patterns, which is vital for both land administration and environmental protection. With the help of both active and passive remote sensing, satellites ensure that essential information about the Earth's surface is obtained. These image improvement techniques, like preprocessing techniques, for example, radiometric and geometric correction, preserve their reliability and make satellite images easier to understand. The aim of change detection analysis is to spot changes over time in how land is used, with classification methods supporting the task of categorizing various types of land surface. Organizations and governments can use the regular monitoring and analysis of satellite data to devise better strategies for managing and protecting the environment. The role of satellite imagery in developing a clear understanding of the Earth's changing landscapes becomes more evident as we grow more skilled in solving challenges related to environmental protection and land management.

Keywords: Change detection, image analysis, land use pattern, remote sensing, satellite

12.1 Introduction

Using remote sensing has made it possible for experts to better understand the world's changing climate and important connections between different parts of our planet. Remote sensing involves the use of sensors and platforms to take pictures and data of Earth from a distance. With this technique, it is possible to collect data on the planet's surface, its atmosphere, and its oceans. Remote sensing is important in many fields such as environmental protection, managing resources, dealing with disasters, and conducting research. Because of many environmental changes occurring in today's scenario, remote sensing is used to track and analyze the changes

Kundan Meshram and Chatrabhuj. A Sustainable Approach to Land Use, (301–332) © 2026 Scrivener Publishing LLC

in the environment, track changes in the land, and detect environmental problems such as deforestation, development on land, and climate change [1–3]. It enables decision-makers to develop plans for nature conservation and resource management using complete and reliable facts from remote sensing. In disasters, remote sensing helps by providing fast information, keeping decision-makers aware of what is happening, and preparing for any emergencies triggered by floods, wildfires, or hurricanes [4]. It also helps in agriculture by aiding crop management, maintaining food supply, and shielding agriculture from negative climate change.

Remote sensing provides information that is needed by scientists to look at the processes of Earth systems—lands with interesting geological features, activities in the atmosphere, and environment's reaction to human behaviours. It provides the information needed for infrastructure planning and development through its data on topography, how land is used, and city development. It aids the plans for cities, building infrastructure, and exploring varieties of resources. Remote sensing is important for planning and developing cities that are sustainable and have strong infrastructure. Remote sensing is used by public health experts to observe factors in the environment that influence the spread of diseases [5–7]. Remote sensing is crucial for tracking diseases, identifying threats, and taking steps to protect

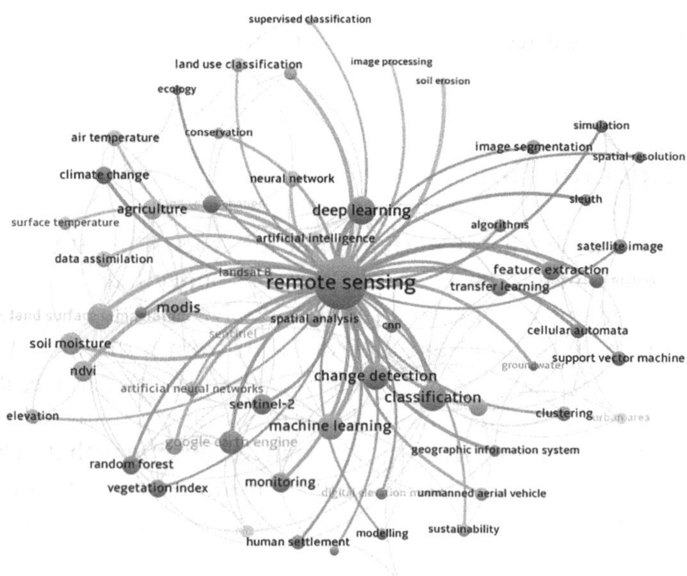

Figure 12.1 Remote sensing and its domain.

people's health. It helps with worldwide issues, promotes sustainability, and broadens scientific knowledge, as can be seen in Figure 12.1.

12.2 Remote Sensing Technique

Remote sensing uses various devices and sensors to collect information about things, locations, or occurrences without any physical interaction. Utilizing sensors and devices allows us to detect and measure the electromagnetic radiation (EMR) that is reflected, emitted, or scattered from the Earth's surface, atmosphere, or the rest of the universe [4, 5, 8]. The gathered data is subsequently processed and analyzed to get significant insights into the subject of interest.

Active and passive remote sensing are the main techniques that help us collect data on the Earth's surface and atmosphere. Table 12.1 compares active with passive remote sensing.

The elements and procedures used in remote sensing are analyzed below.

Energy source:
In remote sensing, one must have some kind of energy source so one can see what is going on down on Earth. Most of the time, sunlight is used because it shines out diverse kinds of electromagnetic radiation, but if there is not enough sunlight, like at night or when there are too many clouds, systems like radar or LiDAR are used, which send their signals and pick them back up after they hit the ground. Passive remote sensing systems need sunlight to get bounced-back signals from the ground. The choice of energy source mostly decides what kind of information can be gathered from remote sensing and what kinds of tasks it can help with.

Interaction with the target:
Once the energy from the source reaches any point on Earth or an object, it interacts with the material in several ways. The light takes part in reflection, absorption, transmission, and scattering when it touches the Earth's surface or another object. The different factors, such as the material, surface, moisture, and angle, affect these interactions. Vegetation absorbs visible light, but it reflects near-infrared waves, while water bodies usually absorb much of the radiation and seem dark in several bands. Such patterns of interaction are called "spectral signatures," and they play a key role in allowing us to differentiate between different objects on Earth with remote sensing.

Table 12.1 Active vs. passive remote sensing.

Aspect	Active remote sensing	Passive remote sensing
Energy Source	Emits its own source of energy	Relies on external natural energy sources (e.g., sunlight)
Energy Emission	Sends out electromagnetic waves	Do not emit energy, only detects reflected or emitted energy
Typical Wavelength	Microwave, Radio waves, Laser (for LiDAR)	Visible, Infrared, Thermal infrared
Energy Reflectance	It measures the amount of energy that is sent back to the sensor by reflection or scattering.	Measures naturally reflected or emitted energy from surfaces
Penetration Capability	Depending on wavelength, it can penetrate through vegetation, soil, and some structures.	Restricted to surface and sub-surface observations
Cost	Generally, more expensive and complex	Typically, less expensive and simpler
Data Aquations	Can collect data day or night, regardless of weather	Only in broad daylight
Operational Time	Operates effectively 24/7	Most effective during the daytime
Weather	Less affected by weather; can penetrate clouds, rain, and fog	Weather-dependent, performance can be hindered by clouds
Examples	Radar (Radio Detection and Ranging), LiDAR (Light Detection and Ranging)	Optical cameras, Radiometers, and Multispectral scanners
Application	Terrain mapping, vegetation structure analysis, surface deformation studies	Land cover classification, oceanography, and climate monitoring

Sensors and platforms:
Sensors are used to detect and measure the electromagnetic signals involving or coming from the Earth's surface. The sensors used for observing the Earth vary in their abilities, the types of energy they detect, and how detailed they are. Optionally, sensors can be deployed on orbiting satellites, manned aircraft, unmanned aerial vehicles (UAVs or drones), or on the ground in stationary units. The use of satellites, manned aircraft, drones, and ground units helps in studying various parts of the Earth in detail. Satellites can repeatedly collect data over a wide area, drones can provide detailed information over small areas, and ground systems are used for detailed close-up work. Sensors working together with the different platforms provide thorough monitoring that helps with both global studies and analyzing local infrastructure.

Data acquisition:
Data acquisition in remote sensing means getting information about a certain place by using the sensors on devices flying over it. This process means taking in how much and which kinds of light get bounced, given off, or just bounce around from the ground, water, or anything else below a satellite. The data collected may show up as photos taken from different light wave ranges, simple radar images, or maps made using bouncing laser pulses. Depending on what the mission wants to know, the sensors can be set to see different types of light, like visible, infrared, or microwave. This flexibility means remote sensing can look at lots of different things, like how healthy plants are, moisture content in soil, land type, and development of cities.

Data transmission and storage:
The data collected by remote sensing is either transmitted in real time or stored onboard for later examination. Thanks to improvements in telemetry and satellite communication, data can now be quickly and safely transmitted, in any case, from space or remote places. These days, large amounts of data are transferred from satellites by improving the speed of communication links. On top of that, storage technology on spacecraft now allows data sizes from new, advanced missions to be stored properly.

Data processing and analysis:
Before the data can be interpreted, raw remote sensing information must undergo a lot of processing. Before the raw data is interpreted, it is important to process it by carrying out geometric correction, radiometric calibration, and atmospheric correction. Once everything is finished in

preprocessing, the raw data is then analyzed using methods such as image classification, change detection, and object-based analysis. Once all the necessary data processing is done, it allows experts to see patterns, trends, and any anomalies, helping them make decisions in various fields.

Interpretation and applications:
The final part of remote sensing involves looking at the processed data to draw useful and relevant conclusions. Analysts or automatic systems based on machine learning can be used to interpret the data obtained in this step. The applications for interpreted data are numerous and cover many areas. A variety of real-world activities benefit from applying data from remote sensing, like management of crops, tracking environmental damage, dealing with emergencies, planning, and urban infrastructure. Remote sensing has become a really important tool that helps government agencies, researchers, and businesses take care of natural resources, prevent environmental problems, and organize development in a good way.

Remote sensing allows scientists to observe and look after the Earth's surface and its atmosphere globally. Remote sensing allows for an organized and efficient way to assess and keep an eye on various sectors. It makes it possible for researchers, policymakers, and land managers to learn about environmental processes, find out about any changes over time, and act wisely for the sake of sustainability and managing resources.

12.3 Necessity and Importance

Many fields use remote sensing, for instance, scientific studies, monitoring the environment, dealing with resources, and disaster management. The following sections give details on the necessity and importance of remote sensing.

12.3.1 Environmental Monitoring

With the help of satellite imaging, remote sensing makes it possible to observe the Earth's surface from above, which greatly helps in environmental monitoring. Remote sensing makes it possible to detect and analyze environmental issues on a large scale, mainly deforestation, urban expansion, desertification, challenges with wetlands, and changes in land cover and land use. Observing the Earth's features over a period gives scientists and policymakers a chance to study the ecosystem and highlight areas at risk of being damaged. Thus, monitoring the environment through neglect helps put conservation policies and sustainable ways to handle land into practice, allowing for the long-lasting stability of the ecology.

12.3.2 Natural Resource Management

Effective management of natural resources relies on remote sensing to obtain the necessary information. It offers information about forest cover, water, land areas for farming, minerals, and other such resources. The information from remote sensing supports activities such as mapping, inventorying, and monitoring of these resources, which helps keep them safe for use in the future. Forest and water resource managers rely on satellite imagery to oversee tree cover and monitor the state of reservoirs and their surrounding areas [9–13]. This technology also helps in restoration projects by showing progress and detecting which areas need further help. As a result, data collected from remote sensing helps with protecting, managing, and distributing the planet's natural resources.

12.3.3 Disaster Management and Response

No matter the type of disaster, remote sensing helps the field of disaster management in planning ahead, reacting, restoring, and preventing future problems. Remote sensing helps identify places in the field of disaster management that are at risk of floods or landslides before a disaster happens. Satellite images are used in disasters to measure the cause, figure out where the people are affected, and help to direct emergency actions. Images taken by satellites are relied upon after disasters to keep watch over the progress of recovery and rebuilding. Whether it is an earthquake, hurricane, flood, wildfire, or volcanic eruption, remote sensing speeds up the process of collecting large-scale and unbiased data to help with disaster relief.

12.3.4 Climate Change Monitoring

When it comes to climate change, remote sensing helps enormously with studying and responding to it. It serves as a way for scientists to detect trends and peculiarities in sea levels, polar ice, and many other changes occurring in climate and the environment. Due to these observations, one can spot changes and unusual events in how the Earth's climate behaves. Also, remote sensing using satellites helps us monitor greenhouse gases like carbon dioxide and methane in the atmosphere. The continuous monitoring data collected using remote sensing supports the development of climate models, assesses climate effects, and helps form worldwide strategies for dealing with climate change. With climate change getting stronger, remote sensing will be vital for keeping an eye on the environment and forming official policies.

12.3.5 Agriculture and Food Security

Remote sensing greatly improves agriculture and food security through accurate tracking of crops and the soil. It collects information about moisture in the soil, crop health, the stages in the crop's growth, where crops fit best, and overall yield over wide agricultural areas. The information helps farmers to find the best approach to irrigation, use fertilizer, control pests, and plan when to harvest their crops. Remote sensing is used by governments and international bodies to predict possible food scarcity and deal with upcoming challenges. In the face of climate change, remote sensing helps farmers adapt and manage their farms effectively. Remote sensing allows for efficient management of farmland and agricultural supplies, making sure there is enough food for everyone.

12.3.6 Infrastructure Planning and Development

It is through remote sensing that spatial data is collected for infrastructure planners to view current land use, topography, population position, and built-up land. They can see what the land is used for, its location relative to topography, where groups of people exist, and developed areas. This information helps in planning roads, bridges, means of public transit, utility lines, and housing projects. When remote sensing is combined with GIS, information on the environment, urban changes, and risks to natural areas can be gathered and identified. Additionally, using remote sensing after construction, planners can ensure infrastructure is still safe and up to date over the years. It allows for more sustainable planning and construction of city infrastructure.

12.3.7 Public Health

Remote sensing helps to improve public health by identifying possible threats in the environment that can affect how diseases spread. For example, it can spot still water where mosquitoes can breed, letting us keep track of how diseases such as malaria and dengue spread. It also checks the air quality, the cleanliness of lakes and rivers, and changes in land use that can affect our health. Health officials use this information to map out where health risks are, so they can find areas that need extra help and protect people and communities before problems start. Health agencies use this data to help find out about possible health threats early, help plan ways to take care of people's health, and make sure they allocate resources where they are most needed.

12.3.8 Scientific Research

In different scientific fields, remote sensing helps a lot to obtain useful results. It supports the creation of detailed elevation models as well as classifications of landforms in topography. It makes it possible to map and analyze biodiversity, model the areas where species live, and check the health of ecosystems. Remote sensing is used in geology for looking for minerals, identifying faults, and studying the evolution of landforms. Satellite data is used by atmospheric scientists for weather prediction, checking air quality, and reviewing climate-related changes. Oceanographers make use of remote sensing to keep track of sea surface temperatures, levels of chlorophyll, and ocean currents. Remote sensing is very helpful because it gives scientists long-term, reliable data needed to learn more about Earth system processes and develop new theories. They can use remote sensing to look at areas that researchers could not otherwise observe.

12.3.9 Policy and Decision-Making

Remote sensing gives us clear, up-to-date maps and information that helps governments, planners, and decision makers make smart choices and plans. Whether for things like planning cities, giving out resources, dealing with the environment, or deciding how land is used, remote sensing helps by giving clear information that makes things fairer and keeps officials honest. It helps people make better decisions by giving up-to-date and long-term data, making sure policies are based on what is actually happening out in the world. For example, deforestation rates from satellite pictures can help make laws about protecting forests, and maps of floodplains can affect building rules and how society gets ready for floods. Thus, remote sensing helps people make better decisions by giving clear information about what is happening in an area.

12.3.10 Global Connectivity and Collaboration

Remote sensing bridges distances and challenges by providing access to information from tough, politically unstable, or environmentally unsafe places. Shared data enables countries and societies to work better together to handle ongoing issues such as climate change, disasters, and environmental threats. Due to GEOSS and UN-supported remote sensing programs, there is more collaboration and research by countries and international agencies. By making information easy to use, it helps scientists and policymakers from around the world strive for sustainable development.

12.4 Satellite Imagery in Identifying Shifts in Land Use Patterns

It is through satellite imagery that users can detect and look into shifts in how land is used. It is needed for urban planning, environmental checks, agriculture, crisis management, and studies about climate change because it provides reliable, unbiased, and detailed information on land usage over vast and long areas [14–17]. Sophisticated techniques and the use of satellite images with other types of spatial data help researchers and policymakers to control and understand the long-term changes in land use across the world.

12.4.1 Tracking the Process of Urbanization

Urbanization refers to the process of making rural or undeveloped regions into cities. Satellite images allow us to accurately follow the growth and expansion of cities. Urban areas can be identified by their special features, which are mainly structures, roads, and all kinds of urban infrastructure.

- Urban expansion in China: In recent years, China has experienced urban growth at a high pace. In a lot of cities in China, including Beijing, Shanghai, and Shenzhen, researchers have observed the documentation of urban expansion using data from Landsat and Sentinel-2 satellites. Looking at images gathered over time from satellites like the Landsat shows more areas are becoming built-up and gives information that is helpful to urban developers and politicians.
- The United States has seen multiple suburban regions around New York, Los Angeles, and Chicago quickly grow in the past decades. The suburban areas of New York, Los Angeles, and Chicago have seen significant expansion. Seeing how the suburbs of these three cities have grown has

been made possible by examining high-resolution satellite data from the Landsat and WorldView satellites. Analysis of this data supports efforts to detangle problems in infrastructure, transportation, and sustainability.

- This phenomenon of urban sprawl has appeared in cities like Mumbai, Delhi, and Bengaluru in India. By using data from Sentinel-2 and Cartosat satellites, researchers are able to study how fast informal settlements and infrastructure in Indian cities have grown. Using this data from satellite images helps with urban planning and solving problems relating to housing, sanitation, and the use of important resources in cities.

12.4.2 Utilization of Agricultural Land

Satellite images make it possible to observe how lands are used for agriculture, as well as changes in crop rotation, more agricultural space, and different crops. It is possible to identify the different types of crops using the spectral properties captured by multispectral and hyperspectral images. With this technology, researchers can map many types of crops and find out how healthy the crops are growing.

- European precision agriculture: For tracking their crops, European farmers make use of satellite images from Sentinel-2 and WorldView. Using data helps in making decisions that boost the efficiency and continued success of farming. U.S. farmers in the Midwest use MODIS and Landsat images to identify and map different crop varieties, like corn, soybeans, and wheat. Many crops are cultivated in the United States in the Midwest area. To support crop variety monitoring in the Midwest, researchers can use data from MODIS and Landsat satellites. They are vital for learning how crop rotation changes over different years.
- Cultivation of rice in Southeast Asian regions is monitored using information from Sentinel-2 and MODIS satellites. Researchers comb through satellite data from Sentinel-2 and MODIS to manage rice field cultivation in Vietnam, Thailand, and Indonesia in Southeast Asia. The tracking of changes in planting and picking seasons, along with the change in size of rice fields, aids in organizing food security and managing water in the region.

12.4.3 Assessment of the Environmental Impact

Satellite images are important for starting EIAs, as the images allow for the monitoring and tracking of modifications in any region that is being developed. This allows for monitoring and assessing the changes caused by such developments in land use.

- Mining operations: A good example is the Appalachian region in the United States, where people monitor the environmental effects of mines and the water quality with satellite imagery. It is possible to measure and watch over time the effects of mountaintop removal mining on the area's plants and land in the Appalachian Mountains region.
- Urban development projects: Researchers have used satellite images to determine the implication of urban development projects on the local environment. To fully understand the effects of major developments on the environment, Environmental Impact Assessments (EIAs) are put into practice. Satellite pictures are used to see how development has changed the earth's surface, allowing researchers to notice any change in land, water, and air quality.
- Impact of infrastructure projects: Scientists rely on satellite images to understand the environmental impacts of infrastructure works like highways and dams. The impact of the Belo Monte Dam in Brazil on forest and river areas has been carefully studied by evaluating satellite images.

12.4.4 Deforestation and Reforestation Analysis

The action of removing plants from an area is called deforestation, and adding new plants to these places is known as reforestation. Deforestation and reforestation both greatly impact the environment and climate levels. Remote sensing from satellites is crucial for checking on how forests are changing and how reforestation and afforestation efforts are going.

- o Boreal forests in Canada: The RADARSAT and Landsat satellites are used to examine and survey Canada's boreal and northern forests. In Europe, these forests might be affected by logging, burning caused by wildfires, or mining. Satellites are helpful in monitoring both the occurrence of deforestation and the success of reforestation and afforestation.

- Forest regeneration in Europe: Sentinel-2 and Landsat are the main satellites used to trace forest restoration in European countries. To tackle climate change and promote diversity, Spain, Italy, and France have put in place projects aimed at enlarging forest regions. Satellite information is beneficial for monitoring the various initiatives made to help forests.
- Amazon rainforest: The rainforest in the Amazon is both important for biodiversity and storehouse of carbon. Data from Landsat has helped track the extent of deforestation in the Amazon rainforest, especially where land was cleared for farming and mining purposes. The success of conservation initiatives in the Amazon Rainforest can be judged by monitoring how much is being replanted.

12.4.5 Coastal Areas Change Analysis

Satellite photography is key for reporting on and checking how the land in coastal areas is being developed, erupted, or covered with coastline infrastructure. It helps explain the impact of people and nature on different coasts.

- Mangrove loss in Southeast Asia: Aquaculture and increasing coastal development are causing the mangroves in Southeast Asia to decline. Landsat and Sentinel-2 satellite images are widely used to see and guard against the loss of mangroves in Southeast Asia.
- Erosion in the Mississippi Delta: The Mississippi Delta's land is being significantly eroded and subsiding. The difference in the delta's coastline because of restoration projects can be seen easily using satellites from Landsat and Sentinel-2.
- Coastal development in Dubai: The growth of the Palm Islands and the world archipelago during the land reclamation efforts in Dubai has been properly documented with the use of high-resolution satellite photos. In Dubai, these land development projects have altered the coastline and affected the living space for marine animals.

12.4.6 Research on Climate Change

Focusing on farming areas, melting glaciers, and changes in the patterns of rainfall helps in analyzing the influence of climate change on land use.

With this information, one can create models and predictions for coming changes in land use, related to different climate conditions.

- Melting glaciers in the Himalayas: Because of the effect on the climate, Himalayan glaciers are shrinking. The images from Landsat and Sentinel-2 are studied to observe the effects on water and land utilization after glaciers start to melt.
- Changing precipitation patterns in Africa: Changes in precipitation brought by climate change on the African continent are affecting the amount of crops grown and the availability of water. The use of satellite data helps to track these changes and supports the formation of plans to manage land and water resources.
- Shifting agricultural zones: Satellite imaging has detected how climate change is causing agricultural zones to shift northward in the Northern Hemisphere, affecting crops and different farming methods. As an example, satellite data can be used to check for growing areas where wheat is more suitable for growing in Canada.

12.4.7 Identification of Illicit Actions

Satellite images are able to point out and identify unauthorized changes to the use of land, such as illegal mining, logging, and building activities. This skill plays an important role for both law enforcement and agencies dealing with environmental protection.

- Unauthorized logging in Indonesia: By studying satellite images, including those from WorldView, law enforcement can detect and manage illegal logging in Indonesian forests. When satellite photographs reveal this kind of crime, they are sent to the authorities, who can charge individuals for the harm they do to the environment and use that data to keep the forests secure.
- Illegal mining in the Amazon: Sentinel-2 is one of the satellites that helps detect and watch illegal gold mining activities in the Amazon. Due to the satellite images, appropriate actions can be taken against illegal mining, and the facilities can be monitored.

12.4.8 Monitoring Biodiversity and Habitats

It helps to track and identify changes in habitats and biodiversity through satellite imaging. It is possible to identify degradation and dispersal of habitats and the ways land is being used that endanger animals.

- The Great Barrier Reef is affected by climate change and people, which makes it vulnerable. Using satellite pictures from Sentinel-2 and Landsat, coral bleaching issues and reef changes are found, thus helping with conservation.
- Deforestation in Borneo: Due to the construction of palm oil plantations, people are clearing out the forests of Borneo. The monitoring of deforestation by Landsat and Sentinel-2 allows for conservation actions to shield orangutans and other endangered animals.
- Habitat loss in the Serengeti: Satellite imagery helps in the recognition and study of how the actions of humans and agriculture influence wildlife and the diversity of species found in the Serengeti. It is important for this type of information to guide conservation methods and protect the habitat.

12.4.9 Emergency Management and Rehabilitation

Satellite imaging helps a lot in determining the impact of natural disasters on the use of land, such as flooding, storms, and wildfires. Thanks to such images, it becomes much simpler to determine where the damage occurred and to monitor the progress of recovery.

- Flooding in Bangladesh: Due to monsoons and cyclones, Bangladesh gets hit by flooding regularly. By using Sentinel-1 and Landsat images, officials can find land that has been submerged, notice the damage to croplands and building, and see how recovery is happening.
- Hurricane impact in the Caribbean: Hurricanes that affect the Caribbean cause a lot of harm and destruction. Satellite imagery from NOAA and NASA is used to observe and measure the damage done to the land, infrastructure, and coastal areas because of hurricanes. It supports responses to disasters and helps plan how to reconstruct the affected areas.

- Wildfire recovery in California: Following wildfires in California, information gathered from MODIS and Landsat satellite images helps determine the extent of damage, check regrowth of plants, and support planning of efforts to reforest the area. This data helps with the recovery by showing where resources should be directed and with the planning of reforestation.

12.5 Satellites System and Their Capabilities

The critical geospatial data obtained from satellites helps improve planning and management of different sectors. Being able to observe the Earth from space, satellites help by observing changes that are happening everywhere on land, at any time. Satellites play a significant role in activities such as watching over the environment, supporting agricultural research, urban planning, building infrastructure, and disaster management. Satellites, orbiting Earth in different ways, have advanced sensors to collect information from the Earth's atmosphere in many forms [18, 19]. With their special features, satellites help improve how land is managed and used to the benefit of both local and global projects. Details about the key capabilities of satellites are given below.

12.5.1 Earth Observation Satellites Capabilities

These satellites are made to keep an eye on the Earth's land and air, providing important data for science, environment, and development activities. These satellites are specifically designed using optical, radar, or thermal sensors to capture a variety of images over different areas and time spans. They use high-resolution cameras to observe and examine the changes on Earth's surface. With these types of imaging, scientists are able to detect even the smallest differences in land use, vegetation, and geological features. By collecting images at regular intervals, the changes on the ground can be easily spotted, which helps with urban setup, land management, and conservation.

Examples:

- Sentinel Series (ESA): Uses high-quality optical and radar images to track changes in vegetation, the soil, water in different regions, urban landscape, and disaster events.

- SPOT (France): Specialized in providing high-resolution optical images used in planning cities, farming, and assessing the environment.
- Landsat Series (USA): Provides pictures with moderate detail useful for studying land use over time and for spotting changes in vegetation.

12.5.2 Navigation Satellites Functionalities

Using these satellites enables clear positioning and timing in many different services, including engineering and agriculture. By these satellites, Global Navigation Satellite Systems (GNSS) are supported to deliver highly accurate and worldwide navigation and tracking to various users.

Geospatial services: They help by creating precise electronic maps, doing land boundary measurements, and collecting geospatial information. They play a role in planning cities, defining property borders, and building different types of infrastructure.

Positioning and navigation: This information allows for improved accuracy and safety in many fields, especially agriculture, traffic management, tracking fleets of vehicles, autonomous transportation, rescue services, and other activities. Their reliable location data helps businesses and public services work better and avoid errors worldwide.

Examples:

- GPS (USA): The most popular GNSS system, providing accurate location and timings all around the Earth.
- Galileo (ESA): Galileo space navigation is a GNSS system that precisely determines locations and times for operations in both Europe and across the globe.
- GLONASS (Russia): GLONASS is similar to GPS in its services and aids in supporting a wide range of civilian and military tasks around the globe.

12.5.3 Weather Satellites Functionalities

Weather satellites are used by agencies to monitor the weather and gather details on atmospheric changes.

Weather Forecasting: Weather satellites capture information about clouds, air pressure, and winds to help predict scheduled weather events that involve storms, cyclones, and heat waves.

Climate Studies: Weather satellites track long-term data that helps scientists analyze the effects of global warming, changes in carbon, higher greenhouse gas concentration, and other activities by people that affect the world's climate.
Environmental Monitoring: Satellites further provide essential information on ocean temperatures, weather conditions across the sea, snow cover, and the air composition (ozone and particles).

Examples:

- GOES (USA): In the United States, GOES satellites monitor the atmosphere and the environment, providing regular updates about weather and storms.
- Meteosat (ESA): Used to observe climate and weather patterns in real time in Europe, Africa, and the Atlantic Ocean.
- Himawari (Japan): Supplies detailed video and photos of the environment for regular updates across Asia and the Pacific.

12.5.4 Radar Satellites Functionalities

Unlike optical satellites, radar satellites use microwaves to take pictures and check for reflections off the Earth. They can see clearly in any kind of weather and at any time of the day.

All-Weather imaging: Thanks to their active remote sensing ability, radar satellites can take pictures that are highly reliable during all weather and at any time of day or night. These satellites are extremely useful where clouds are a regular problem or in time of emergencies.

Surface and subsurface monitoring: These satellites have the power to find small variations in soil and water levels, as well as in plant life. With SAR (synthetic aperture radar), the monitoring of ground deformations, sinking land, and building stability is possible, and it can even see through vegetation to reveal surface features.

Examples:

- RADARSAT (Canada): RADARSAT provides high-quality SAR imagery that supports work in agriculture, forestry, hydrology, and emergency situations.
- TerraSAR-X (Germany): Used for mapping cities, checking on infrastructure, and studying the environment.
- Sentinel-1 (ESA): Allows for the use of SAR imagery to survey land, oceans, find floods, and assess the risks of disasters.

12.5.5 Communication Satellites Functionalities

Communication satellites act as middlemen in space to send voice, video, and data signals far apart on Earth. They are a vital part of how words, pictures, and online information move across long distances for TV, phones, the internet, and other services.

Telecommunications: Thanks to these satellites, people living in far and isolated locations can use television, satellite phones, mobile services, and broadband internet.

Data relay: Communication satellites also help send data from Earth observation satellites to ground stations so it can be used by scientists or others who need it. These satellites are vital for furthering science goals, fighting in armed conflicts, and exploring outer space.

Examples:

- Iridium provides worldwide voice and data connectivity using a network of satellites in low Earth orbit.
- SES provides satellite communication services for the media, fixed data, and mobility markets.
- Intelsat offers satellite communication services for television, internet, and data transmission.

12.6 Satellite Series

12.6.1 Landsat Satellite Series

The Landsat satellite series, under the joint management of NASA and the U.S. Geological Survey (USGS), has been a fundamental component of Earth observation since the early 1970s. Every satellite in the series has supplied useful data for monitoring and comprehending alterations in land use and the environment. Table 12.2 below is a chronological sequence of the Landsat satellites, along with their respective characteristics and purposes:

12.6.2 Sentinel Satellite Series

Sentinel satellites play a crucial role in gathering essential data for many purposes, such as environmental monitoring, land use analysis, disaster response, and climate change studies. Their contribution is vital in global Earth observation, which is shown in Table 12.3.

Table 12.2 Details of Landsat satellite series.

Satellite	Launch date	Decommission date	Key instruments	Resolution	Bands	Functions
Landsat 1	July 23, 1972	January 6, 1978	Multispectral Scanner System (MSS)	79 meters	4 spectral bands (visible to near-infrared)	First to provide multispectral images, used for land cover mapping and environmental studies
Landsat 2	January 22, 1975	February 25, 1982	Multispectral Scanner System (MSS)	79 meters	4 spectral bands (visible to near-infrared)	Continued Landsat 1's mission, enhanced global coverage and data continuity
Landsat 3	March 5, 1978	September 7, 1983	Multispectral Scanner System (MSS), Return Beam Vidicon (RBV)	40 meters (RBV), 79 meters (MSS)	4 spectral bands (MSS), 3 spectral bands (RBV)	Introduced RBV camera for higher resolution; limited success due to technical issues

(Continued)

Table 12.2 Details of Landsat satellite series. (*Continued*)

Satellite	Launch date	Decommission date	Key instruments	Resolution	Bands	Functions
Landsat 4	July 16, 1982	June 15, 2001	Thematic Mapper (TM), Multispectral Scanner System (MSS)	30 meters (TM), 79 meters (MSS)	7 spectral bands (TM), 4 spectral bands (MSS)	Introduced Thematic Mapper, improved spatial resolution and additional spectral bands
Landsat 5	March 1, 1984	June 5, 2013	Thematic Mapper (TM), Multispectral Scanner System (MSS)	30 meters (TM), 79 meters (MSS)	7 spectral bands (TM), 4 spectral bands (MSS)	Nearly three decades of continuous data, critical for long-term environmental monitoring
Landsat 6	October 5, 1993	Failed to reach orbit	Enhanced Thematic Mapper (ETM)	30 meters	7 spectral bands	Intended to continue Landsat 5's legacy with enhanced imaging; mission failed during launch

(*Continued*)

Table 12.2 Details of Landsat satellite series. (*Continued*)

Satellite	Launch date	Decommission date	Key instruments	Resolution	Bands	Functions
Landsat 7	April 15, 1999	Operational (with reduced capabilities since 2003)	Enhanced Thematic Mapper Plus (ETM+)	15 meters (panchromatic), 30 meters (multispectral), 60 meters (thermal)	8 spectral bands	Provided high-resolution imagery, introduced panchromatic band; SLC failure in 2003 reduced effectiveness
Landsat 8	February 11, 2013	Operational	Operational Land Imager (OLI), Thermal Infrared Sensor (TIRS)	15 meters (panchromatic), 30 meters (multispectral), 100 meters (thermal)	11 spectral bands	Improved imaging capabilities, additional spectral bands, higher radiometric resolution
Landsat 9	September 27, 2021	Operational	Operational Land Imager 2 (OLI-2), Thermal Infrared Sensor 2 (TIRS-2)	15 meters (panchromatic), 30 meters (multispectral), 100 meters (thermal)	11 spectral bands	Continuation of Landsat program with enhanced data quality and reliability

Table 12.3 Details of Sentinel satellite series.

Satellite	Launch date	Status	Key instruments	Resolution	Bands
Sentinel-1A	April 3, 2014	Operational	C-SAR (Synthetic Aperture Radar)	5 meters by 20 meters (range by azimuth)	Radar bands
Sentinel-1B	April 25, 2016	Operational	C-SAR (Synthetic Aperture Radar)	5 meters by 20 meters (range by azimuth)	Radar bands
Sentinel-2A	June 23, 2015	Operational	Multispectral Instrument (MSI)	10 meters (visible/NIR), 20 meters (red edge/SWIR), 60 meters (coastal/aerosol/water vapor)	13 spectral bands
Sentinel-2B	March 7, 2017	Operational	Multispectral Instrument (MSI)	10 meters (visible/NIR), 20 meters (red edge/SWIR), 60 meters (coastal/aerosol/water vapor)	13 spectral bands
Sentinel-3A	February 16, 2016	Operational	OLCI (Ocean and Land Colour Instrument), SLSTR (Sea and Land Surface Temperature Radiometer), SRAL (SAR Altimeter)	300 meters (OLCI), 500 meters to 1 km (SLSTR), 300 meters (SRAL)	Multiple spectral bands (optical, thermal)
Sentinel-3B	April 25, 2018	Operational	OLCI, SLSTR, SRAL	300 meters (OLCI), 500 meters to 1 km (SLSTR), 300 meters (SRAL)	Multiple spectral bands (optical, thermal)

(Continued)

Table 12.3 Details of Sentinel satellite series. (*Continued*)

Satellite	Launch date	Status	Key instruments	Resolution	Bands
Sentinel-5P	October 13, 2017	Operational	TROPOMI (Tropospheric Monitoring Instrument)	3.5 km by 7 km (nadir)	UV, visible, near and short-wavelength infrared
Sentinel-6 Michael Freilich	November 21, 2020	Operational	Poseidon-4 Radar Altimeter, AMR-C (Advanced Microwave Radiometer)	300 meters (Poseidon-4), 1.1 km (AMR-C)	Multiple bands (microwave, radar)

Applications:
Sentinel-1:
It enables the provision of high-resolution, all-weather, day-and-night pictures through the utilization of synthetic aperture radar. Monitoring of land cover, agriculture, forestry, mapping of soil moisture, and emergency response can be executed. When Sentinel-1B is used in conjunction with Sentinel-1A, the revisit time and coverage are both improved.

Sentinel-2:
A multispectral device that is capable of capturing thirteen different spectral bands is included. Optical images with a high resolution that can be utilized for monitoring vegetation, soil, water cover, inland waterways, and coastal areas. In addition to enhancing coverage and revisit frequency, Sentinel-2B is a useful complement to Sentinel-2A.

Sentinel-3:
Multiple devices are included to conduct extensive monitoring of the topography of the sea surface, the temperature of the sea and land surface, and the colour of the ocean and land. The Sentinel-3B satellite is designed to supplement the Sentinel-3A satellite, which is used for monitoring the environment on land and in the ocean.

The Sentinel-5P:
Monitoring the composition of the atmosphere, including trace gases, aerosols, and cloud characteristics, is the primary focus of this activity. Provides assistance with high geographic resolution for monitoring efforts pertaining to climate and air quality.

Sentinel-6 Michael Freilich:
Provides measurements of the height of the sea surface with a high degree of precision. Microwave radiometry and radar altimetry are utilized to provide support for oceanographic and climate monitoring initiatives.

12.6.3 SPOT Satellite Series

The SPOT (Satellite Pour l'Observation de la Terre) satellite series offers high-resolution optical imagery for environmental monitoring, land use, and natural disaster management. The following Table 12.4 is a comprehensive summary of the primary properties and functions of the SPOT satellite series.

Table 12.4 Details of SPOT satellite series.

Satellite	Launch date	Decommission date	Key instruments	Resolution	Bands	Functions
SPOT 1	February 22, 1986	December 31, 1990	High Resolution Visible (HRV)	10 meters (panchromatic), 20 meters (multispectral)	3 spectral bands (green, red, near-infrared), 1 panchromatic	Land use and cover monitoring, environmental studies, agricultural mapping
SPOT 2	January 22, 1990	July 2009	High Resolution Visible (HRV)	10 meters (panchromatic), 20 meters (multispectral)	3 spectral bands (green, red, near-infrared), 1 panchromatic	Similar functions as SPOT 1, providing continuity and enhanced coverage
SPOT 3	September 26, 1993	November 14, 1997	High Resolution Visible (HRV)	10 meters (panchromatic), 20 meters (multispectral)	3 spectral bands (green, red, near-infrared), 1 panchromatic	Continued mission of SPOT 2; operational life cut short due to satellite malfunction
SPOT 4	March 24, 1998	July 2013	High Resolution Visible and Infrared (HRVIR)	10 meters (panchromatic), 20 meters (multispectral)	4 spectral bands (green, red, near-infrared, shortwave infrared), 1 panchromatic	Added shortwave infrared band for improved vegetation and soil monitoring, water resource management

(Continued)

Table 12.4 Details of SPOT satellite series. (*Continued*)

Satellite	Launch date	Decommission date	Key instruments	Resolution	Bands	Functions
SPOT 5	May 4, 2002	March 31, 2015	High-Resolution Geometric (HRG), High-Resolution Stereoscopic (HRS)	2.5 meters (supermode panchromatic), 5 meters (panchromatic), 10 meters (multispectral)	4 spectral bands (green, red, near-infrared, shortwave infrared), 1 panchromatic	Improved resolution and stereoscopic capabilities for 3D mapping, urban planning, and infrastructure monitoring
SPOT 6	September 9, 2012	Operational	High-Resolution Optical Instrument (HRSI)	1.5 meters (panchromatic), 6 meters (multispectral)	4 spectral bands (blue, green, red, near-infrared), 1 panchromatic	Enhanced resolution and agility for detailed mapping, land use planning, and natural disaster assessment
SPOT 7	June 30, 2014	Operational	High-Resolution Optical Instrument (HRSI)	1.5 meters (panchromatic), 6 meters (multispectral)	4 spectral bands (blue, green, red, near-infrared), 1 panchromatic	Complements SPOT 6, offering increased revisit frequency and coverage for similar applications

12.6.4 MODIS Satellite Series

The MODIS (Moderate Resolution Imaging Spectroradiometer) satellite series, which is carried by NASA's Terra and Aqua satellites, offers extensive data on a wide range of environmental and climatic factors. Table 12.5 below is a comprehensive, concise overview of the essential characteristics.

Terra (MODIS):
This was launched as a component of NASA's Earth Observing System (EOS). The device is equipped with 36 spectral bands that cover a wide range of wavelengths, from visible light to thermal infrared. It offers three main spatial resolutions: 250 meters, 500 meters, and 1 Km, and its uses cover the observation of changes in land cover, recognition of fires, measuring the strength of vegetation, and counting temperatures on land and sea. It studies aerosols, clouds, and energy transfer as part of efforts to understand climate change.

Aqua (MODIS):
This device was launched under NASA's EOS program, with a specific emphasis on investigating the Earth's hydrological cycle. Additionally, it is equipped with 36 spectral bands that are similar to the MODIS instrument on Terra. It gathers data with consistent spatial resolutions of 250 meters, 500 meters, and 1 Km, and it enhances Terra's observations by offering a more inclusive perspective of the Earth's processes. The primary uses of this technology are to observe and track water vapour levels, cloud characteristics, ocean colour, and sea surface temperatures as shown in Table 12.6.

Table 12.5 Details of MODIS satellite series.

Satellite	Launch date	Status	Key instruments	Resolution	Bands
Terra (MODIS)	December 18, 1999	Operational	Moderate Resolution Imaging Spectroradiometer (MODIS)	250 meters (bands 1-2), 500 meters (bands 3-7), 1 km (bands 8-36)	36 spectral bands (visible, NIR, SWIR, thermal)
Aqua (MODIS)	May 4, 2002	Operational	Moderate Resolution Imaging Spectroradiometer (MODIS)	250 meters (bands 1-2), 500 meters (bands 3-7), 1 km (bands 8-36)	36 spectral bands (visible, NIR, SWIR, Thermal)

Table 12.6 Functionalities of MODIS satellite series.

Functions	Sub-area
Land Monitoring	Land cover changes
	Vegetation health
	Deforestation
Ocean Monitoring	Measurement of ocean colour
	Phytoplankton concentration
	Sea surface temperature
Atmospheric Monitoring	Observation of aerosols
	Cloud characteristics
	Atmospheric water vapor
Climate Research	Climate change
	Greenhouse gas emissions
Disaster Response	Identifies and tracks natural calamities such as wildfires, droughts, and floods

It enhances comprehension of the interplay of the ocean, atmosphere, and land.

12.7 Conclusion

Understanding changes in how land is being used is essential for land management and saving the environment, and this is aided by satellite imagery and remote sensing. Using satellites and remote sensing methods enables observation of Earth's surface from high above in space. A variety of preprocessing methods, including radiometric and geometric corrections, and image enhancement, ensure that satellite images are both accurate and easier to interpret. Temporal changes in land use can be identified using change detection techniques, and classification techniques allow different types of land cover to be grouped. The fact that stakeholders can use satellite systems to keep an eye on land use changes helps them make good decisions and contribute to better land management. With a greater understanding of satellite imagery analysis, one can handle tough problems related to looking after the Earth and its land use. Because of this, it becomes clear that satellite imagery plays an important role in teaching us about Earth's dynamic lands.

References

1. Kumar, S. and Arya, S., Change detection techniques for land cover change analysis using spatial datasets: A review. *Remote Sens. Earth Syst. Sci.*, 4, 3, 172–185, 2021, https://doi.org/10.1007/s41976-021-00056-z.
2. Lasaponara, R., Abate, N., Fattore, C., Aromando, A., Cardettini, G., Di Fonzo, M., On the use of sentinel-2 NDVI time series and Google Earth Engine to detect land-use/land-cover changes in fire-affected areas. *Remote Sens.*, 14, 19, 4723, 2022, https://doi.org/10.3390/rs14194723.
3. Hermosilla, T., Wulder, M.A., White, J.C., Coops, N.C., Hobart, G.W., Disturbance-informed annual land cover classification maps of Canada's forested ecosystems for a 29-year Landsat time series. *Can. J. Remote Sens.*, 44, 1, 67–87, 2018, https://doi.org/10.1080/07038992.2018.1437719.
4. Almutairi, A. and Warner, T.A., Change detection accuracy and image properties: a study using simulated data. *Remote Sens.*, 2, 6, 1508–1529, 2010, https://doi.org/10.3390/rs2061508.
5. Aly, A.A., Al-Omran, A.M., Sallam, A.S., Al-Wabel, M.I., Al-Shayaa, M.S., Vegetation cover change detection and assessment in arid environment using multi-temporal remote sensing images and ecosystem management approach. *Solid Earth*, 7, 2, 713–725, 2016, https://doi.org/10.5194/se-7-713-2016.
6. Fichera, C.R., Modica, G., Pollino, M., Land cover classification and change-detection analysis using multi-temporal remote sensed imagery and landscape metrics. *Eur. J. Remote Sens.*, 45, 1, 1–18, 2012, https://doi.org/10.5721/EuJRS20124501.
7. Kusetogullari, H. and Yavariabdi, A., Unsupervised change detection in Landsat images with atmospheric artifacts: a fuzzy multiobjective approach. *Math. Probl. Eng.*, 2018, 24–26, 2018, https://doi.org/10.1155/2018/7274141.
8. Wulder, M.A., *et al.*, Current status of Landsat program, science, and applications. *Remote Sens. Environ.*, 225, 127–147, 2019, https://doi.org/10.1016/j.rse.2019.02.015.
9. Ahmed, O.S., Wulder, M.A., White, J.C., Hermosilla, T., Coops, N.C., Franklin, S.E., Classification of annual non-stand replacing boreal forest change in Canada using Landsat time series: a case study in northern Ontario. *Remote Sens. Lett.*, 8, 1, 29–37, 2017, https://doi.org/10.1080/2150704X.2016.1233371.

10. Zhang, P., Lv, Z., Shi, W., Local Spectrum-Trend Similarity Approach for Detecting Land-Cover Change by Using SPOT-5 Satellite Images. *IEEE Geosci. Remote Sens. Lett.*, 11, 4, 738–742, April 2014, doi: 10.1109/LGRS.2013.2278205.
11. Tan, K.C., Lim, H.S., MatJafri, M.Z., Abdullah, K., Landsat data to evaluate urban expansion and determine land use/land cover changes in Penang Island, Malaysia. *Environ. Earth Sci.*, 60, 1509–1521, 2010, https://doi.org/10.1007/s12665-009-0286-z.
12. Amiro, B.D., Todd, J.B., Wotton, B.M., Logan, K.A., Flannigan, M.D., Stocks, B.J., Mason, J.A., Martell, D.L., Hirsch, K.G., Direct carbon emissions from Canadian forest fires, 1959-1999. *Can. J. For. Res.*, 31, 3, 512–525, 2001, https://doi.org/10.1139/x00-197.
13. Coppin, P., Jonckheere, I., Nackaerts, K., Muys, B., Lambin, E., Review ArticleDigital change detection methods in ecosystem monitoring: a review. *Int. J. Remote Sens.*, 25, 9, 1565–1596, 2004, https://doi.org/10.1080/0143116031000101675.
14. Belgiu, M. and Drăguţ, L., Random forest in remote sensing: A review of applications and future directions. *ISPRS J. Photogramm. Remote Sens.*, 114, 24–31, 2016, https://doi.org/10.1016/j.isprsjprs.2016.01.011.
15. Franklin, S.E., Ahmed, O.S., Wulder, M.A., White, J.C., Hermosilla, T., Coops, N.C., Large area mapping of annual land cover dynamics using multitemporal change detection and classification of Landsat time series data. *Can. J. Remote Sens.*, 41, 4, 293–314, 2015, https://doi.org/10.1080/07038992.2015.1089401.
16. Hermosilla, T., Wulder, M.A., White, J.C., Coops, N.C., Hobart, G.W., Regional detection, characterization, and attribution of annual forest change from 1984 to 2012 using Landsat-derived time-series metrics. *Remote Sens. Environ.*, 170, 121–132, 2015, https://doi.org/10.1016/j.rse.2015.09.004.
17. Hermosilla, T., *et al.*, Mass data processing of time series Landsat imagery: pixels to data products for forest monitoring. *Int. J. Digital Earth*, 1–20, 2016, https://doi.org/10.1080/17538947.2016.1187673.
18. Millard, K. and Richardson, M., On the importance of training data sample selection in random forest image classification: A case study in peatland ecosystem mapping. *Remote Sens.*, 7, 8489–8515, 2015, https://doi.org/10.3390/rs70708489.
19. Wulder, M.A., White, J.C., Gillis, M.D., Walsworth, N., Hansen, M.C., Potapov, P., Multiscale satellite and spatial information and analysis framework in support of a large-area forest monitoring and inventory update. *Environ. Monit. Assess.*, 170, 1–4, 417–433, 2009, https://doi.org/10.1007/s10661-009-1243-8.

13

Satellite Image Analysis

Abstract

Using a satellite image processing system helps reveal critical insights into Earth systems by collecting, analyzing, and interpreting satellite images. Data is gathered over a range of wavelengths using multiple types of sensors during image collection. Preprocessing removes imperfections and enhances images to make them more readable. A wide range of land cover types and other features can be accurately classified using the latest classification techniques that often involve machine learning and artificial intelligence. Changes such as deforestation and urban growth are detected using change detection techniques. Analyzing the spectral reflectance of materials on Earth's surface is possible with spectral analysis. Satellite image processing impacts efforts in environmental monitoring, agriculture, urban planning, disaster management and natural resource management. Satellite images combined with multiple geographic data sets improve decision-making and management of resources. Advancements in technology will improve the accuracy and detail offered by satellite image processing to better understand Earth's ever-changing systems.

Keywords: Environmental monitoring, machine learning, processing, satellite image

13.1 Introduction

The discipline of satellite image processing involves obtaining, examining and understanding images that satellites gather as they orbit around our planet. These images provide a wealth of data on Earth's features and conditions, enabling us to make informed decisions for purposes such as mapping, changes in environment, supporting growth, development, monitoring crops, natural disasters and much more. Essential phases involved in satellite image processing are described below.

Kundan Meshram and Chatrabhuj. A Sustainable Approach to Land Use, (333–358) © 2026 Scrivener Publishing LLC

13.1.1 Image Acquisition

Image acquisition initiates the satellite image processing and involves satellites capturing images of the Earth from above. The sensor that is mounted on the satellite greatly influences the quality and features of the captured data. There are two primary classifications of sensors: passive and active. Passive sensors can measure radiation coming from the Earth that is either reflected or emitted by various materials on the ground. Sunlight is generally the most common energy source detected using passive sensors. Optical sensors and thermal sensors form part of the group of passive sensors [1–4]. Alternatively, active sensors transmit energy to the Earth, then measure how much of that energy is reflected back from various points on the ground. Active sensors, such as radar systems and LiDAR, can send out signals and analyze the returns even in adverse environmental conditions. Many sensors measure information at different spectral bands spanning from visible light to infrared and microwave. This allows for an in-depth analysis of the various features and events occurring on the Earth.

13.1.2 Data Preprocessing

The first step in analyzing images from a satellite is often to preprocess them to correct for distortions and enhance their quality. During preprocessing, radiometric, geometric and atmospheric corrections are all taken into account. Radiometric correction reduces noise from the data and corrects the sensor's response so that the pixel values show the correct light reflected from the ground. Geometric adjustment realigns images with accurate map coordinates and alters any abnormalities caused by the sensor's movement and Earth's curvature [5–7]. Atmospheric correction accounts for the impact that light from air particles and gases may have on image quality, resulting in accurate surface information in the data.

13.1.3 Image Enhancement

Image enhancement is performed after preprocessing to make features in the image easier to observe, which improves the interpretation of the data for analysis. Standard approaches to enhance images involve using contrast stretching, filtering techniques and edge detection. Contrast stretching is an important method for making objects in the image stand out clearly from each other. Filters are used to either suppress unimportant noise or increase the prominence of particular details in an image [8–10].

Edge detection methods make the boundaries between different regions more prominent, allowing an improved analysis of distinctive features.

13.1.4 Classification and Examination

State-of-the-art algorithms and machine learning methods are used to identify and categorize the characteristics in the image. This often involves identifying and separating various types of land cover, including forests, cities and water bodies. Tools for change detection allow analysis of variations that take place over time, including cases like deforestation and urbanization. Machine learning techniques, such as supervised and unsupervised classification, automatically organize large datasets, improving the efficiency and precision of the subsequent analysis.

13.1.5 Data Integration

Frequently, satellite imagery is combined with information from geographic information systems (GIS) to gain a better understanding of a particular region. The adoption of multiple sources results in more accurate and useful analysis as it provides an added level of insight into the subject. Combining satellite imagery with topographic data improves understanding of the nature and features of a particular area. Using demographic data can be instrumental in making more effective decisions about urban development and management of natural resources.

13.1.6 Visual Representation and Analysis

Processed images and information are showcased visually *via* maps, charts and various types of images. Visualizations play a crucial role in interpreting data and guiding people to make informed choices using the gathered knowledge. Visualization techniques highlight important patterns and trends that are accessible to specialists like scientists, scholars, political leaders and the community as a whole. Visualization transforms complex statistical data into clear, meaningful and useful information that facilitates improved understanding and decision-making.

13.2 Applications of Satellite Image Processing

Different uses of satellite image processing are vital for understanding and managing both the planet's resources and the environment. Satellites play a vital role in environmental surveillance by measuring ecosystem changes, following deforestation trends, analyzing water quality and detecting the impacts of climate change. The analysis of satellite images enables officials to make wise decisions that protect and preserve natural habitats. Satellite imagery supports urban planning by providing insight into how urban areas grow, the status of infrastructure projects and adjustments made to land use patterns. The information obtained is used to create responsible approaches to city planning and development. Having instant and accurate information from satellite imagery proves invaluable for evaluating the scale of damage after events such as hurricanes, floods and earthquakes. These images help with efforts to swiftly respond to and recover from disasters. Moreover, satellite images are particularly valuable in advancing the field of natural resource management by enabling the discovery and administration of mineral, oil and gas deposits. This helps to ensure that resources are used in the most responsible and effective way. Satellite imagery helps us to better protect and manage the Earth's resources in a variety of fields.

13.3 Image Acquisition

A wide variety of procedures and cutting-edge technology are employed in satellite imaging to collect highly accurate information about the ground. Each technique features unique sensors, captures different wavelengths and uses specialized processes to obtain and transmit the information. There are mainly three techniques for gathering images.

13.3.1 Optical Imaging

Optical sensors play a crucial role in remote sensing by allowing for the measurement of surface information on Earth by detecting reflected light energy from various regions of the electromagnetic spectrum. Multispectral imaging and hyperspectral imaging are the two primary modalities found in optical remote sensing.

Multispectral imaging: Multispectral sensors transmit and measure radiation in several discrete spectral bands covering various wavelength regions in the visible, near-infrared and shortwave infrared portions of the spectrum. This enables distinguishing and categorizing different materials on the Earth's surface as well as monitoring the health of vegetation. Popular multispectral instruments found on satellites such as Landsat 8's

Operational Land Imager and Sentinel-2's MSI are known for their ability to collect high-quality, continuously updated images needed for long-term monitoring of the planet.

Hyperspectral Imaging: Hyperspectral imagery is achieved by sensors detecting radiation in a large number of very narrow bands that provide detailed spectral information about different materials. This extremely high spectral resolution makes it possible to accurately discriminate various types of materials, including minerals, vegetation and water. EO-1 was an important space mission equipped with the Hyperion sensor, equipped with 220 spectral bands ranging from 400 to 2500 nanometers. Hyperspectral images are essential for cutting-edge uses in precision farming, environmental surveillance and mineral ore survey.

13.3.2 Thermal Imaging

Thermal imaging allows for the analysis of temperature changes within the Earth's surface and the thermal properties of various objects and locations. Thermal sensors detect heat in real time from items, so they can be used at all times. The Thermal Infrared Sensor (TIRS) is installed on Landsat 8 which detects radiation in the thermal infrared spectrum, approximately between 10 and 12 micrometers. The remote sensing of temperature allows small changes to be found on land and water. Data from weather observations can be used to explore why cities are frequently much warmer than the countryside nearby. Because there is a way to monitor temperature changes, this technology helps with the management of aquatic ecosystems by spotting thermal pollution. By using thermal imaging, it is easier to understand how severe a drought is and to monitor soil moisture by noticing signs of its increased heating due to evaporation. Remote sensing is useful for following natural disasters, such as finding wildfires and monitoring volcanoes. Thermal imaging is important in both checking environment, city planning and seeing how climate change influences our planet's surface.

13.3.3 Microwave and Radar Imaging

Microwave and radar imaging can work through clouds and in almost any weather, and it has greatly improved advancements in remote sensing. SAR, which is a type of remote sensing, takes microwave images bounced off the planet to produce detailed maps and photos. SAR takes its data from broad antenna paths through signal processing, creating accurate images. Radar is commonly employed for drawing topographic landscapes and discovering soil moisture, analyzing changes in the ground layer and locating earth subsidence. Collectively, the Sentinel-1 series (supported by the European Space Agency) and the Radarsat series (from Canada) have made a significant difference in environmental monitoring, handling disasters and managing land use planning. Both the transmission and reception of radar pulses are carried out with a stationary array or a non-synthetic antenna.

13.3.4 Light Detection and Ranging (LiDAR)

Pulsed laser beams are used by LiDAR to exactly find the distance of the ground or other nearby items from the device. The LiDAR instrument continuously launches laser pulses that bounce off features on the ground and return to the LiDAR, which uses those timings to generate elevation data critical for detailed 3D models of the land. LiDAR's ability to go into tree canopies allows quick and accurate mapmaking of forests and things hidden under trees. LiDAR technology is used widely in forests for counting biomass, modeling animal habitats and tracking changes on the ground. Satellite-attached LiDAR is used for particular worldwide studies, for example, keeping an eye on the ice caps. ICESat from NASA has taken advantage of satellite LiDAR to track alterations in Earth's globally distributed environments. Reports on topography, various land surfaces and climate monitoring are made possible by LiDAR in environmental science.

13.4 Preprocessing Techniques

Using preprocessing makes it possible to prepare the original image data for further processing and study. These procedures help fix many problems and raise the accuracy and dependability of the data's outcomes [11–14]. Almost all groundwork in preprocessing involves radiometric correction, geometric correction and removing noise.

13.4.1 Radiometric Correction

Removing flaws in the sensors and effects on pixels from the atmosphere is possible with radiometric correction. This is done so that the Digital Number found by the sensor accurately reflect both incoming and reflected radioactivity from the landscape. It makes satellite images obtained from different sensors and dates more similar in quality. Radiometric correction relies heavily on calibration, which helps adjust for any problems with the sensor and the sensor's changing response. Using the satellite's coefficients, absolute radiometric calibration can change Digital Numbers into corresponding values of the sensor's radiation on board. Through that approach, scientists, researchers and analysts can make precise comparisons using various types of satellite sensors and mission data sets. By relative radiometric calibration, unequal measuring by different sensor detectors is eliminated, and the whole image radiates an even brightness and radiance for similar results.

13.4.2 Atmospheric Correction

It is important to put atmospheric correction at the beginning of satellite image processing to correct mistakes caused by absorption or light scattering. As a result, the retrieval of surface reflectance is closer to reality. Atmospheric gases, aerosols and water vapor all change how sunlight arrives on Earth. Dark Object Subtraction (DOS) has been used extensively because dark areas in a picture, including deep water and shadows, are not exposed. The resulting diffuse radiance is used to treat every pixel in the image and decrease effects from scattering. Highly refined atmospheric corrections are possible through models including the 6S Model (Second Simulation of the Satellite Signal in the Solar Spectrum). The 6S Model effectively handles precise correction of atmospheric data because it accounts for different layers of the atmosphere, including aerosol, water and ozone. FLAASH is specifically designed for the atmospheric correction of hyperspectral imaging. Working with the way light connects with particles and gases in the air, FLAASH corrects atmospheric influences and provides useful data about surface reflectance in a variety of spectral bands. Satellite research relies on these techniques to gather trustworthy information that supports environmental studies, land-use analysis and climate research.

13.4.3 Geometric Correction

Geometric correction changes the orientation of satellite images according to a known location framework. The system takes into account many factors, among them the shape of the instruments used, Earth's movement, heights involved and changes in the satellite's place. With geometric correction, the planet's true spatial relationships become clear in the imagery. The main step in geometric correction is to choose Ground Control Points (GCPs) that line up image pixels so they appear in the proper places on the Earth's surface. These landmarks can be spotted either by manually comparing features on the image with well-known points or through automated approaches that identify matches between picture features and known datasets. It is important to georeference the image by using mathematics to align its features with a map projection. Polynomial transformations allow us to change the appearance of an image using the coordinate system to shape the pixels. Remote sensing relies on geometric correction to support work in GIS and to improve results in monitoring, city planning and managing resources.

13.4.4 Reduction of Noise

During satellite image preprocessing, noise reduction is necessary to get rid of questionable data in the images that come from sensor errors, atmosphere or other external causes. Interruptions of this kind may cause trouble with data understanding and damage the overall quality of images captured by remote sensing. Median filtering is one of the most often used methods. It involves finding the median value of pixels in a square region and placing it over each pixel; usually, this square area has 9 or 25 pixels, depending on the kernel used. The method is well known for cutting out noise without making the image borders look blurry. Sometimes, Gaussian filtering is done to lessen high-frequency noise and boost the whole image's attractiveness without eliminating notable details along edges. Wavelet transform methods are used for advanced and adaptive noise reduction. DWT breaks an image into a series of approximation and detail sub-bands, enabling noise to be targeted and minimized by dividing the detail coefficients into meaningful bands. Multiscale decomposition allows analyzing images at different resolutions, providing an effective way to detect and remove noise from different spatial frequencies. Employing these noise reduction methods leads to improved visual presentation and analytical accuracy of satellite imagery, allowing for more precise extraction, classification and quantification in remote sensing tasks.

13.5 Image Enhancement Techniques

Image enhancement techniques play an important role in satellite image analysis because they improve the visual clarity and ease of understanding of the images. The techniques encompass contrast enhancement to enhance the differentiation between features, spatial filtering to decrease noise and amplify edges, frequency domain transformations for sophisticated filtering, colour enhancement for improved visualization of multi-spectral data, and advanced techniques such as adaptive filtering and morphological processing to refine the enhancement based on local image characteristics [15–17]. By implementing these methodologies, satellite imagery becomes more valuable for diverse applications such as environmental surveillance, land utilization assessment, and catastrophe mitigation.

13.5.1 Contrast Enhancement

- **Histogram equalization:** Changes the image's intensity settings to make the contrast better. The image's pixel intensity values are changed so that the final image's histogram (frequency distribution) looks a lot like a uniform distribution. This method spreads out the most common intensity values, which makes the difference stronger in places where a lot of the intensity values are the same. When the background and subject are both bright or dark, this method works well.
- **Adaptive Histogram Equalization (AHE):** Improves contrast by making changes locally instead of globally. The picture is broken up into small areas called tiles, and histogram normalization is used on each tile. Each local area has more contrast, which makes local features stand out, but can make noise louder.
- **Linear contrast stretching:** Increases the range of intensity numbers so that they cover the whole spectrum. It is known what the lowest and highest pixel numbers are. Then, these numbers are pushed to the very end and very beginning of the range that was chosen, 0 to 255 for an 8-bit image. To keep proportional relationships, intermediate numbers are changed as needed.
- **Logarithmic and exponential transformations:** Boosts contrast in either darker or lighter parts of the picture. It uses two methods:
  - Exponential Transformation: Makes the image's dynamic range bigger. This method brings out the details in parts of the picture that are bright.
  - Logarithmic Transformation: The image's dynamic range is shrunk. This method brings out details in the dark parts of a picture.

13.5.2 Spatial Filtering

- **Low-pass filtering (Smoothing):** By averaging pixel values, it lowers noise and makes the picture smoother. It is done by
  - Averaging Filter: Changes the value of each pixel to the average of its peers within a certain kernel. (e.g., 3x3, 5x5).
  - Gaussian Filter: Uses a Gaussian function to figure out how much nearby pixels are worth, with closer pixels getting more weight. A simple average filter does not keep edges as well as this does.
- **High-pass filtering (Edge Enhancement):** By boosting high-frequency parts, it improves edges and fine features. This is achieved by the following methods.
  - Sobel Filter: Finds edges by calculating changes in the image's strength. To find edges, it uses both a horizontal and a vertical gradient.
  - Laplacian Filter: Uses the second derivative to draw attention to areas where the strength changes quickly (edges).
- **Band-pass filtering:** Some spatial frequencies are boosted while others are lowered. Combines low-pass and high-pass filtering to keep features within a certain frequency range, making certain patterns or textures stand out.

13.5.3 Transformations and Filtering in the Frequency Domain

- **Fourier transform:** Looks at the picture in the frequency domain and splits it into sine and cosine parts. Changes the picture from the spatial domain to its frequency domain representation. In the frequency domain, filtering processes are carried out. Next, the Inverse Fourier Transform is used to change the image back to the spatial domain. One can get rid of irregular noise or boost certain frequency patterns with this method.
- **Wavelet transform:** It breaks the picture down into different scales and analyzes it at multiple resolutions. It breaks the picture down into approximation and detail coefficients at different levels using wavelets. Changes the wavelet coefficients before reconstructing the picture to highlight or hide certain scales, such as edges or textures.

13.5.4 Colour Enhancement

- **Colour stretching:** Enhances colour contrast in multi-band images. Adjusts the colour values in each band individually to make the colour contrast better as a whole. Often used in RGB (Red, Green, and Blue) composites to make natural colour pictures look better.
- **Principal Component Analysis (PCA):** Reduces the number of dimensions and makes multispectral pictures easier to understand. Changes the original bands that were correlated into a new set of components that are not correlated (principal components). Most of the information in the data is usually in the first few main components, which can be stretched and shown to make the data easier to understand.
- **False colour composite:** Improves some things that are hard to see in true colour pictures. Assigns various spectral bands to the RGB channels to make an image with coloured highlights for certain elements, like water and plants.

13.5.5 Advanced Enhancement Techniques

- **Adaptive filtering:** It improves certain features while adapting to the features of the local picture. Changes the filtering process based on local statistical traits. As an example, adaptable smoothing filters can cut down on noise in areas with a lot of the same thing while keeping the edges in areas with a lot of different things.
- **Morphological processing:** Uses mathematical morphology to improve picture structures. Uses operations like dilation, erosion, opening, and closing to bring out or hide certain parts of a picture. It can be used to improve linear features (like roads and rivers) or get rid of small noise items.

13.6 Classification Techniques

The classification of satellite imagery is an essential operation in remote sensing since it allows for the extraction of vital data regarding land cover, land use, and environmental changes. Diverse classification strategies are utilized to categorize pixels within satellite photos into significant classes [18–20]. The techniques can be classified into two main categories:

Table 13.1 Differences between supervised and unsupervised classification.

Aspect	Supervised classification	Unsupervised classification
Training Data	Requires labeled training data, with known class labels.	Clusters are created by identifying inherent patterns in the data without the need for pre-labeled training data.
Purpose	Classifies data into predetermined categories based on existing knowledge.	Clusters data based on intrinsic similarities without prior knowledge of classes.
Input	Require known class labels for training.	Processes input data without any pre-existing class information.
Decision-Making Process	Based on similarities to known training data.	Based on natural groupings and distances between data points.
Inter-pretability	Results are interpreted based on predefined classes.	Clusters do not have predefined labels, requiring further analysis to interpret results.
Use Cases	Highly suitable for applications that use known and named classes, such as picture classification.	Valuable for conducting exploratory data analysis, identifying patterns, and uncovering latent structures within data.
Algorithm Examples	Maximum Likelihood Classification (MLC), Support Vector Machines (SVM), Neural Networks.	K-means Clustering, ISODATA Clustering, Hierarchical Clustering.

supervised and unsupervised methods, each with unique benefits and uses as shown in Table 13.1.

13.6.1 Supervised Classification

A common technique in remote sensing called supervised classification uses information from labelled training samples to categorize features in satellite images. The process is aided by the use of labelled reference data, enabling the algorithm to identify the distinctive spectral signatures for different land cover classes. A range of methods is used in supervised classification, highlighting their different areas of advantage. The Maximum Likelihood Classifier estimates the mean and variance (covariance) for each class and classifies pixels to the most probable class based on these estimates. SVMs find the plane that separates classes best and allows their use for linearly and non-linearly separable data. ANNs are able to address complex and non-linear associations in the data by replicating connected computational units that adjust their performance by self-tuning mechanisms. ANNs are particularly effective when classifying land use/land cover requires the analysis of complicated associations between spectral features. Supervised classification excels when comprehensive, accurately labelled training samples exist, enabling precise thematic maps for detailed land use and land cover analysis, as well as for agricultural monitoring and detection of environmental changes.

13.6.2 Unsupervised Classification

Unsupervised classification organizes pixels into groups without requiring prior information about the class assignment. It uses the inherent characteristics of the data to organize the data into clusters. Algorithms like K-means Clustering and ISODATA Clustering use calculations of pixel-by-pixel similarity to identify clusters in the data. Clusters are formed by K-means Clustering by routinely adjusting the location of cluster centers to reduce the maximum distance between each data point and its closest center. ISODATA Clustering iteratively divides and combines clusters to achieve the highest possible classification, and it regularly adjusts the number of clusters to improve results. Using unsupervised categorization allows for exploratory analysis, identification of unforeseen patterns and the production of themed maps without required labels for the groups.

Techniques combining supervised, unsupervised and advanced approaches are often employed to enhance the precision of classifications and resolve unique problems. These approaches may involve the combination of restricted labelled data with unsupervised methods, known as semi-supervised approaches. Another method, called object-based classification, takes into account the spatial relationships between surrounding pixels. Advanced machine learning methods, such as Random Forests (RF) and Convolutional Neural Networks (CNNs), are being used more and more for satellite image classification. This is because they are able to effectively capture intricate spatial and spectral patterns.

13.7 Change Detection Techniques

Change detection techniques are essential for monitoring environmental changes over time, which is critical for applications such as land cover analysis, urban expansion monitoring, and disaster assessment [18–21]. Change detection methods play a crucial role in remote sensing for tracking changes in the Earth's surface over time, which helps with applications like land cover change analysis, urban growth monitoring and post-disaster impact assessment. They allow for the identification, measurement and spatial representation of variations observable between sequential satellite imagery. Image differencing brings about the calculation of the discrepancy between two distinct images by removing their mutual pixel values. Many change detection techniques return a difference image that shows where reflectance levels have changed for individual pixels. However, image differencing suffers from strong dependency on radiometric and atmospheric factors that differ between image acquisition times, making it necessary to carry out careful preprocessing to guarantee useful results. Post-classification comparison compares classified images from two time points in order to detect and categorize transitions in land cover. This method allows for a precise assessment of change in categorical land cover, but its accuracy tends to be influenced by the initial classifications' accuracy.

Change Vector Analysis (CVA) is capable of analyzing the degree and direction of appearance, disappearance and transformation in a multidimensional space based on spectral features. CVA estimates the magnitude and direction of change for each pixel, allowing for accurate identification of both the extent and type of land cover transformation. Setting thresholds on vector magnitudes allows the identification of notable transitions from more subtle fluctuations. The effectiveness of CVA is determined greatly by the quality of the spectral data, temporal resolution and knowledge of the study area and the particular sensor being used.

13.8 Spectral Analysis Techniques

Spectral analysis techniques help reveal significant details contained within remote sensing data through evaluation of the various spectral signatures present in the data. The purpose of these techniques is to enhance specific features or to separate combined pixels into their isolated elements, which in turn enables a better understanding of remote sensing imagery.

13.8.1 Spectral Indices

By examining the spectral signature of each observed event, spectral analysis methods help us find useful details in remote sensing data. The purpose of these techniques is to optimize specific properties or separate blended pixels into their constituent parts, making the interpretation and analysis of remote sensing photographs easier. Several frequently used spectral indices include those described below.

Normalized Difference Vegetation Index (NDVI):
The NDVI has grown to be a recognizable spectral index used to observe vegetation on the ground, plant vigour and biomass. The NDVI is boosted as plants absorb mainly the red wavelength and reflect extra NIR light. Values are generally located within the bounding values of 0.2 to 0.8. An NDVI number very close to zero or in the negative suggests the area lacks vegetation, and high NDVI values indicate a strong presence of vegetation. NDVI keeps track of farming, forestry and natural areas by spotting times of vegetation growth and noticing how drought affects them.

Normalized Difference Moisture Index (NDMI):
NDMI is created to monitor leaf moisture and discover cases of moisture stress in plants. NDMI is calculated by using this equation: (NIR – SWIR) ÷ (NIR + SWIR). It examines vegetation water content by noticing that it reflects a lot of NIR radiation and absorbs a large amount of SWIR radiation. Sites with a higher NDMI typically contain more moisture in their vegetation than spots with a low NDMI. It is useful for predicting wildfire hazard, irrigation water management and spotting dry spell onset in agricultural lands.

Soil Adjusted Vegetation Index (SAVI):
SAVI was developed to correct for the effect of soil on NDVI measurements in regions with little or no vegetation cover. It corrects for the effects of soil brightness on NDVI by including a scaling term, L, so that the formula becomes: [(NIR – Red)/(NIR + Red + L)] × (1 + L) L is usually set to 0.5 to take both soil and vegetation signals into consideration. When adopted for monitoring vegetation in drylands and semi-arid areas, SAVI adjusts for the impact of soil reflectance and better detects changes in the vegetation cover. It enables studies on desertification, rangeland monitoring and dryland crop management.

Normalized Difference Built-up Index (NDBI):
NDBI distinguishes urban and built-up regions in satellite images. The value is obtained by dividing the difference between SWIR and NIR by the sum of these two bands. Built-up areas tend to exhibit strong reflectance in the SWIR band and lower reflectance in the NIR band, leading to high NDBI values over those surfaces. NDBI distinguishes urban areas from vegetation and water-filled regions clearly on satellite images. It supports urban planning, analysis of developments and built-up areas, land use monitoring and evaluating how urbanization affects the environment.

Normalized Difference Water Index (NDWI):
NDWI distinguishes water from land by computing the difference between NIR and green bands and dividing that value by the sum of both bands. NDWI values are greatly elevated over water bodies while remaining negative over land because water tends to absorb NIR radiation and reflect green wavelengths. NDWI can significantly enhance the difference between visible water and other surfaces in remote sensing data and is frequently utilized in flood management, hydrology and surveillance of water level fluctuations.

Normalized Burn Ratio (NBR):
NBR quantifies vegetation loss in areas following a fire based on comparisons of pre- and post-fire imagery. It is calculated by dividing NIR and SWIR band differences by their sum. Areas damaged by fire often reflect less NIR and more SWIR, leading to lower NBR values. Comparing pre- and post-fire NBR images produces ΔNBR, a measure of the intensity of a fire event. This index plays a crucial role in effectively managing, planning recovery from and monitoring how well ecosystems respond to wildfires.

Normalized Difference Bareness Index (NDBAI):
NDBAI locates land areas with scarce vegetation as a result of deforestation or desertification. It often combines NDBI with SAVI or uses a ratio of SWIR to NIR reflectance. NDBAI allows the differentiation of barren land from areas with vegetation or human development which is useful for tracking desertification, investigating soil degradation and assessing the potential of sites for afforestation or developmental purposes.

Enhanced Vegetation Index (EVI):
EVI reduces the impact of atmospheric conditions, shadows and soil background radiation on the estimation of vegetation vigor. The addition of the blue band in the EVI formula helps correct for light scattering in the

atmosphere. The formula for EVI is G × [(NIR – Red)]/[NIR + C_{1} × Red – C_{2} × Blue + L], in which G stands for a gain term and C_{1}, C_{2} and L are several coefficients. EVI performs better than NDVI in areas of abundant biomass where NDVI often becomes limited. It plays a key role in monitoring changes in vegetation, analyzing carbon storage and studying environmental biodiversity.

Green Chlorophyll Index (GCI):
GCI is an index that responds to changes in the level of chlorophyll, determined by subtracting 1 from the ratio of NIR compared to Green. GCI helps evaluate crop plant health and detect nutrient shortages. A high GCI value and healthier plants are usually found in areas with intense green colour. In precision farming, GCI can decide the best amount of fertilizer for crops and gauge their development.

Atmospherically Resistant Vegetation Index (ARVI):
With the help of the blue band and reduced atmospheric effects, ARVI improves on NDVI. NIR – (Blue+2*Red))/(NIR + (Blue+2*Red)). The method delivers reliable results for vegetation monitoring, as it can handle changes in fog and other aerosols in the atmosphere. It functions well in environments typical of tropical and subtropical regions, so it is a great device for monitoring the environment.

Tasseled Cap Transformation (TCT):
TCT splits the reflectance readings into the three parameters titled Brightness, Greenness and Wetness. Certain combinations of these variables help determine how a soil acts, how strong the vegetation is and its level of dryness or water saturation. The greenness component is used as a replacement for NDVI. Using the wetness index, one can recognize wetlands and floods. TCT supports activities such as mapping land cover, tracking how land is utilized, and research focused on hydrology.

13.8.2 Spectral Unmixing

Spectral unmixing is an advanced way to investigate mixed pixels in remote sensing by separating out their different spectral components and classifying them. A common consequence of sensors with low spatial resolution is mixed pixels when different land cover types are not clearly distinguished in remote sensing. For this reason, the reflected light contains various substances which make it hard to interpret or sort out features in the data.

Using spectral unmixing helps create a better understanding of what is happening on the land from the data.

Linear Spectral Unmixing (LSU):
This approach is now preferred, as it is easy to understand and needs little computation. Reflectance at a particular pixel is claimed to be the sum of contributions from each of the existing pure materials, with the weightings given by how often they are present. This problem can be expressed as a system of linear equations subject to conditions that result in realistic spectra with positivity and complete coverage of the spectrum. LSU excels in areas with clearly identifiable materials and limited interference between them, including agricultural fields, bare soil patches and urban mosaics. LSU is commonly applied to tasks like mineral identification, vegetation analysis and land use classification where discernible endmember materials are present.

Non-linear unmixing:
This is appropriate for situations where light propagation between materials is highly complicated and cannot be explained by simple linear mixing models. As a result, the measured reflectance is shaped by nonlinear processes such as multiply scattering and close intermixture. Non-linear models use a wide range of mathematical functions, including bilinear, polynomial and kernel approaches and are sometimes improved with machine learning methods such as neural networks or support vector regression. Non-linear unmixing is computationally challenging but produces increased accuracy in areas where the spectral response is highly non-linear.

Applications and benefits of spectral unmixing:
Spectral unmixing finds important applications in a wide range of Earth observation fields. It allows precise characterization of mixed land use by estimating the proportion of different vegetation, soil and water types across each pixel. Spectral unmixing helps identify different mineral types by their unique spectral signatures, often allowing for the detection of low abundances of those minerals. Spectral unmixing proves valuable for identifying built-up areas, vegetation and different types of materials that characterize cities. As a consequence, spectral unmixing helps us identify and measure the influence of deforestation, land degradation or water contamination on ecosystems from high-resolution satellite pictures. By using unmixing, more accurate details are separated, which gives remote sensing

data better thematic and spatial qualities and leads to more reliable geospatial decision-making.

13.9 Machine Learning and Artificial Intelligence Techniques

Analysis and interpretation of satellite imagery data depend greatly on AI and ML approaches. They find valuable meaning and patterns in large datasets. Because of them, computers find features and identify patterns in data from Earth observation, which helps in detailed analysis of these data.

13.9.1 AI and ML

With new advancements in AI, satellite image analysis is evolving because of the automated and improved feature analysis and real-time processing. They greatly help us process a lot of satellite data, which is useful for farming, expanding cities, analyzing the environment and reacting to disasters. Studies of this kind become much faster when AI and ML tools are in use to automatically identify lands such as buildings, roads and plant cover. Because of their abilities to pick out trends, they provide useful information about issues such as deforestation and expanding cities. AI systems can quickly sort land use and land cover classes in images and identify different objects as well. These ways increase the image quality with effective Super Resolution technology. AI and ML are shaking up satellite image analysis by offering many advantages, better managing resources and enhancing decision making in several sectors.

13.9.2 Deep Learning

With deep learning, massive data is analyzed by using neural networks with several layers, so that the main structure and significance are discovered. Because of its capability to learn on its own from a wide variety of data, there have been great advances in computer vision, speech recognition, natural language processing and remote sensing. Deep learning is crucial in geospatial work because it makes image analysis automatic, improves classification results and facilitates quicker and timelier choices based on satellite and aerial images.

Convolutional Neural Networks (CNNs):
CNNs are neural networks designed to study how different features are placed next to each other in two dimensions, such as pixels in an image. Many tasks in remote sensing such as fetching features, spotting objects, analyzing scenery and deciding on land use and cover rely on CNNs. CNNs feature layers where filters look at the input image and draw out information about basic edges, different textures and increasingly harder patterns as the structure becomes deeper. The structure of features hierarchically in CNNs allows them to recognize cities, coverage from plants, water bodies, roads and many other surfaces. Using semantic segmentation methods, you can sort every pixel in an image and produce maps and monitor features in the environment in detail.

Recurrent Neural Networks (RNNs):
Analyzing satellite imagery over time requires RNNs, as the pictures are frequently recorded at different times to track changes in land, plants, cities and weather. By examining vegetation indices during the crop's development, RNNs alert farmers to possible problems early. Land use, forest cover and urban growth are all patterns that RNNs are skilled at discovering and monitoring. They can recall earlier information and spot repeated seasonal cycles, growing trends and striking anomalies.

Long Short-Term Memory (LSTM):
Using LSTM networks has allowed scientists to study and forecast changes in remote sensing data through time. LSTMs include a special memory cell and adjustable gates to handle and keep information in the network throughout a sequence. It is easy to analyze changes over time with remote sensing, since it uses the same image repeatably. LSTMs are applied in monitoring plants by watching their vegetation indices so as to detect changes in cropping, estimate yields and pick out signs of plant stress. A lot of the time, LSTMs are applied to satellite statistics to predict sea surface temperature, probability of drought and rain levels. LSTMs are able to process both space and time data when added to neural networks such as ConvLSTM. Spatial analysis is required in such cases as flood forecasting and tracking wildfires because the information being used changes as the event unfolds. LSTMs handle well the challenging situations, noise and sparsity in satellite image data. Besides, their strong point in showing how things develop over time allows them to detect changes and warn about dangers in environmental monitoring.

Autoencoders and Variational Autoencoders (VAEs):
To represent data effectively and accurately, autoencoders use unsupervised learning by bringing the differences between the original and rebuilt data as close together as they can. By using autoencoders, dimensions in remote sensing data can be reduced, unusual patterns can be found, and noise can be removed from hyperspectral and multispectral images. Their capacity to model various forms of data means they work well for making artificial imagery, modeling terrain and recovering missing parts in satellite records. They help produce better data and make remote sensing information easier to understand.

Generative Adversarial Networks (GANs):
To create realistic results, GANs train a generator and a discriminator in a coordinated manner. It is evident from tests that Generative Adversarial Networks help create realistic photos and enhance weak satellite images. Using GANs, researchers can clear out clouds, mix up optical and SAR data and reproduce observations from missing sensors in remote sensing. They are especially useful when extending datasets can significantly improve how the model performs across unobserved data points.

3D Deep Learning and Advancements in Point Cloud Processing:
Deep learning is now being used to process 3D data from sources such as LiDAR, digital elevation models and photogrammetric point clouds. Methods like 3D CNNs, Graph Neural Networks and PointNet models enable the extraction of information about the structure, objects and terrain of the environment in three dimensions. These models can be used for applications such as tracking changes in cities, assessing the amount of vegetation in forests and calculating volumes of materials in the Earth's crust.

Transfer Learning and Pretrained Models:
Transfer learning applies deep learning models that have been initially trained on general image data (e.g., ImageNet) to specific problems in remote sensing. This becomes especially important when there is a shortage of labelled data in the geospatial field. Training neural networks that have been pretrained on satellite imagery generally leads to faster, more accurate and more reliable models. Transfer learning is commonly applied in monitoring unauthorized transformation of land use, illegal quarrying and changes to the built environment following disasters.

Combining Deep Learning with Geographic Information Systems (GIS):
Connecting deep learning models to GIS systems improves the analysis and interpretation of spatial datasets. Integrating model predictions with spatial data enables the generation of maps that show urbanization patterns, crop growth and risk areas for various natural disasters. Increasingly, deep learning tools are being integrated into GIS platforms such as Google Earth Engine, ArcGIS and QGIS for real-time classification, forecasting and visualization of remote sensing data.

Challenges and Future Directions:
Deep learning for remote sensing encounters numerous obstacles that hinder its general applicability and efficiency. A key problem is that training and deploying deep learning models usually impose significant demands on computational resources and can slow down real-time or computationally limited applications. Furthermore, collecting sufficient, accurate and representative training data for remote sensing is challenging, making it difficult to develop effective deep learning models. The capacity of deep learning models hinders their adoption in crucial tasks such as remote sensing, because it becomes hard to corroborate and understand the underlying reasons for their output. Furthermore, moving trained models to new regions, sensors, or time periods frequently causes performance to drop because of unexpected changes in the input data. Researchers are now concentrating on developing methods that make it possible to interpret and understand how models arrive at decisions, thereby improving their reliability and boosting users' confidence in their outputs. Efforts are being made to develop lean, quick models that can operate on edge devices and provide rapid processing for remote or low-bandwidth environments.

Innovative learning approaches, such as self-supervised and unsupervised models, are being explored to achieve more accurate predictions with limited need for annotated data and take full advantage of massive unlabelled remote sensing imagery. Progress in transfer learning and domain adaptation improves a model's ability to handle variations in geographic location and imaging sensors. Using multiple data modalities by fusing optical, radar and LiDAR signals is becoming increasingly popular because it enhances feature representations and enables stronger performance under various environmental conditions. Addressing these challenges and adopting these innovative techniques will allow deep learning models to realize their full potential for environmental monitoring, agriculture, urban planning and many other fields of remote sensing.

Random Forests:
Random forests are commonly used to perform classification as well as regression. Combining several decision trees in an ensemble helps boost forecast reliability and minimize the risks of overfitting. Random forests are built by training a number of decision trees on different subsets of the training data and features, all separately. The results from each individual tree are either averaged or acted on by other methods to form the final decision. Generalization performance becomes better when an ensemble of trees is used rather than a single decision tree. Remote sensing uses random forests for tasks such as land cover identification, predicting crop output and observing forests. It is possible to analyze challenging spatial data smoothly and gain useful insights with the help from this technology.

Improvements in deep learning and random forests using machine learning and AI have made a big difference in how people look at remotely sensed data. It is now simpler to find the most important points within large and intricate datasets. They play a key role in land cover discovery, detecting changes, checking the environment and assessing possible natural dangers. Researchers and professionals use ML and AI techniques to uncover image traits and identify patterns in remote sensing photos. After doing these calculations, they can apply the results to problems or decisions in research, conservation and daily living.

13.10 Conclusion

Using advanced techniques, satellite image processing takes satellite images of the Earth to gain useful information. The process is initiated when images are collected from sensors to observe the land and atmosphere. Imperfections in data are corrected, and significant improvements are made with preprocessing before analysis. When images are enhanced, identifying marks become easier to spot, and the data is simpler to review closely. In most cases, artificial intelligence and machine learning are applied to sort out and arrange a variety of land feature categories. Change detection tools allow specialists to trace the evolution of the Earth's surface over time, such as the development of urban areas and the destruction of forests. Spectral analysis methods make use of multispectral data collected from satellite images to provide accurate data about the materials and conditions present on Earth's surface. Satellite image processing finds widespread use in numerous sectors such as environmental monitoring, agriculture, urban planning, disaster management and natural resource management. Satellite image processing allows for thorough analysis and well-informed decision-making in several sectors by integrating image data with other geographical information. With the progression of technology, satellite sensors and image processing techniques are always improving, allowing us to better comprehend and efficiently oversee the Earth's resources and habitats.

References

1. Shafaey, M.A., Salem, M.A.-M., Ebied, H.M., Al-Berry, M.N., Tolba, M.F., Deep learning for satellite image classification, in: *International Conference on Advanced Intelligent Systems and Informatics*, Springer International, Cham, pp. 383–391, 2018, https://doi.org/10.1007/978-3-319-99010-1_35.
2. Ouchra, H., Belangour, A., Erraissi, A., Machine learning for satellite image classification: A comprehensive review, in: *2022 International Conference on Data Analytics for Business and Industry (ICDABI)*, IEEE, pp. 1–5, 2022, doi: 10.1109/ICDABI56818.2022.10041606.
3. Rahman, A., Abdullah, H.M., Tanzir, M.T., Hossain, M.J., Khan, B.M., Miah, M.G., Islam, I., Performance of different machine learning algorithms on satellite image classification in rural and urban setup. *Remote Sens. Appl.: Soc. Environ.*, 20, 100410, 2020, https://doi.org/10.1016/j.rsase.2020.100410.
4. Ferdous, H., Siraj, T., Setu, S.J., Anwar, M.M., Rahman, M.A., Machine learning approach towards satellite image classification, in: *Proceedings of International Conference on Trends in Computational and Cognitive Engineering: Proceedings of TCCE 2020*, Springer Singapore, pp. 627–637, 2021, https://doi.org/10.1007/978-981-33-4673-4_51.
5. Adegun, A.A., Viriri, S., Tapamo, J.R., Review of deep learning methods for remote sensing satellite images classification: experimental survey and comparative analysis. *J. Big Data*, 10, 93, 2023, https://doi.org/10.1186/s40537-023-00772-x.
6. Phiri, D., Simwanda, M., Salekin, S., Nyirenda, V.R., Murayama, Y., Ranagalage, M., Sentinel-2 data for land cover/use mapping: A review. *Remote Sens.*, 12, 14, 2291, 2020, https://doi.org/10.3390/rs12142291.
7. Sarker, H., Deep learning: A comprehensive overview on techniques, taxonomy, applications and research directions. *SN Comput. Sci.*, 2, 420, 2021, https://doi.org/10.1007/s42979-021-00815-1.
8. Zhu, X., Helmer, E.H., Gao, F., Liu, D., Chen, J., Lefsky, M.A., A flexible spatiotemporal method for fusing satellite images with different resolutions. *Remote Sens. Environ.*, 172, 165–177, 2016, https://doi.org/10.1016/j.rse.2015.11.016.
9. Benz, U.C., Hofmann, P., Willhauck, G., Multiresolution, object-oriented fuzzy analysis of remote sensing data for GIS-ready information. *ISPRS J.*

Photogramm. Remote Sens., 58, 3-4, 239–258, 2004, https://doi.org/10.1016/j.isprsjprs.2003.10.002.
10. Zhang, X. and Hu, J., Remote sensing image classification based on deep learning: A review. *Remote Sens.*, 12, 1, Article 166, 2020, https://doi.org/10.3390/rs12010166.
11. Zhong, P. and Wang, R., Learning Conditional Random Fields for Classification of Hyperspectral Images. *IEEE Trans. Image Process.*, 19, 7, 1890–1907, Jul. 2010, https://doi.org/10.1109/TIP.2010.2045034.
12. LeCun, Y., Bengio, Y., Hinton, G., Deep learning. *Nature*, 521, 436–444, May 2015, https://doi.org/10.1038/nature14539.
13. Chen, X., Xiang, S., Liu, C.-L., Pan, C.-H., Vehicle Detection in Satellite Images by Hybrid Deep Convolutional Neural Networks. *IEEE Geosci. Remote Sens. Lett.*, 11, 10, 1797–1801, 2014, https://doi.org/10.1109/LGRS.2014.2309695.
14. Cheng, G., *et al.*, Object detection in remote sensing imagery using a discriminatively trained mixture model. *ISPRS J. Photogramm. Remote Sens.*, 85, 32–43, 2013, https://doi.org/10.1016/j.isprsjprs.2013.08.001.
15. Aytekın, O., Erener, A., Ulusoy, I., Düzgün, S., Unsupervised building detection in complex urban environments from multispectral satellite imagery. *Int. J. Remote Sens.*, 33, 7, 2152–2177, 2011, https://doi.org/10.1080/01431161.2011.606852.
16. Qi, W., Huang, C., Wang, Y., Zhang, X., Sun, W., Zhang, L., Global-Local Three-Dimensional Convolutional Transformer Network for Hyperspectral Image Classification. *IEEE Trans. Geosci. Remote Sens.*, 61, 1–20, 2023, https://doi.org/10.1109/TGRS.2023.3272885.
17. Haroun, F.M.E., Deros, S.N.M., Din, N.M., Detection and monitoring of power line corridor from satellite imagery using RetinaNet and K-mean clustering. *IEEE Access*, 9, 116720–116730, 2021, https://doi.org/10.1109/ACCESS.2021.3106550.
18. Ni, Y., Li, X., Ye, Y., Li, Y., Li, C., Chu, D., An investigation on deep learning approaches to combining nighttime and daytime satellite imagery for poverty prediction. *IEEE Geosci. Remote Sens. Lett.*, 18, 9, 1545–1549, 2020, https://doi.org/10.1109/LGRS.2020.3006019.
19. Zhang, G., Xu, P., Xu, H., Hsu, L.T., Prediction on the urban GNSS measurement uncertainty based on deep learning networks with long short-term memory. *IEEE Sens. J.*, 21, 18, 20563–20577, 2021, https://doi.org/10.1109/JSEN.2021.3098006.
20. Sengan, S., Karn, A.L., Pustokhin, D.A., Pustokhina, I.V., Alharbi, M., A hybrid learning model for efficient classification of Land Use and land change from satellite images. *Phys. Chem. Earth Parts A/B/C/*, 128, 103284, 2022, https://doi.org/10.1016/j.pce.2022.103284.
21. Benbriqa, H., Abnane, I., Idri, A., Tabiti, K., *Deep and ensemble learning based land use and land cover classification*, Springer International, pp. 588–604, 2021, https://doi.org/10.1007/978-3-030-86970-0_41.

14

Artificial Intelligence and Machine Learning in Land Feature Extraction

Abstract
Artificial Intelligence (AI) and Machine Learning (ML) in land use analysis are getting noticed because they could change how one understands and uses land resources. AI and ML are demonstrated in this chapter for analyzing land use, emphasizing their advantages, disadvantages and what lies ahead. It introduces the subject area and looks at how AI and ML are used to study land use trends, forecast alterations, and affect choices made on environmental planning. Using AI and ML, a lot of spatial data can be processed, detailed patterns and relationships can be identified, and many regular activities can be automated, increasing the efficiency and accuracy of analyzing land use. In land use studies, there are other problems with AI and ML, such as the quality of data, how easy it is to understand the algorithm, and possible biases. Issues in these fields are best addressed when multiple specialists collaborate, approaches are made clear, and verification methods are effective. It was determined that further studies and breakthroughs in AI and ML are required for land use analysis to fully contribute to sustainability, protect the environment and improve decision-making on different scales.

Keywords: Artificial intelligence, land use pattern, machine learning, sustainability

14.1 Introduction

Advancements in AI and ML have significantly changed the scene by allowing computers to process much information, such as satellite pictures, maps and changes in the environment, with greater speed, correctness and ability to react to changes. In the past, people used to survey sites by hand, interpret what they saw using expertise and rely on data that was not regularly updated. AI and ML help detect minute changes in land cover, anticipate urban development and check how humans and the environment interact at various places and over time [1–4]. Consequently, these new

Kundan Meshram and Chatrabhuj. A Sustainable Approach to Land Use, (359–376) © 2026 Scrivener Publishing LLC

tools are helping planners, researchers, and policymakers to rely on taking initiative with reliable and intelligent decisions supported by continued learning and the use of advanced technology.

When planning cities, analyzing land use is very important for sustainability and conserving resources and the environment. It studies the ways land is being used, as well as the influence of what humans do on the environment. Studying how land is used can indicate where nature is sensitive and aid in planning new projects, inclusive of strategies to cope with climate change and support growth. The use of AI and ML is able to get meaningful insights and identify small patterns from a wide range and huge amount of data. AI and ML can process and understand the world's geography quickly and precisely when they receive information from high-resolution satellites, Geographic Information Systems, remote control sites and social and economic data [5–9]. With this, land use planning can be done in real time, with predictions and scenario testing, all important for planning responsibly. They assist in identifying where cities are being built, observing changes in forests, determining land erosion and gauging the distribution of natural resources. Therefore, AI and ML are important for sustainable urban development and caring for the environment because they support data use, flexible planning and foresight when making key decisions.

By using AI and ML, information about land use, infrastructure, the environment and human actions can be studied by planners and city administrators at any time. It, in turn, promotes better, quicker and longer-lasting development in cities. AI and ML are most valuable for their capacity to reduce congestion on roads, better handle utilities and help with setting directions for new development projects. With these systems, traffic can be redirected based on how many cars are on the roads, alternative routes may be suggested, and upcoming traffic congestion can be noticed early. AIs are able to examine the projected impacts of new building projects on traffic, energy, noise and pollution in cities. The ability to predict issues enables development to support sustainability and adhere to all relevant regulations. Additionally, using information on people and their jobs, AI can help locate less-served neighbourhoods and support fair development. AI is playing a key role in improving both safety and city administration. If one uses data gathered from different sensors, cameras and IoT devices in a city, it becomes possible to notice and react to unlawful activities, run public utilities better, ensure sites are safe and verify the timely completion of large projects. AI and ML are helping cities become more efficient, inclusive, sustainable and resilient to face the issues of our time.

AI and ML are playing a major role in bringing about precision cultivation in agriculture. Using satellite and drone photos, producers get timely information to raise their crop yield, reduce damage from pests, estimate yields and monitor the condition of their crops. AI also examines the condition of the soil, the climate and whether enough water is available, picking crops that best fit the land.

AI and ML play a major role in helping with the conservation and management of the environment. With the help of satellites, these devices spot deforestation instantly and assist in managing and stopping illegal cutting of trees. ML techniques help determine where wildlife lives, meaning that habitats can be recognized and areas set aside for their defence. AI and ML also play an important role in handling disasters. Using old weather records, the shapes of the land and land utility, AI can estimate the risk of floods and support planning for any resulting disasters. When there is an earthquake, AI determines the amount of damage to buildings and structures by studying satellite pictures, which supports quick management of resources for disaster relief.

Research on climate change is being affected by advances in AI and ML. The development of ways to reduce climate change greatly depends on these technologies due to their ability to gauge the quantity of carbon in various land cover types. Moreover, ML strategies help to understand the effects of global warming by analyzing the motions and disintegration of glaciers. AI and ML help with building and maintaining transportation and infrastructure by studying satellite and GPS data, which aids in road network mapping and updating [10–13]. It allows experts to maintain the current infrastructure and plan future road developments. Tracking, scheduling and monitoring goods movements by rail are safer and more efficient due to the use of AI systems on railway aerial imagery and sensor data.

To determine the value of land and property, ML models are used in real estate and property management by analyzing its location, what is done with the land and recent trends in the industry. By doing this, it helps to decide where and for how much to invest. AI technologies also check land use to confirm that rules and regulations for zoning are followed, avoiding any illegal land use changes and guaranteeing proper planning of activities.

By using AI and ML on satellite photos, forestry management can detect the initial outbreak of forest fires and quickly put them out. Moreover, with the help of ML models, it is easier to plan for sustainable timber harvesting. Using AI and ML technology, experts observe how land use occurs, which supports better management of watersheds, helps manage water resources, and guarantees that materials and water are available.

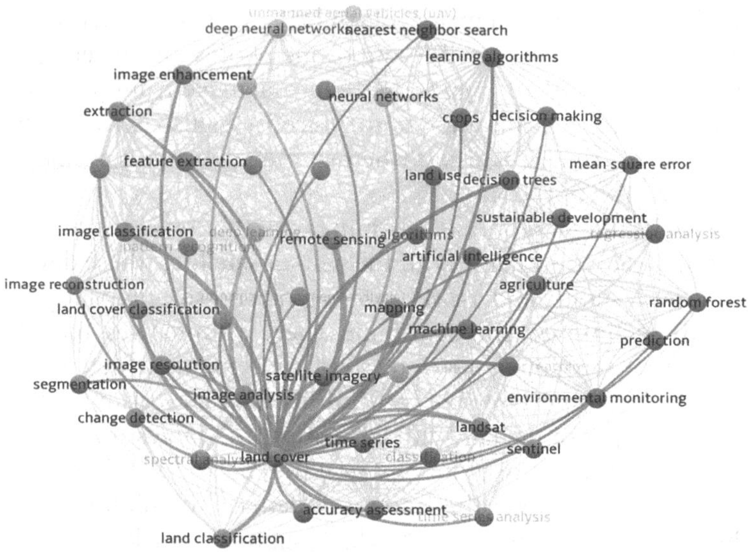

Figure 14.1 Land cover.

AI enables farmers to adjust their watering schedules based on knowledge of the weather, moisture in the soil and the needs of the crops [14–18]. By using AI and ML, the selection of suitable land for solar and wind farms can consider land usage, surrounding climate and the topography involved. Furthermore, they are used in planning energy infrastructure by estimating energy needs as affected by how the population and land use are changing. Through understanding land usage, managing resources well and focusing on sustainability, AI and ML are revolutionizing many different industries.

All these systems prove the major impact AI and ML are having on land use analysis, as is clear from Figure 14.1. Such technologies are greatly valued nowadays because they can identify useful patterns from diverse and large urban and environmental datasets. AI and ML are key to the progress made in land use planning, creating policies and developing strategies for sustainability today.

14.2 Role of AI and ML in Land Use Analysis

Using AI and ML makes land use analysis more effective. Due to their wide array of applications, land management and understanding have been transformed [19–23]. Here, the main uses of AI and ML in land use analysis are summarized.

14.2.1 Automated Land Cover Classification

AI and ML are used to separate and classify land into categories like forests, open, bare, urban, water bodies and farms. The approach relies on reviewing satellite photos and other remote sensing information. It is faster and more accurate than doing the job manually. Data acquisition and preprocessing begin first, where satellite data is collected from places such as Landsat, Sentinel or through commercial sources, and all corrections are carried out before analyzing the data. Thereafter, each segment of the image is tagged with a label, indicating its matching land cover. In many cases, these labels come from real data, existing land cover maps or by handwriting their meaning over the images. The use of rotation, flipping, scaling and removing noise from data aims to increase the quality and variety of the training dataset to display all land cover classes.

Convolutional neural networks (CNNs) are preferred because they can recognize spatial features in images without manually finding and adding features. Unlike decision trees, random forests or SVMs, CNNs use the arrangement of pixels in images, which allows them to be effective in handling a battery of remote sensing questions. Researchers apply CNNs to study patterns and light-reflecting traits of various things on the ground, such as plants, urban settlements, rivers and exposed land, by processing many annotated images from satellites. The model's accuracy for classification can be improved by using SGD, Adam or RMSprop to iteratively alter its parameters. Once trained, CNN knows how to identify and assign categories to every pixel in satellite images or certain regions (in object-based classification). CNNs can also return the likelihood of an image being a member of every class. Using this kind of output is useful when one needs to consider uncertainty in a particular process. To boost the success of the results, many post-processing approaches are regularly used. A validation dataset is used, apart from the data for training, to measure the quality of the model's performance. To measure how accurate a classification is, several evaluation metrics are applied, for example, overall accuracy, precision, recall, F1-score and the kappa coefficient. With the help of this matrix, researchers are able to see the mistakes made in classifying various types of data.

Many areas, including the environment, cities, farming and emergency plans, make use of automated land cover classification. It supports the tracking of changes in nature, creation of maps for city planning, inspection of farmland and crops and evaluation of spots damaged by disasters. Users can easily call upon proficient models in cloud-based services, at the edge or through Geographic Information System (GIS) maps to make decisions with instant results, regardless of their experience. Remotely sensing land cover has improved with the introduction of AI and ML. This way of working allows prompt and precise monitoring of land cover changes. Tools that enable remote sensing are crucial when managing the environment, planning cities, monitoring farms and reacting to disasters.

14.2.2 Change Detection and Monitoring

To manage the environment, urban areas and resources well, one must analyze the change in land use and cover. Time-series data are often analyzed by ML algorithms to uncover and track differing trends. As a result, one can understand the ways humans influence the ecosystem and design methods to manage land sustainably.

Initially, images from Landsat, Sentinel and MODIS are collected month-by-month in a scheduled manner. Due to these satellite systems, scientists can monitor and observe changes on the Earth's surface for many decades. For consistency over the years, the images collected are first corrected for their position and light intensity using different methods. With standardization, one avoids mistakes in which ML models label certain differences as land cover changes when they are only variations in how the images were taken.

Recurrent Neural Networks (RNNs) and Long Short-Term Memory (LSTM) networks are commonly used to analyze data sets that change over time. They use sequential information and can find relationships that last just a little time or by looking further back in the past. RNNs keep track of changes from one frame to another using feedback loops, while LSTMs remember and update important details as the data comes in sequence, allowing them to recognize various shifts occurring in land over time. Some examples include deforestation, city expansion, various crop rotation cycles, damaged land or turning wetlands into other uses.

When taught using past images, these algorithms identify unexpected changes in land cover in the latest pictures. Alterations in the data are pointed out and recorded, then the changes are grouped and displayed to observe consistent patterns and places where those patterns occur. Temporal-spatial analysis is used to highlight, with a map, the areas that are undergoing the most changes. These ideas are needed in many fields, for example, to follow deforestation rates in tropical forests, review the state of wetlands and coastal areas, measure progress with afforestation or conservation and map changes in urban areas to support area planning.

Furthermore, mixing ML with GIS and Google Earth Engine (GEE) allows for easier and faster processing of data. The platforms make it possible to track, present and monitor extensive geospatial data across various locations. Combining AI, ML, and cloud computing makes it possible for stakeholders to access tools that quickly and successfully monitor land cover. Time-series analysis of Land Use and Land Cover (LULC) data greatly supports decisions made by policymakers, agencies focused on the environment and urban planners. With them, decisions are made using evidence, land is managed in advance, and strategies are developed to withstand pressures from humans and the environment. As a result, AI and ML change how one oversees the environment, by giving quick, precise and useful insights to help with the safe management of our environment.

14.2.3 Predictive Modeling

Using data from the past and the present through predictive modeling, land use analysts try to estimate future changes in land usage. It is largely helpful for those planning urban areas, managing the environment and making government policies, who have to foresee development results and plan to maintain land sustainably. Essentially, predictive modeling uses past experiences to provide guidance for future actions. To begin, all key historical land use datasets must be gathered and studied. Usually, these are combined with many additional details such as population growth, economic conditions, housing requirements, land features, distant bodies of water and various regulations. This information is used to analyze how a range of factors have helped shape land use throughout history.

To show how these relationships exist, ML algorithms are used, such as decision trees, random forests, support vector machines and other techniques like gradient boosting machines (examples include XGBoost and LightGBM). Analyzing the importance of each feature also helps recognize the main predictors behind land transformation. Training the models takes place only after identifying important features. The models analyze data with information on land use to find the link between input variables and results. Mostly, datasets are split into training and validation segments to judge if the model can generalize. Models are checked using cross-validation or holdout testing to avoid fitting too closely to the training data. Model performance is checked by measuring accuracy, precision, recall, F1-score and the AUC-ROC.

Once successful training is finished, the predictive models start being used to project future land use changes according to the present situation and future strategies. They can also predict when cities may expand, farming fields could be turned into industries, or buildings may reach areas protected by nature. Hence, models allow experts to estimate how populations in cities surroundings will expand when roads or airports are established and spot locations prone to deforestation because of economic reasons. Planning and policymaking rely heavily on the findings of these predictive models. Environmental regulators can search for upcoming problems in the environment and introduce action plans to address them. This technology can also be used for analyzing the effects of different strategies in various scenarios. Converting both old and current data into useful predictions for land use, effective predictive modeling adds significant certainty to land use planning. Through the use of ML algorithms, the system can provide accurate forecasts in a range of cultures and places. As a result, this approach leads to lasting strategies that address how the land is developed and deserves to be used without harming the environment for those using the land now and in the future.

14.2.4 Incorporating Socioeconomic Data

Socioeconomic data makes it possible to fully understand what controls land use in a certain area. There is a significant effect of socioeconomic factors on the usage of land. By using AI and ML, it becomes convenient to gather information from various datasets and see how land is used over time. At the beginning of the process, information about the population, the economy, development and environment is combined into one organized system. GIS is regularly applied to work with and represent the combination of data sets, giving the study a spatial perspective. After all the data is gathered, multivariate analytic methods are employed to explore how people's economic status and how they use their land affect different communities. By using machine learning, scientists can understand complicated ways that factors interact and influence land use, as well as their impact on the environment.

Land-use research that considers socioeconomic indicators provides more evidence for guiding future public policies. Reviewing changes in human trends, income, jobs and markets, along with land cover, reflects how human actions affect the environment. It supports the creation of approaches that can improve the economy and conserve our environment. Having gained this understanding, governments can make their zoning rules more effective, develop areas close to transport hubs and establish protected green areas where more homes can be built. Similarly, using maps of land use together with information about the livelihoods or incomes of people at the household level can identify rural locations prone to either damaging the environment or losing the ability to feed the population. To increase yields and avoid expanding the farmland area, payments for services that protect the environment, technology for farming, and more can be offered to encourage climate-smart practices. Socioeconomic–land-use integration shows that certain areas are not treated equally.

14.3 Benefits of AI and ML in Land Use Analysis

There are several benefits of AI and ML in analyzing land use; among them are automation, better efficiency, more accurate results, scalability, accurate predictions and cost-efficient ways to achieve the best decisions. These technologies give individuals the tools needed to make sense of geospatial data. As a result, experts can foresee what is ahead and decide wisely on how best to manage sustainable land development, which is given in Table 14.1.

Table 14.1 AI/ML applications in land use planning and management.

Broad area	Sub-area	Benefits
Automation and Efficiency	Automated Feature Extraction	- Fast extraction of land features from satellite and drone data - Reduces human error and improves consistency - Enables large-area assessment
	Scalable Workflows	- Processes large datasets efficiently *via* cloud platforms - Adapts to growing data volume dynamically - Supports real-time and long-term land monitoring
	Real-Time Data Processing	- Instant ingestion and analysis of remote sensing and sensor data - Enables timely decision-making in emergencies - Facilitates dynamic land change detection

(*Continued*)

Table 14.1 AI/ML applications in land use planning and management. (*Continued*)

Broad area	Sub-area	Benefits
	Customizable Models	- Domain-specific model tuning (e.g., urban zoning, soil erosion) - Allows expert involvement for relevancy - Enhances model adaptability
	Resource Optimization	- Automates repetitive tasks, saving time - Reduces workforce requirement - Improves cost and time efficiency in planning workflows
Precision and Accuracy	Advanced Data Analytics	- Deep learning for pattern recognition in land cover - High predictive accuracy for planning decisions - Detects subtle environmental changes
	Multi-Source Fusion	- Combines optical, LiDAR, radar, and socioeconomic data - Reduces uncertainty in mapping - Offers comprehensive situational analysis
	High-Resolution Classification	- Identifies fine-scale land use classes - Helps in urban planning, vegetation mapping, and water body delineation - Generates usable geospatial layers
	Semantic Segmentation	- Recognizes meaningful objects (e.g., roads, rivers, buildings) - Enhances map readability - Useful in object-based image analysis (OBIA)
Managing Extensive Datasets	Distributed Computing	- Uses parallel frameworks (e.g., Spark, Hadoop) for fast image processing - Handles petabytes of satellite and sensor data - Enables continent-scale land analysis in hours

(*Continued*)

Table 14.1 AI/ML applications in land use planning and management. (*Continued*)

Broad area	Sub-area	Benefits
	Cloud-Native Infrastructure	- No physical servers needed - Enables remote collaboration and model deployment - Offers real-time access to AI tools and APIs
	Multi-Modal Remote Sensing	- Fuses radar, optical, thermal, and hyperspectral data - Ensures coverage in cloudy or low-light conditions - Enhances land use classification reliability
	Temporal Monitoring	- Detects land use change over time - Assesses sustainability indicators (e.g., NDVI trends) - Supports long-term environmental tracking
Predictive Insights	Long-Term Forecasting	- Simulates future scenarios (urban growth, deforestation) - Informs proactive policy-making - Supports SDG-linked decision frameworks
	Scenario Simulation	- Tests "what-if" policies and land use plans - Helps visualize consequences of zoning or infrastructure projects - Encourages participatory planning
	Early Warning Systems	- Predicts floods, droughts, landslides using real-time data - Aids in timely evacuation and disaster preparedness - Supports climate-resilient planning
	Vulnerability Analysis	- Identifies at-risk zones from climate or urban pressures - Guides sustainable adaptation and resilience strategies - Supports ecological prioritization

(*Continued*)

Table 14.1 AI/ML applications in land use planning and management. (*Continued*)

Broad area	Sub-area	Benefits
Economic Efficiency	Cost-Effective Data Collection	- Selects optimal remote sensing tools (satellites, UAVs) - Reduces redundant data acquisition - Enhances monitoring ROI
	Efficient Workflows	- Streamlines data processing pipelines - Reduces manual intervention - Supports reproducible and open-source workflows
	Evidence-Based Decisions	- Integrates AI outputs into policy dashboards - Facilitates actionable insights from geospatial data - Improves governance in land management
	Cost-Benefit Assessment	- Quantifies trade-offs (e.g., development vs. conservation) - Informs prioritization of land investments - Encourages transparent environmental accounting

14.4 Challenges and Considerations

Bringing AI and ML into how land is used can bring enormous benefits, but it also creates a range of serious problems. To make AI and ML-based systems beneficial for all, one must handle these issues when making decisions on land use.

➢ Feature selection and representation:
AI and ML methods run into trouble when sorting through significant aspects of land use data and showing detailed patterns of land use. Building better and clearer predictive models is made possible by creating effective techniques and methods to mine relevant data from complex and multi-dimensional data.

➢ Complexity and interpretability:
AI and ML models, particularly deep learning models, are complex, and it is often tough to understand the important factors that direct land use decisions.

To ensure the right decisions are made, AI and ML systems for land use must be built to be open about their methods while remaining simple to understand.

➢ Temporal and spatial dynamics:
The temporal and spatial changes in land use show that it is important to use specialized models that can detect space and time relations among land use categories. Because land use pattern analysis is complex, one needs to design spatio-temporal models and incorporate geospatial analysis methods into AI and ML tools.

➢ Data quality and availability:
Good data and plenty of availability of data are necessary for AI and ML to work well. Still, training models that work accurately and dependably can be challenged by the lack of, unevenness in or outdatedness of land use data. Guaranteeing AI and ML work well for land use pattern analysis means filling in data gaps and making data more precise by collecting, combining, filtering and checking the data.

➢ Robustness and uncertainty:
As a result of algorithmic constraints, model assumptions, data variability, various restrictions and traits of the models, as well as the variability in the data, AI and ML models are likely to be uncertain. Achieving informed decisions and managing the risks requires understanding model prediction uncertainty, implementing techniques for propagating it and enhancing the model's stability as input varies.

➢ Long-term monitoring and evaluation:
Since land cover and use can change and interventions keep happening, it is important to keep monitoring and evaluating all the time. Land use planning and management improve as a result of feedback loops during decision-making, the use of trustworthy indicators and regular, sustained monitoring.

➢ Ethical and social consequences:
The use of AI and ML in examining land use gives rise to ethical issues about who owns and uses data, the consent involved, monitoring practices and responsibility. Rules should be set in place to guarantee people's rights, increase transparency and maintain good ethical practices in data-based systems. When stakeholders are involved in making these systems, trust and socially responsible ideas are promoted.

➢ Bias and fairness:
When training data contains inequality and prejudice, the use of AI and ML by decision-makers can cause land use decisions that are unfair or marked by bias. Failing to represent areas of marginalized communities or specific land types can cause decisions that continue to widen inequality. Dealing with algorithmic fairness calls for early identification of biases, the employment of balanced algorithms and a wide range of datasets. All land use planning should include fairness across groups, areas and economic status.

➢ Capacity building and expertise:
One must help land use planners, policymakers, and stakeholders understand AI and ML to make the best use of them and avoid the risks that come with them. By teaching, training and assisting practitioners with relevant skills and tools, land use pattern analysis using AI and ML technologies can benefit from greater adoption. Knowledge of AI and ML by practitioners in land use planning makes the effective use of these technologies possible. Now, both public agencies and regions that are developing face a skills gap. Helping stakeholders obtain knowledge through training, workshops, various training courses, and shared technologies allows for better use of AI and ML in sustainable development.

➢ Data bias and representativeness:
AI and ML performance rely on land use data, and any favourable depiction of certain land cover classes or regions can affect it. By processing the data with unbiased sampling, image augmentation and using corrections suited to the type of land use, one can improve the performance of land-use models. Some land use models rely on data that is not evenly spread or well-representative, mainly in regions that are not fully surveyed. This situation makes it hard for the model to solve problems in new conditions. Improving the information present in data by sampling wisely, creating artificial data and leveraging existing models guarantees models can work reliably in a wide range of contexts.

➢ Transferability and generalization:
Generalizations of AI and ML models built just for a specific place or period can fail in other situations because of differences in environmental factors, how land is used, and social and economic settings. To make models transferable and usable in many regions and timeframes, people use cross-validation, domain adaptation and transfer learning methods. Because environmental, socioeconomic and policy situations differ

between regions and over time, AI and ML models do not usually apply well to them. Model adaptability is improved by using domain adaptation, fine-tuning and meta-learning approaches. For land use planning practices to be applicable in various settings and to succeed multiple times, models must be generalized.

➢ Interdisciplinary collaboration:
Analysis of land use patterns using AI and ML calls for scientists from geography, ecology, computer science and social sciences to work together. One can increase the use of data-driven knowledge and improve land use understanding by encouraging interdisciplinary collaborations, sharing information among various groups and crossing the lines between different specialities. Because of its diverse nature, land use analysis requires teamwork between environmental scientists, computer specialists, urban planners, economists and social researchers. Using different sets of knowledge, methods for solving issues and understanding stakeholder needs, interdisciplinary teams improve how AI and ML models are made and used. A mix of disciplines provides better insights into land use and increases the usefulness of what the model shows.

➢ Policy and regulatory frameworks:
The use of AI and ML in planning and managing land use raises problems with data governance, protecting privacy and holding people responsible for any actions. Setting guidelines, standards and regulations for bringing AI and ML into land use analysis is necessary, so the interests of law, those involved and public trust are protected. Banking, finance, insurance, land use planning and manufacturing sectors must follow strong rules and guidelines when using AI and ML. They must deal with topics like managing, protecting and using data, holding people accountable and ensuring compliance with set standards, local and worldwide. To use these technologies well and ethically, one must create laws that support responsible progress while still helping the wider public.

➢ Resource constraints:
AI and ML are not always used in land use pattern analysis because some regions have limited resources for equipment, storing data and qualified experts. Open-access data, cloud-based solutions and strengthening institutional knowledge can make AI and ML available to many and solve problems related to sustainable land use management, overcoming troubling resource issues. Because of difficulties with technology, data and engineering in low-resource places, adopting AI and ML is more difficult.

Some solutions are using lightweight models, cloud computing, data shared openly and group platforms that help everyone access advanced technologies. For digital divides to be fixed and land use innovations to benefit many, and need to improve infrastructure and build stronger institutions.

➢ Public engagement and trust:
To make sure AI and ML land use analyses are accepted, one needs to explain them and build trust among the public. Support for using data in sustainable land use planning can grow if the government is transparent and accountable, involves the public in decisions and explains both the benefits and drawbacks of AI and ML technologies to the public. To achieve successful AI and ML use in land use planning, the public needs to have trust and be engaged. When people are given honest explanations about models, how they work and what they can do, it earns those systems acceptance. When local communities, civil society and officials work together to develop plans, technological solutions fit well with local customs and priorities.

14.5 Conclusion

AI and ML have brought about a new way of handling land and habitat analysis, organization and planning. As a result of using these tools, it has become much faster and more accurate to work with huge and complex data. The use of socioeconomic information helps to better explain the connections between human actions and the environment. Even though AI and ML offer a lot of value, people still have to deal with certain key challenges, including issues with data quality, accessing data, understanding models and deciding what is right. Resolving these issues is crucial to guarantee that land use and environmental planning make good use of such technologies. The integration of data and technology leads to better environmental management and urban planning, which can deal with major global issues such as unplanned urban development, degradation and climate change. These developments can direct societies toward greater sustainability, equality and the ability to withstand challenges.

References

1. Huang, B., Zhao, B., Song, Y., Urban land-use mapping using a deep convolutional neural network with high spatial resolution multispectral remote

sensing imagery. *Remote Sens. Environ.*, 214, 73–86, 2018, https://doi.org/10.1016/j.rse.2018.04.050.
2. Chen, X.L., Zhao, H.M., Li, P.X., Yin, Z.Y., Remote sensing image-based analysis of the relationship between urban heat island and land use/cover changes. *Remote Sens. Environ.*, 104, 2, 133–146, 2006, https://doi.org/10.1016/j.rse.2005.11.016.
3. Hu, F., Xia, G.S., Hu, J., Zhang, L., Transferring deep convolutional neural networks for the scene classification of high-resolution remote sensing imagery. *Remote Sens.*, 7, 11, 14680–14707, 2015, https://doi.org/10.3390/rs71114680.
4. Huang, Z., Qi, H., Kang, C., Su, Y., Liu, Y., An ensemble learning approach for urban land use mapping based on remote sensing imagery and social sensing data. *Remote Sens.*, 12, 19, 3254, 2020, https://doi.org/10.3390/rs12193254.
5. Rogan, J. and Chen, D., Remote sensing technology for mapping and monitoring land-cover and land-use change. *Prog. Plann.*, 61, 4, 301–325, Dec. 2004, https://doi.org/10.1016/S0305-9006(03)00066-7.
6. Li, F., Yigitcanlar, T., Nepal, M., Nguyen, K., Dur, F., Machine learning and remote sensing integration for leveraging urban sustainability: A review and framework. *Sustainable Cities Soc.*, 96, 104653, 2023, https://doi.org/10.1016/j.scs.2023.104653.
7. Alem, A. and Kumar, S., Deep Learning Methods for Land Cover and Land Use Classification in Remote Sensing: A Review. *2020 8th International Conference on Reliability, Infocom Technologies and Optimization (Trends and Future Directions) (ICRITO)*, Noida, India, pp. 903–908, 2020, doi: 10.1109/ICRITO48877.2020.9197824.
8. Zhang, Y., Li, Q., Huang, H., Wu, W., Du, X., Wang, H., The combined use of remote sensing and social sensing data in fine-grained urban land use mapping: A case study in Beijing, China. *Remote Sens.*, 9, 9, 865, 2017, https://doi.org/10.3390/rs9090865.
9. Wang, J., Bretz, M., Dewan, M.A.A., Delavar, M.A., Machine learning in modelling land-use and land cover-change (LULCC): Current status, challenges and prospects. *Sci. Total Environ.*, 822, 153559, 2022, https://doi.org/10.1016/j.scitotenv.2022.153559.
10. Chaturvedi, V. and de Vries, W.T., Machine learning algorithms for urban land use planning: A review. *Urban Sci.*, 5, 3, 68, 2021, https://doi.org/10.3390/urbansci5030068.
11. Yu, Y., Cao, Y., Hou, D., *et al.*, The study of artificial intelligence for predicting land use changes in an arid ecosystem. *J. Geogr. Sci*, 32, 717–734, 2022, https://doi.org/10.1007/s11442-022-1969-6.
12. Aburas, M.M., Ahamad, M.S.S., Omar, N.Q., Spatio-temporal simulation and prediction of land-use change using conventional and machine learning models: A review. *Environ. Monit. Assess.*, 191, 205, 2019, https://doi.org/10.1007/s10661-019-7330-6.

13. Liao, X., Huang, X., Huang, W., ML-LUM: A system for land use mapping by machine learning algorithms. *J. Comput. Lang.*, 54, 100908, 2019, https://doi.org/10.1016/j.cola.2019.100908.
14. Yang, H., Nguyen, T.N., Chuang, T.W., An integrative explainable artificial intelligence approach to analyze fine-scale land-cover and land-use factors associated with spatial distributions of place of residence of reported dengue cases. *Trop. Med. Infect. Dis.*, 8, 4, 238, 2023, https://doi.org/10.3390/tropicalmed8040238.
15. Hossain, Md.T., Zarin, T., Sahriar, M.R., Haque, M.N., Machine learning based modeling for future prospects of land use land cover change in Gopalganj District, Bangladesh. *Phys. Chem. Earth Pt. A/B/C/*, 126, 103022, 2022, https://doi.org/10.1016/j.pce.2021.103022.
16. Fonseca, L.M.G., *et al.*, Pattern recognition and remote sensing techniques applied to land use and land cover mapping in the Brazilian Savannah. *Pattern Recognit. Lett.*, 148, 54–60, 2021, https://doi.org/10.1016/j.patrec.2021.04.028.
17. Xu, Q., Wang, Q., Liu, J., Liang, H., Simulation of land-use changes using the partitioned ANN-CA model and considering the influence of land-use change frequency. *ISPRS Int. J. Geo-Inf.*, 10, 5, 346, 2021, https://doi.org/10.3390/ijgi10050346.
18. Woldemariam, G.W., *et al.*, Machine-learning algorithms for land use dynamics in Lake Haramaya Watershed, Ethiopia. *Model. Earth Syst. Environ.*, 8, 3719–3736, 2022, https://doi.org/10.1007/s40808-021-01296-0.
19. Samardžić-Petrović, M., Kovačević, M., Bajat, B., Dragićević, S., Machine learning techniques for modelling short term land-use change. *ISPRS Int. J. Geo-Inf.*, 6, 12, 387, 2017, https://doi.org/10.3390/ijgi6120387.
20. Jayanandini, G., Kathiravan, P., Janani, V., Land use suitability analysis using artificial intelligence and machine learning for developing urban areas, in: *2023 Annual International Conference on Emerging Research Areas: International Conference on Intelligent Systems (AICERA/ICIS)*, pp. 1–8, 2023, Doi 10.1109/AICERA/ICIS59538.2023.10420204.
21. Liu, X., Ou, J., Li, X., Ai, B., Combining system dynamics and hybrid particle swarm optimization for land use allocation. *Ecol. Modell.*, 257, 11–24, 2013, https://doi.org/10.1016/j.ecolmodel.2013.02.027.
22. Feizizadeh, B., Omarzadeh, D., Kazemi Garajeh, M., Lakes, T., Blaschke, T., Machine learning data-driven approaches for land use/cover mapping and trend analysis using Google Earth Engine. *J. Environ. Plann. Manage.*, 66, 3, 665–697, 2021, https://doi.org/10.1080/09640568.2021.2001317.
23. Abdi, A.M., Land cover and land use classification performance of machine learning algorithms in a boreal landscape using Sentinel-2 data. *GISci. Remote Sens.*, 57, 1, 1–20, 2019, https://doi.org/10.1080/15481603.2019.1650447.

15
Land Use Pattern of Cities

Abstract
The chapter investigates the rate of urbanization and change in land use trend from the year 2000 to 2020. This research includes an analysis of land use patterns for cities like Bhopal and Raipur for the years 2000, 2010 and 2020. Firstly, satellite images from Landsat were obtained for the regions, and their analysis was done on QGIS 3.30. The results from the investigation show that, due to development, the urban areas and agricultural areas are expanding at the cost of natural and open spaces. The finding shows the need for an intelligent, sustainable, and environmental land management system. It stresses the role of advanced city planning and efforts to increase land use sustainability in metropolitan centers. Comprehending land use patterns helps understand urban trends and supports efforts in sustainability, protecting nature, sustainable agriculture, climate resilience, and effective urban governance.

***Keywords*:** Agricultural expansion, land use patterns, population growth, sustainability, urbanization

15.1 Introduction

The arrangement of residential, commercial, industrial, recreational and public areas within a city or metropolitan area is called land use in urban areas. Such patterns do not simply appear; instead, they develop through a complicated interaction of several factors playing a part in urban growth and design. Older cities are usually built with mixed uses, while newer ones are more often divided into areas for specific purposes. Industries, business areas and commercial areas being concentrated in few places greatly impact how much and what kind of land is needed in those urban areas. The suitability and accessibility of land are greatly influenced by physical geography such as topography, climate, water bodies and natural barriers. Meanwhile, lifestyle, artistic choices, different family structures and demographic features help determine both the design of homes and the layout

Kundan Meshram and Chatrabhuj. A Sustainable Approach to Land Use, (377–392) © 2026 Scrivener Publishing LLC

of community spaces. Besides, land is managed according to rules and frameworks developed at both community and national scales. Such policies are zoning ordinances, building codes and investments in infrastructure, which all join to ensure that how land is used coincides with what is best for society [1–7]. In order to analyze, plan and manage cities, urban land use is generally divided into several major classes, as described below.

- Residential land use: These areas are mainly intended for people to live and for housing. There are many types of housing in these zones, including large apartment buildings, mid-sized condos, secure communities and small detached homes. In addition, they might include residential areas or housing colonies that feature facilities for everyone to use, like public centers, parks and safety services. How residential areas are arranged and how many homes are found there depends on economic status, accessible land and distance to employment and other services.
- Commercial land use: Business activities and economic transactions happen primarily in the commercial zones. Retail stores, shopping malls, office buildings, restaurants, hotels, entertainment venues and service center businesses like banks or salons fall into these areas. Placed mostly in CBDs, sub-centers or along main roads, commercial land use is critical for city economies, since it hires workers, helps carry out trade and provides necessary services for residents and travellers.
- Industrial land use: Industrial land use covers places where goods and materials are made, stored and moved. The zones found in manufacturing are factories, manufacturing plants, warehouses, logistics centers and workshops. One must likely find them outside of cities or along routes like highways, train lines or ports, to ensure less traffic, fewer effects on people's daily lives and better deliveries. Zoning and environmental laws help control pollution, noise and safety in areas used for industry.
- Institutional and public facilities: They are places reserved for organizations that give necessary services to the public. Examples in this group are schools, colleges, universities, hospitals, clinics, government offices, administrative centers, police and fire stations and religious institutions. Institutional land use ensures that cities have learning,

health, administrative and community facilities which benefit the public.

- Recreational and open spaces: Urban quality of life greatly depends on having recreational and open spaces. They consist of public parks, botanical gardens, green belts, sports facilities, playgrounds, community spaces and city woods. People can use these spaces for relaxing, getting active and interacting with their community. Besides helping the neighborhood look better and supporting public health, they contribute to the environment by cleaning the air, keeping cities more comfortable climate-wise and managing rainwater.
- Transportation and infrastructure: Land devoted to transport systems and urban movement is called transportation land use. Among the types of transport included are roads, highways, railways, metro networks, bus stations, airports and paths for pedestrians. The range also extends to bridges, tunnels, places to park and transportation-focused building areas known as TODs (Transit-Oriented Development). Other parts of urban infrastructure, which are just as essential, are water supply networks, sewage systems, power lines and telecommunication towers.
- Mixed-Use developments: Mixed-use projects combine housing, business and, sometimes, light industry in one building or neighbourhood. The goal of these changes is to make communities more enjoyable to walk around, avoiding lengthy journeys to and from work and home and improving sustainable living in cities. For example, many apartments are developed with stores on the ground floors, some allow residents to mix work and home life and city centers may have homes, workplaces and shops near each other. Zoning areas for mixed use encourages people to live, work and play together, while also saving land.

15.2 Need for Identifying Areas

A major part of developing a sustainable city is planning and land use zones in cities systematically. By doing this, communities promote equal use of space for different activities and follow environmental, economic and social goals. When residential, commercial, industrial, institutional and green areas are clearly separated, growth and urban functions can be organized, land use becomes more optimal and wastage is reduced [8–15].

Distinct sections for different land uses in areas of cities are vital for achieving environmentally friendly, useful and strong communities. With the help of strategic zoning, land and infrastructure, it helps reduce costs for buildings and resources, strengthens access to important services and boosts people's standard of living. Managing space encourages residents to use nearby services, reducing traffic, lessening emissions and supporting the use of walking, biking and public transport. Having distinct commercial and industrial zones means investors can find what they need, new employment opportunities are created and the economy grows as businesses grow. Land use planning helps to save natural areas, control growth in cities, look after our natural resources and assists living creatures by protecting nature and including "green" projects in the plans. Detached industrial operations and well-placed emergency and healthcare services protect both public health and safety. Equity in land sharing increases social inclusion because all groups get the chance to use public areas and facilities. This, in turn, limits the differences in wealth between communities, bringing people together. Cities are able to keep their traditions and attract tourists thanks to preserving important places and sites in their midst. Making a plan helps limit building in dangerous locations and ensures vital services are set up away from risks, which makes things safer and better prepares communities for disasters. By identifying land and using zoning, complying with urban planning rules, and avoiding conflicts when using land, development projects fit into overall policies. Using comprehensive land use identification is essential for building communities that are livable, fair, active, environmentally stable and able to handle new challenges.

For cities to become sustainable, habitable and resilient, land use areas must be identified and planned. Optimal resource distribution, enhanced quality of life, reductions in congestion and pollution, economic progress, respect for the environment, public health and safety are all ways cities can prosper. Urban planning that integrates social fairness, cultural preservation, catastrophe resilience, and legal compliance [2, 16–18] is crucial for achieving balanced and harmonious urban development.

15.3 Land Use Pattern for Bhopal and Raipur

Bhopal, commonly known as the "City of Lakes," is a significant metropolitan hub located in the central region of India. The region spans from around 23°12′N to 23°24′N latitude and from 77°24′E to 77°50′E longitude, covering an area of approximately 3,055 km^2. Bhopal is renowned as a cultural and historical centre in Central India, with a population of around 1.8 million inhabitants and a population density of approximately 590 individuals per square kilometer, according to the 2011 Census of India. These tranquil bodies of water not only improve the city's visual attractiveness but also function as vital water sources and recreational areas for the local population. Bhopal has a subtropical climate characterized by an annual temperature range of 24.2°C to 34.6°C. The relative humidity levels demonstrate seasonal fluctuations, ranging from 36.9% to 71.6%.

Raipur, is a prosperous city located in the central region of Chhattisgarh state in India. Raipur is situated among the geographical coordinates of about 21°23′N to 21°50′N latitude and 81°38′E to 82°00′E longitude. It has an area of approximately 226 km^2. The city has a crucial role in regional governance, education, business, and culture, promoting development and progress within the state. The city has a tropical wet and dry climate, with an average annual rainfall of about 127.4 cm. Raipur has substantial seasonal variations, with temperatures ranging from 5°C to 47°C in winter and summer, respectively. The relative humidity levels vary between 20% and 80%, contingent upon the season.

Figure 15.1 and Figure 15.2 depict a gradual magnification sequence, starting from a national level and gradually zooming in to a small scale within India. The initial map (a) displays the entirety of India, with a particular region emphasized by a red square. The second map (b) focuses on the highlighted region, offering more precise borders within a state or region. The third map (c) provides a more detailed view by zooming in on a specific study area.

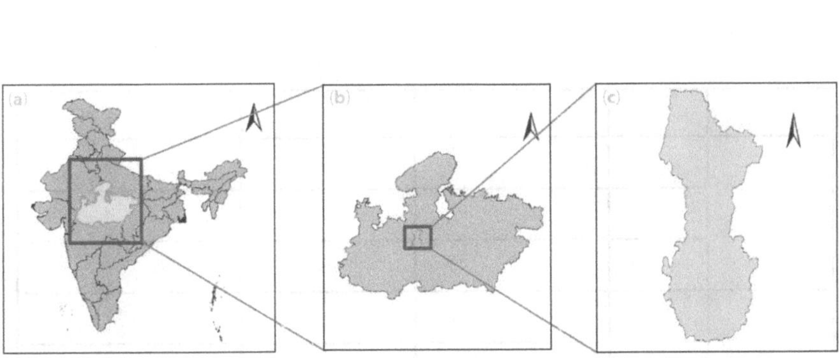

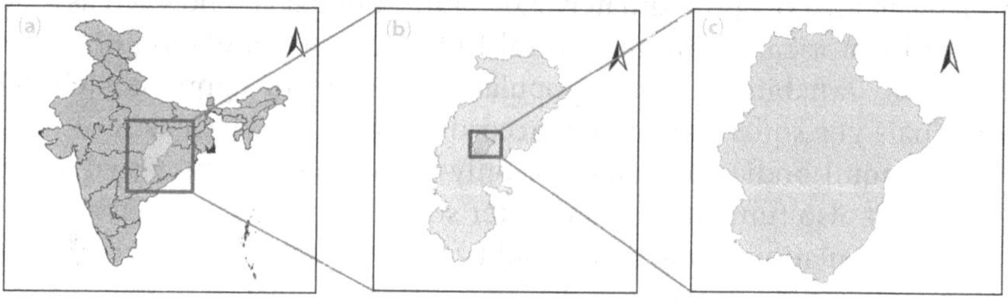

15.3.1 Source of Data

For analysis of land use patterns and their trend in change over time, remote sensing techniques were used. Several images from satellites were studied and understood. Remotely sensed data from Landsat 7 & 8 satellites is available at a 30 m resolution. Along with the satellite photos, users can see the path and row data so they know the purpose of each feed area. USGS Earth Explorer was used to assess the study area data. Users can access the USGS Datasets at this website: https://earthexplorer.usgs.gov/

As Table 15.1 shows, USGS Data Centre provided a set of satellite images.

15.3.2 Methodology

In this analysis, a detailed approach was used to assess changes in land use and land cover (LULC) occurring during three time periods (2000, 2010 and 2020) using satellite images from the USGS Earth Explorer.

Photos from Landsat 7 Enhanced Thermal Mapping (ETM+) and Landsat 8 Operational Land Imager (OLI) were taken by the Landsat series satellite system at a 30-meter resolution. The satellite pictures were extensively pre-processed using the Quantum GIS (QGIS) program. After applying radiometric and atmospheric corrections, the scan line errors from the Landsat 7 series for the year 2010 were fixed. Afterward, many bands were layered to form composite images. Subsequently, the pertinent portions, specifically the maps matching the desired administrative region, were isolated. The classification approach employed unsupervised classification techniques. After the classification process, the land use patterns were discovered and evaluated for correctness using ground truthing. By surpassing a 75% accuracy threshold, the established land use patterns were utilized for conducting change detection analysis [2, 16–18].

Land use classifications as shown in Table 15.2, including water bodies, forest cover, urban areas, cultivated land, and open spaces, were assigned unique colours for differentiation. This facilitated a distinct vision and comprehension of the categorized land. The analysis spanned three decades in order to identify substantial alterations in development. The change detection percentage was computed for the time intervals 2000-2010, 2010-2020 and 2000-2020.

Table 15.2 Descriptions of land use class.

Land use type	Description
Water Bodies	The collection encompasses various bodies of water such as ponds, streams, rivers, lakes, riverine areas, and river banks. It primarily consists of surface water.
Urban Area	It encompasses residential dwellings, industrial facilities, roadways, public edifices, constructions, and human habitation areas.
Agriculture land	Land suitable for cultivating both annual and perennial crops.
Forest cover	It refers to the expanse of land that is encompassed by indigenous woodlands, trees, cultivated areas, and other local forested regions.
Open area	The area is devoid of any woodland, flora, or man-made structures. Additionally, it comprises prominent mountains and exposed rock formations.

To test how accurate the classification system is, the results were checked on Google Earth. With the accuracy of 500 data points by seeing how well they matched with the categories on the map. The results from the accuracy assessment depended on the chosen size of the training set and were expressed using an error matrix. With satellite images, application of pre-processing methods, several classification approaches, assessing accuracy and change detection, the chosen urban regions' land use patterns could be tracked within the given time period.

15.3.3 Results

1. Land use pattern

Three graphs in Figure 15.3 display information on how land use in Bhopal, India, has shifted from 2000 to 2020. Each map identifies open land (pink), forests (green), urban regions (red), agriculture land (yellow) and water bodies (blue).

Map (a) from 2000 represents Bhopal's national land use, with lots of forest and farmland, some spread-out cities and open zones. The changes seen from 2000 to 2010 are clearly shown in the second map (b). The expansion and development seen in urban areas are evident through the noticeable rise they are going through. Although parts of the forest and agricultural fields have been changed, some are transitioning to city development. Changes in where open spaces are found point to new choices about how land is being used. On the third map (c), Bhopal's land use in 2020 reflects a rise in urban development. The urban areas have experienced a substantial expansion, as indicated by the larger red patches, demonstrating the ongoing urbanization and population growth in the city. Forest areas have had decreases or fragmentation as a result of urban expansion, although they remain significant. There has been increase in agricultural fields, with

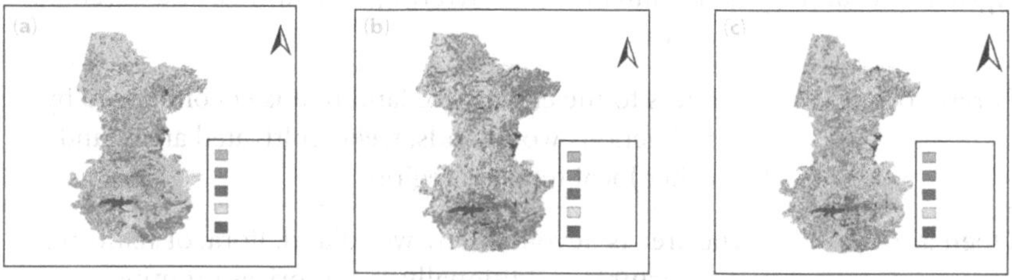

certain sections being transformed into urban areas, reflecting shifts in land use and development methods.

Over the course of twenty years, the maps demonstrate a significant and changing pattern in the way land is utilized in Bhopal, mostly due to the process of urbanization. The open spaces and bodies of water also undergo alterations, possibly due to the construction of infrastructure and adaptations to the environment. This sequence emphasizes the substantial influence of urbanization on natural and agricultural environments, emphasizing the intricate relationship between expansion and sustainability in the area.

The land use pattern of Bhopal experienced notable and dynamic changes between 2000 and 2020 as depicted in Figure 15.4, mostly marked by major urban growth. Urban regions had a substantial expansion between 2000 and 2010, indicating considerable development endeavors. In 2020, the trend of urbanization persisted, with urban areas expanding and occupying a significant amount of the landscape. This expansion is a result of economic growth and an increasing population. By 2020, forested regions had a certain degree of fragmentation, indicating that urban expansion has encroached to some extent into these areas while overall remaining reasonably steady. The agricultural regions had a significant amount of variation. Although they were expansive in 2000, certain portions of these lands underwent a transition to urban usage by 2010 and continued to do so until 2020. This shift in land use priorities indicates a change from emphasizing agriculture to prioritizing urban development. Open landscapes and water bodies experienced alterations, with certain

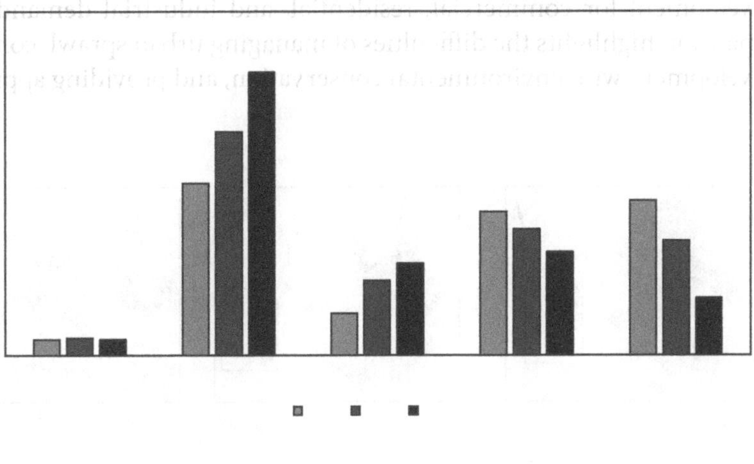

open areas potentially being developed or transformed, while water bodies remained mostly unchanged, suggesting efforts to conserve these natural characteristics amidst urban expansion. In summary, the land use pattern in Bhopal has shown a consistent increase in urban development at the cost of agricultural and forested areas. This emphasizes the importance of implementing balanced development strategies that incorporate sustainable practices to minimize the negative effects on natural and agricultural landscapes.

Due to fast growth, Raipur has changed its land use patterns, which are depicted in Figure 15.5. From 2000 to 2020, the land use patterns in the investigated area have exhibited clear tendencies. Over time, the coverage of water bodies has steadily expanded, whereas vegetation has shown varying patterns with a noteworthy decline. Urban areas have continuously grown, indicating ongoing urbanization. The availability of open space, which serves a variety of uses, has decreased, possibly as a result of its conversion for urban and developmental operations. During the study period, these changes emphasize the dynamic transformation of the environment, urban areas, and development, driven by various variables.

From Figure 15.6 it is evident that water bodies, including rivers and lakes vital to ecological health and human use, increased from 1.7% in 2000 to 3.2% in 2010, then decreased to 2.7% by 2020. Agricultural land, needed to feed the city's rising population, climbed from 25% in 2000 to 36% in 2010 and 45% by 2020. The rise of farming in Raipur may be due to technological advances, agricultural regulations that boost productivity, and rising food demand. Raipur's urban areas grew from 8.1% in 2000 to 10.2% in 2010 and 16% by 2020. The city's increasing urbanization is due to economic expansion, rural population migration, and infrastructure development for commercial, residential, and industrial demands. Such expansion highlights the difficulties of managing urban sprawl, combining development with environmental conservation, and providing appropriate

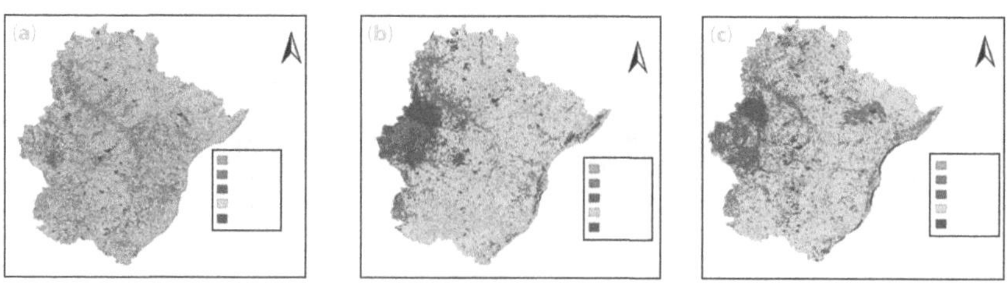

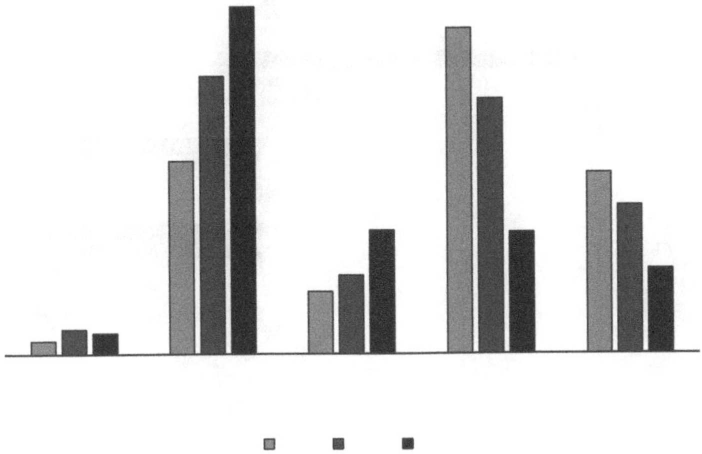

infrastructure and services to urban populations. Open land fell from 42.1% in 2000 to 33% in 2010 and 15.7% by 2020. Raipur's natural habitats, biodiversity, and ecosystem services may be threatened by this reduction, which signals growing urban and agricultural conversion. Raipur's forest cover fell from 23.4% in 2000 to 19.2% in 2010 and 11% by 2020. This decrease highlights forest conservation issues such as deforestation, land use changes for agriculture or infrastructure, and natural disturbances. To protect Raipur's forest ecosystems, conservation and sustainable land management are essential.

2. Change in trend analysis

The below Figure 15.7 shows the changes in land types in Bhopal during the past twenty years indicating substantial transformations in many categories. Water bodies exhibited variations, suggesting endeavours in water management and preservation in response to urban expansion and environmental concerns. The expansion of cultivated/agricultural areas has been consistently increasing, which indicates the intensification of agriculture and the economic motivations behind the drive for food production. Bhopal had rapid urbanization, resulting in its transformation into a vibrant metropolitan hub and a substantial decrease in available open land. The forest cover had a significant decrease, which emphasizes the difficulties in preserving forests in the face of urban growth and the repercussions

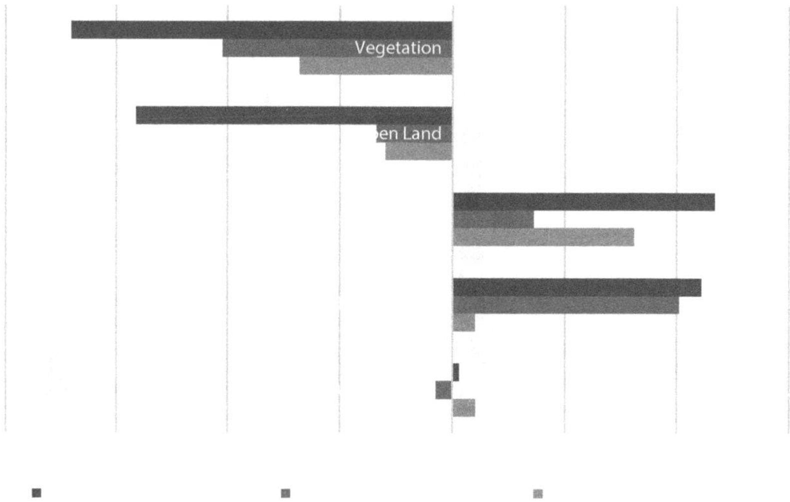

of climate change. The aforementioned modifications highlight the intricate equilibrium required between the growth of metropolitan areas, the advancement of agriculture, and the preservation of the environment in Bhopal's developmental path.

Raipur water bodies showed a positive but moderate net change from 2000 to 2020, as shown in Figure 15.8. Maintaining ecosystem health,

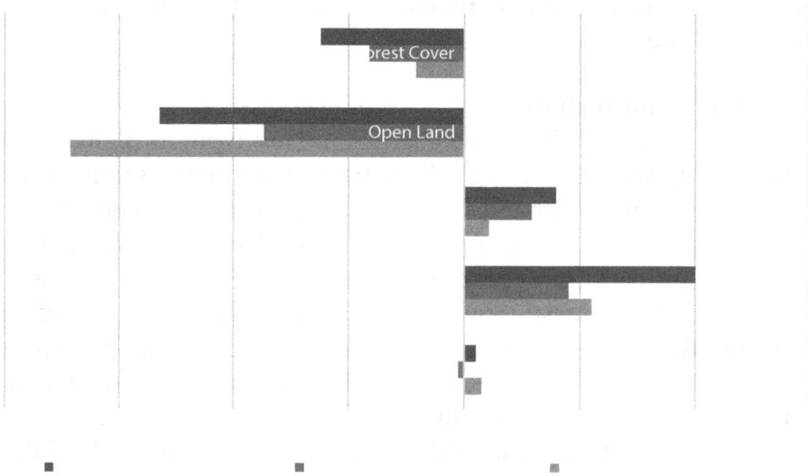

biodiversity, and urban and rural water needs requires sustainable water resource management in the face of urbanization and agricultural intensification. Between 2000 and 2010, there was a large increase in cultivated/agricultural land. Government policies promoting agricultural growth, technological advances in farming, and population increase driving food consumption have contributed to this expansion. Farmers may have expanded land use to boost productivity, causing a rise in cultivated areas. Although slower, urban growth continued from 2010 to 2020. The increase was mainly driven by city growth, special efforts by the government and expenses on real estate and infrastructure. There is a need for sustainable urban planning to maintain growth and keep open areas, land used for farming and natural habitats safe. There has been a major decrease in Raipur's open land over the past two decades, with most of it happening between 2000 and 2010. Mainly, this loss is due to urbanization and increased farming which change open areas into cities or fields. Growing numbers of people, increased economic growth, improved infrastructure and new policies encouraging urban and agricultural growth cause this change. During the past two decades, forest cover in Raipur has decreased because human actions have increased pressure on its forest ecosystem. In the years 2000 to 2010, people cleared forests for farming, building and city building. These practices hurt the local environment by removing forest habitats, lowering biodiversity and lessening the ability of the ecosystem to recover from damage. Between 2010 and 2020, forest cover dropped because of land use adjustments and certain environmental factors. Forest ecosystems suffered when wood was taken illegally, forest fires broke out, climate change occurred, and conservation was neglected. Activities to protect forests should be carried out to reduce forest loss, improve those that are damaged and support green management in Raipur to maintain both biodiversity and ecosystem benefits. It is clear from the Raipur land use changes that urban progress, wider farming activities and protecting the environment all have an impact on each other. To create a city that can last and be strong, designers, farmers and stakeholders must focus on urban planning, farming and managing environmental risks.

15.4 Conclusion

In each of the years 2000, 2010 and 2020, there was rapid urban growth and population rise in Bhopal and Raipur, leading to informal and unofficial settlements. Urbanization is increasing quickly because of population growth and economic reasons. The loss of natural habitats is due to a steady decrease in forests and green areas because of the development of urban and farm areas. Over the years, there has been increasing pressure on land for farming so food can be grown closer to city centers. These cities are showing the strain on land resources caused by growth in the population. Managing land in a wise and environmentally responsible way helps meet the requirements of residential, agricultural, industrial and infrastructure. Coverage of waterbodies has not been affected by urbanization, suggesting people are trying to preserve these precious natural areas. Clear patterns reveal that planning cities and the environment is needed to keep up with progress while protecting nature. Encouraging people to take action and learn about it ensures urban and rural ecosystems thrive into the future.

References

1. Rawat, J.S., Biswas, V., Kumar, M., Changes in land use/cover using geospatial techniques: A case study of Ramnagar town area, district Nainital, Uttarakhand, India. *Egypt. J. Remote Sens. Space. Sci.*, 16, 111–117, 2013, doi: 10.1016/j.ejrs.2013.01.006.
2. Almutairi, A. and Warner, T.A., Change detection accuracy and image properties: a study using simulated data. *Remote Sens.*, 2, 6, 1508–1529, 2010, https://doi.org/10.3390/rs2061508.
3. Pande, A.B., Moharir, K.N., Khadri, S.F.R., Assessment of land-use and land-cover changes in Pangari watershed area (MS), India, based on the remote sensing and GIS techniques. *Appl. Water Sci.*, 11, 6, 2021, doi: 10.1007/s13201-021-01425-1.
4. Tan, K.C., Lim, H.S., MatJafri, M.Z., Abdullah, K., Landsat data to evaluate urban expansion and determine land use/land cover changes in Penang Island, Malaysia. *Environ. Earth Sci.*, 60, 1509–1521, 2010, https://doi.org/10.1007/s12665-009-0286-z.
5. Cao, M., *et al.*, Multi-Scenario Simulation of Land Use for Sustainable Development Goals. *IEEE J. Sel. Top. Appl. Earth Obs. Remote Sens.*, 15, 2119–2127, 2022, doi: 10.1109/JSTARS.2022.3152904.
6. Aburas, M.M., Ahamad, M.S.S., Omar, N.Q., Spatio-temporal simulation and prediction of land-use change using conventional and machine learning models: A review. *Environ. Monit. Assess.*, 191, 205, 2019, https://doi.org/10.1007/s10661-019-7330-6.
7. Liao, X., Huang, X., Huang, W., ML-LUM: A system for land use mapping by machine learning algorithms. *J. Comput. Lang.*, 54, 100908, 2019, https://doi.org/10.1016/j.cola.2019.100908.
8. Chen, Z., *et al.*, Integrating Remote Sensing and a Markov-FLUS Model to Simulate Future Land Use Changes in Hokkaido, Japan. *Remote Sens.*, 13, 13, 2021, doi: 10.3390/rs13132621.

9. Yang, H., Nguyen, T.N., Chuang, T.W., An integrative explainable artificial intelligence approach to analyze fine-scale land-cover and land-use factors associated with spatial distributions of place of residence of reported dengue cases. *Trop. Med. Infect. Dis.*, 8, 4, 238, 2023, https://doi.org/10.3390/tropicalmed8040238.
10. Yusuf, Y., *et al.*, An Assessment of Land Use Land Cover Change Detection in Misau Local Government Area, Bauchi State, Nigeria, in: *5th International New York Academic Research Congress*, pp. 418–427, April 23-24, 2022.
11. Hossain, Md.T., Zarin, T., Sahriar, M.R., Haque, M.N., Machine learning based modeling for future prospects of land use land cover change in Gopalganj District, Bangladesh. *Phys. Chem. Earth Pt. A/B/C/*, 126, 103022, 2022, https://doi.org/10.1016/j.pce.2021.103022.
12. Fonseca, L.M.G., *et al.*, Pattern recognition and remote sensing techniques applied to land use and land cover mapping in the Brazilian Savannah. *Pattern Recognit. Lett.*, 148, 54–60, 2021, https://doi.org/10.1016/j.patrec.2021.04.028.
13. Kumar, S. and Arya, S., Change detection techniques for land cover change analysis using spatial datasets: A review. *Remote Sens. Earth Syst. Sci.*, 4, 3, 172–185, 2021, https://doi.org/10.1007/s41976-021-00056-z.
14. Xu, Q., Wang, Q., Liu, J., Liang, H., Simulation of land-use changes using the partitioned ANN-CA model and considering the influence of land-use change frequency. *ISPRS Int. J. Geo-Inf.*, 10, 5, 346, 2021, https://doi.org/10.3390/ijgi10050346.
15. Woldemariam, G.W., *et al.*, Machine-learning algorithms for land use dynamics in Lake Haramaya Watershed, Ethiopia. *Model. Earth Syst. Environ.*, 8, 3719–3736, 2022, https://doi.org/10.1007/s40808-021-01296-0.
16. Samardžić-Petrović, M., Kovačević, M., Bajat, B., Dragićević, S., Machine learning techniques for modelling short term land-use change. *ISPRS Int. J. Geo-Inf.*, 6, 12, 387, 2017, https://doi.org/10.3390/ijgi6120387.
17. Fichera, C.R., Modica, G., Pollino, M., Land cover classification and change-detection analysis using multi-temporal remote sensed imagery and landscape metrics. *Eur. J. Remote Sens.*, 45, 1, 1–18, 2012, https://doi.org/10.5721/EuJRS20124501.
18. Jayanandini, G., Kathiravan, P., Janani, V., Land use suitability analysis using artificial intelligence and machine learning for developing urban areas, in: *2023 Annual International Conference on Emerging Research Areas: International Conference on Intelligent Systems (AICERA/ICIS)*, pp. 1–8, 2023, Doi 10.1109/AICERA/ICIS59538.2023.10420204.

16

Geospatial Analysis of Land Use Patterns for Enhancing Sustainable Development Initiatives

Abstract

The United Nations has established 17 Sustainable Development Goals (SDGs) focusing on economic, social, and environmental objectives. Goal 11, "Sustainable Cities and Communities," showcases the importance of utilizing land resources properly for development in a resource-limited world. Bad planning has led to serious threats to many communities. This study aims to analyse the change in land use patterns of urban areas, water bodies, agricultural land, open areas, and forest land, with different time durations, 2000, 2010, and 2020, in Indore, Madhya Pradesh, India. Satellite images were obtained from USGS Earth Explorer. Unsupervised classification was carried out to determine change. Based on changes in land use patterns in percentage, one can predict whether a city is trending toward sustainability by analyzing parameters such as increased green spaces, mixed-use development, and the preservation of natural areas. Results show that forest cover has decreased while urban areas have increased. An examination of the rate of urbanization and the decrease in water bodies and forest cover shows the need for important steps to conserve resources, policymaking and look toward sustainability.

Keywords: Land use patterns, remote sensing, sustainable development goals, urban areas, water bodies

16.1 Introduction

In the present scenario, population outbursts and land demand have led to alarming conditions. The global population is in an increasing phase while the majority of the rural population is moving to urbanized areas for better opportunities [1, 2]. This creates threatened situations in urbanized areas regarding finding resources, including the basic needs for living. In addition, some urban areas are being converted to urbanized areas

Kundan Meshram and Chatrabhuj. A Sustainable Approach to Land Use, (393–408) © 2026 Scrivener Publishing LLC

to accommodate population growth. Therefore, sustainable management of land resources is one of the important aspects of the world today and in the future [3]. However, this management is complex due to various unpredictable scenarios like natural disasters. Climate change has directly impacted the severity of natural disasters [4–6]. Recent floods, wildfires, and landslides happen in many countries and verify the impact of climate change.

Nevertheless, technology plays a vital role in dealing with land use resource management. One of the most dominant technologies in monitoring land use is remote sensing technology. With rapid urbanization, these technologies are essential for promoting sustainable land use. This technology allows the local bodies to make informed decision-making based on data supported by scientific evidence. It will create a balance between resource consumption and environmental preservation [7]. For city growth, it is important to optimize infrastructure, minimize resource waste, better planning, and use land use effectively. Geospatial techniques help local governments in supporting sustainable land use, making better, responsible, and progressive practices. It supports the mapping of natural resources, identification of susceptible land, and preservation of natural ecosystems [8].

Remote sensing and geospatial tools are widely used for collection, processing, analysis, and visualization. They are used for effective planning and management of land, resources, forests, agriculture, water bodies, and land use [9, 10]. By considering these data planners, policymakers will make well-informed decisions. This is in support of the preservation of biodiversity, prevention of overexploitation, and utilization of resources effectively.

Analysis of land use patterns is very important to achieve sustainability in areas like forest fires, coastal area erosion, water bodies mapping, urban area cover, agricultural monitoring, crop health management, irrigation planning and many more. It is used to map changes in river lines [11] and monitor land use pattern change due to urbanization [12–14]. It finds its application in hydrological impact [15, 16] in measuring catchment area and discharge in rivers. It can be used in spatial and temporal analysis [17, 18] to predict future land use patterns with the help of satellite images [19] and machine learning techniques to categorize Land Use and Land Cover (LULC) [20, 21].

According to [22], spatial analysis allows the modeling of complex problems which can also be applied to forecasting and monitoring. According to [23], the availability of public services should be viewed as a critical component that influences social engagement and quality of life, which in turn plays a crucial role in understanding socio-spatial inequality [24]. As a result, the government needs an effective system that will aid in analyzing the situation of services today, like education and its advancement. Large tracts of forest cover in India's Himalayan states are being converted to various land uses due to population growth and urbanization, which is having the effect of extreme degradation of the soil [25].

Land Use Functions (LUFs) have gained popularity as a means of encouraging sustainable regional development. Researchers like [26] emphasize the role of land use regulations in regional development plans. For this study seven variables that represent the sustainability of land use were chosen. The variations in the LUF over time show how different areas respond differently to policies and allocate resources differently. From the standpoint of land use, LUF coordination draws attention to the compromises between the economy, society, and ecology. [27, 28] concluded that land use significantly affects the socioeconomic and environmental systems of an area. LULC is determined by ecological and social conditions along with geographical conditions like elevations, steep slopes, and unstable land. Population growth, heavy industrialization and urbanization have changed LULC [29].

Land use in Saudi Arabia was developed [30] featuring land use maps by using remote sensing technology for the years 1990 and 2023. A temporal study was done to detect changes in bare land, built-up areas, rocks, and vegetation cover. Upon analysing satellite images, it was concluded that most of the urbanization occurs at the outside edge of the city. [31] integrated object-oriented categorization techniques with maximum likelihood; the four classifications that were recognized were vegetation, rocky terrain, arid land, and an urban setting. The classification has a high accuracy rate of 88.9% for the years 1986, 1990, 2000, and 2013. It is identified that the percentage of growth of the urban area in Makkah is 174% and for Al-Taif it is nearly 113%. Figure 16.1 shows the relations between remote sensing, agriculture, urbanization and land use, sustainability, and biodiversity conservation.

The interconnection illustrates the multidisciplinary approach in various fields. The application of geospatial tools is critical in shaping policy in support of land use pattern conservation. This data-driven approach facilitates spatial analysis in the analysis of ecological areas, land use regulation supporting conservation efforts, and environmentally friendly policies. Additionally, it assists in creating land use laws and determining zoning areas, helping policymakers to define areas for urban growth, agricultural land, and conservational areas. With time it helps in monitoring land use trends and detecting unlawful activities. It helps in mitigating climate by

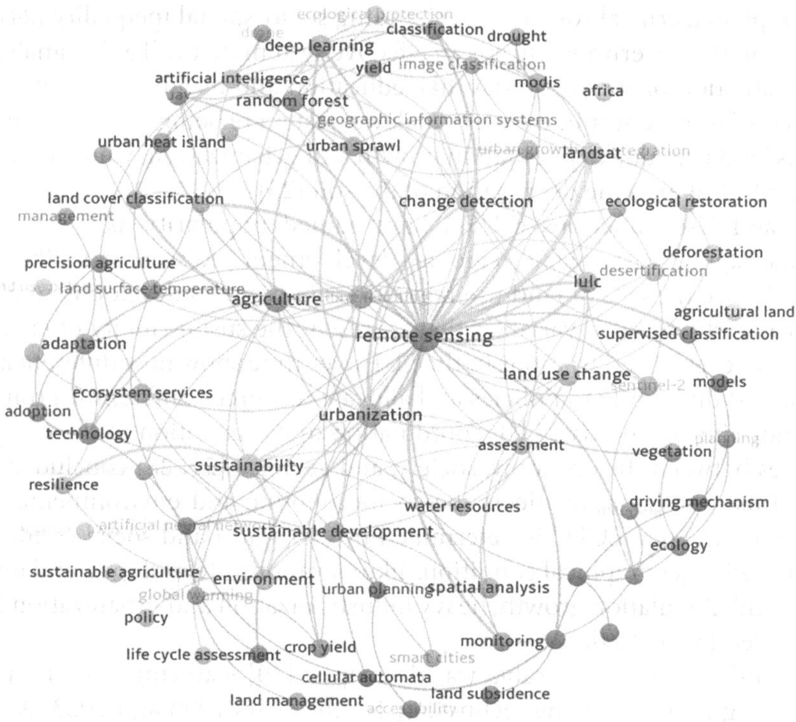

Figure 16.1 Remote sensing and its interconnections.

analysing land degradation and deforestation for sustainable land use. It is essential for sustainable development, offering vital tools for planning and decision-making that drive sustainable growth in diverse sectors [32, 33].

From the above literature, the importance of analysing the land use change patterns in India can be understood. However, such a study has not yet been conducted in Indore, India. Therefore, to overcome the identified research gap, a target area was selected which covers Indore, India, and multi-temporal satellite images of Landsat-7 (TM) and Landsat-8 (OLI) with a resolution of 30 m were extracted. These images were analysed and assessed and are presented in the following sections.

16.2 Study Area and Data

16.2.1 Study Area

Indore is often referred to as the commercial capital of Madhya Pradesh. This thriving urban hub sprawls across the geographical coordinates of

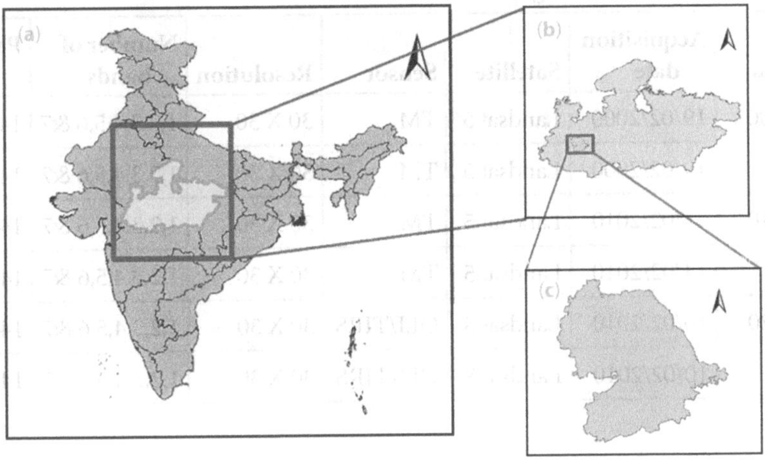

approximately 22°42′N latitude to 22°48′N latitude and from 75°50′E longitude to 76°12′E longitude, covering an area of approximately 530 km^2. Indore holds a distinctive place in the cultural, economic, and educational landscape of the region, with a population of approximately 3.3 million residents, as per the Census of India in 2011. The city experiences a diverse climate, with an average annual rainfall of approximately 1016 mm, with temperature variations ranging from 5.6°C during chilly winter nights to 42.5°C during the peak of summer. Humidity levels range from 31% to 78% during the year. Figure 16.2 shows the study area map of Indore, Madhya Pradesh, India.

16.2.2 Source of Data

To create a land use pattern of Indore, USGS Earth Explorer is the main source of satellite images. To obtain images for the years 2000 and 2010 Landsat TM (Thematic Mapper) 5 were used, and for 2020 Landsat OLI/TIRS (Operational Land Imager/Thermal Infrared Sensor) were used. (The link to USGS is https://earthexplorer.usgs.gov/).

All satellite images were taken in same season (February) to avoid a seasonal effect. It falls in the dry season in Indore, reducing cloud cover and thus ensuring better image quality for analysis. Table 16.1 shows the year, satellite series, resolution, path and row of images used in this study.

Table 16.1 Satellite images for Indore.

Year	Acquisition date	Satellite	Sensor	Resolution	Number of bands	Path/ row
2000	19/02/2000	Landsat 5	TM	30 X 30	1,2,3,4,5,6 &7	146/044
	19/02/2000	Landsat 5	TM	30 X 30	1,2,3,4,5,6 &7	146/045
2010	14/02/2010	Landsat 5	TM	30 X 30	1,2,3,4,5,6 &7	146/044
	14/02/2010	Landsat 5	TM	30 X 30	1,2,3,4,5,6 &7	146/045
2020	10/02/2010	Landsat 8	OLI/TIRS	30 X 30	1,2,3,4,5,6 &7	146/044
	10/02/2010	Landsat 8	OLI/TIRS	30 X 30	1,2,3,4,5,6 &7	146/045

16.3 Methodology

Figure 16.3 showcases the overall methodology of the research carried out. It illustrates the selection of Landsat satellite series 5 and 8 for image acquisition for the years 2000, 2010 and 2020. The acquired images underwent atmospheric and radiometric correction processes, followed by the creation of composite images through the fusion of multiple bands, from

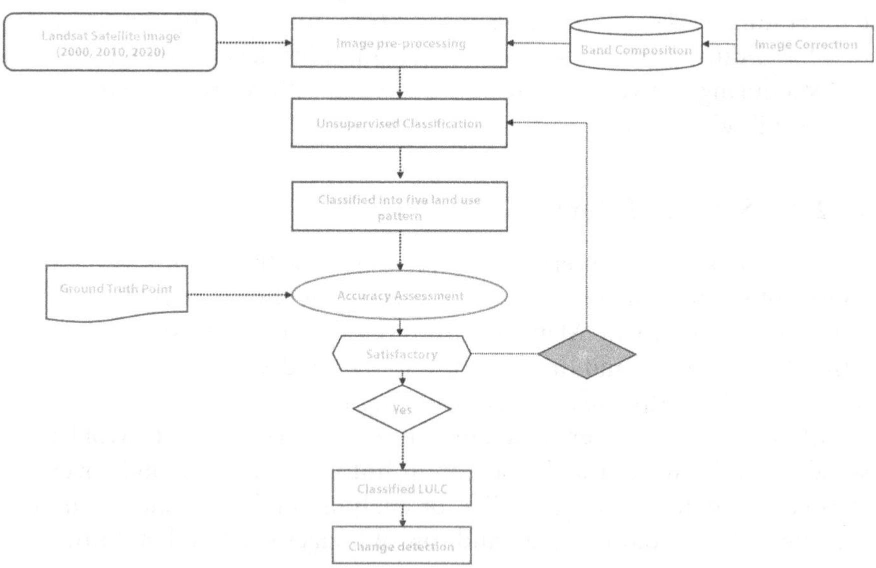

1 to 7. Then the composite images were trimmed as per the administrative shape area of Indore. Subsequently, the composite band was subject to land use pattern classification using an unsupervised ISO classification system. Once the images were classified into land use, they were broadly categorized into five types: water bodies, urban land, agricultural land, open areas and vegetation cover. To validate the classification accuracy, Google Earth imagery was utilized, and if the accuracy met the desired threshold, the classification was approved; otherwise, the entire process was repeated. The classified LULC map was employed to assess change detection for each class and year. Policy decisions were then based on the observed changes in land use.

For accuracy assessment, 500 random points were generated in software then a KML file was made which was exported to Google Earth Pro. Then each point is visually confirmed leading to formation of Kappa coefficient.

$$OA = \frac{\sum D_{ii}}{N} \tag{16.1}$$

Where, OA is overall accuracy, $\sum D_{ii}$ is the total number of correctly classified pixels and N is the total number of pixels in the error matrix.

$$PA = \frac{\sum D_{ij}}{R} \tag{16.2}$$

Where PA is Producers accuracy, D_{ij} is the Number of correctly classified pixels in row and R is the total number of pixels in row

$$UA = \frac{\sum D_{ij}}{C} \tag{16.3}$$

Where UA is the user's accuracy, D_{ij} is the Number of correctly classified pixels in Column and C is the total number of pixels in column.

$$K = \frac{OA - P_e}{1 - P_e} \tag{16.4}$$

Where k is kappa coefficient, OA is user accuracy and Pe is the proportion of agreement expected by chance, calculated based on row and column totals of the confusion matrix.

The change in Sq-Km for different LULC classes were calculated for each period. The percentage of LULC change (trend) for individual LULC classes was calculated using the following formula eq 16.5.

$$\text{Percentage of LULC change} = \frac{(A_2 - A_1)}{A_1} X100 \qquad (16.5)$$

Where, $A1$ and A2 area of LULC type in the initial year and final year, respectively.

16.4 Results and Discussion

16.4.1 Land Use Pattern

Land use was further classified into five types: open area, vegetation, urban, agricultural and water bodies, which are indicated with a particular colour matrix. The blue colour is for water bodies, yellow is for agriculture/cultivated land, red shows urban land/build-up/constructed area, green denotes vegetation cover/forest and purple shows open land/barren/bare land. This classified map serves as a visual narrative of the ever-changing landscape of the study area, offering insights into how urbanization, environmental factors, and agricultural activities have evolved over time. Different land use patterns of the Indore city are given in Figure 16.4 for the years 2000, 2010 and 2020.

From the classified land use pattern of Indore, it is clear that water bodies are very low. Vegetation cover is diminishing with time and agricultural land is most predominant, followed by urban land which is increasing with each decade, which is at the eastern, western, and northern parts of the city. A noticeable trend is that urban areas are expanding at the outskirts of the urban area in the form of a cluster, in the central part of the city. Open lands, which are at the southern region of the city, have a net decrease from 2000 to 2020. So, policy should be inclined towards conservation of water bodies and minimizing urban growth rate. These findings underscore the dynamic nature of land use changes in the studied area, with implications for policy decisions and future planning.

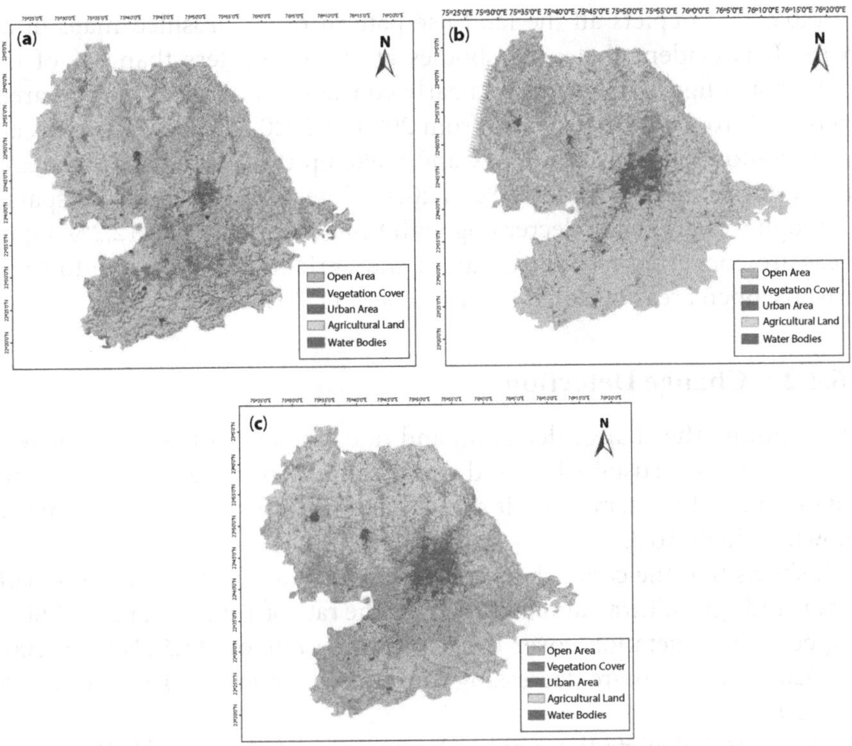

Figure 16.4 Classified map of Indore (a) 2000 (b) 2010 (c) 2020.

Table 16.2 Land use pattern.

Land use class	2000		2010		2020	
	Area (sq-km)	Area (%)	Area (sq-km)	Area (%)	Area (sq-km)	Area (%)
Water bodies	109.94	2.82	113.39	2.91	112.67	2.89
Urban Area	459.96	11.8	549.52	14.1	666.26	17.1
Agriculture Land	1873.04	48.1	2072.14	53.2	2185.08	56.1
Vegetation Cover	705.94	18.1	529.73	13.6	475.56	12.2
Open Areas	748.42	19.2	611.49	15.7	455.97	11.7
Total	3898.00	100	3898.00	100	3898.00	100

Table 16.2 depicts all the land use pattern of all classified maps of all years. It is evident that water bodies are very low, less than 3% of the total area, a figure that remains nearly constant for all years. Urban areas increased from 11.8% to 17.1% from 2000 to 2020. Agricultural areas are most predominant in Indore city and make up the majority of the area—nearly 48.1% in 2000 and 56.1% in 2020. Vegetation cover is very sparse, although it is gradually decreasing with time, from 18.1% to 12.2%. Open areas are converted to built-up areas and agricultural areas due to these reasons; open areas declined from 19.2% to 11.7%.

16.4.2 Change Detection

Determining the change detection and rate of change of land use pattern is important for sustainable land use planning, urban development, and environmental conservation. It shows how land use changes over time, as shown in Table 16.3.

It shows that the cover change is nearly constant for water bodies, with urban and agricultural land increasing at the rate of +10.32% and +15.60%, respectively. Vegetation cover is declining at a rate of -11.52% from 2000 to 2020. Open areas are decreasing over time at a rate of -14.6% from 2000 to 2020.

Figure 16.5 depicts the graphical represents of change of LULC of the city. It is evident that for urban areas there is rapid expansion of 19.47% from 2000 to 2010 and 21.24% from 2010 to 2020, resulting in a 44.85% increase over two decades. This is due to population growth, migration,

Table 16.3 Change in land cover analysis.

LULC	Cover change (%)			Rate of change gain (+)/ loss (-)		
	2010-2000	2020-2010	2020-2000	2010-2000	2020-2010	2020-2000
Water bodies	3.14	-0.63	2.48	0.35	-0.07	0.14
Urban Area	19.47	21.24	44.85	8.96	11.67	10.32
Agriculture Land	10.63	5.45	16.66	19.91	11.29	15.60
Vegetation Cover	-24.96	-10.23	-32.63	-17.62	-5.42	-11.52
Open Areas	-18.3	-25.43	-39.08	-13.69	-15.55	-14.62

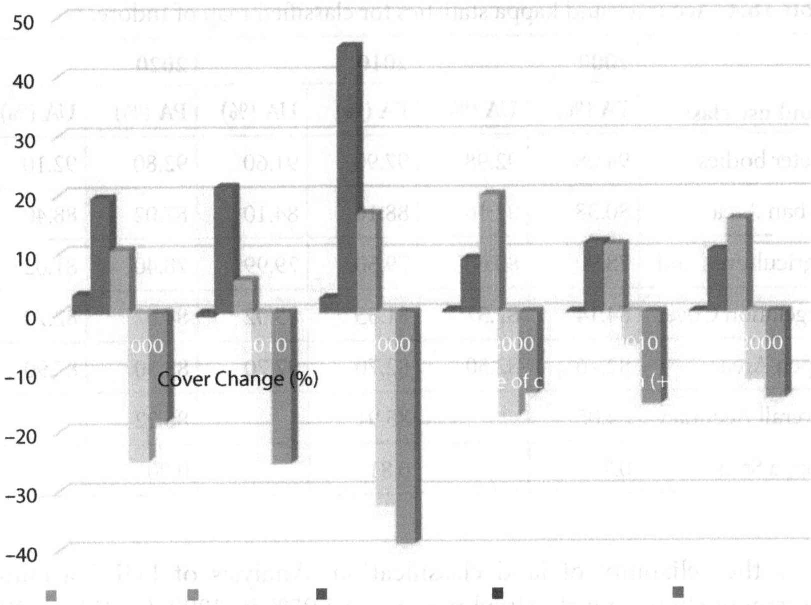

and industrialization. This growth is due to the loss of agricultural land and vegetation cover. In agricultural land there is a decline of 19.91% per decade (2000–2010) to 11.29% per decade (2010–2020), indicating a slowdown in agricultural expansion. Initial expansion shows the high rate of conversion of forest and open area to cultivated land; apparently the slowdown indicates urbanization pressure affecting agriculture. Vegetation cover shows a loss of -17.62% per decade (2000–2010) to -5.42% per decade (2010–2020), but overall negative. Deforestation is a major issue, due to agriculture and infrastructure development; urban sprawl and rate of loss is slowing, affecting local biodiversity and ecosystem. Open areas are declining by a rate of -18.3% from 2000 to 2010 and even more, by -25.43%, from 2010 to 2020, making an overall -39.08% loss over 20 years. Open space in Indore city are barren lands which are rapidly disappearing due to urbanization and agricultural expansion.

16.4.3 Accuracy Assessment

Table 16.4 shows the Producer's Accuracy (PA) and User's Accuracy (UA) for different land use classes, along with overall accuracy and Kappa statistics for classified maps of Indore over three decades. These metrics help

Table 16.4 Accuracy and kappa statistics for classified map of Indore.

	2000		2010		2020	
Land use class	**PA (%)**	**UA (%)**	**PA (%)**	**UA (%)**	**PA (%)**	**UA (%)**
Water bodies	94.08	92.98	97.90	91.60	92.80	92.10
Urban Area	80.38	82.16	88.10	84.10	87.02	88.40
Agriculture Land	78.90	80.60	79.50	79.99	78.40	81.02
Vegetation Cover	84.14	82.37	81.35	85.02	84.62	82.70
Open Areas	82.76	82.50	82.70	87.70	88.30	87.90
Overall Accuracy	84.05		85.91		86.22	
Kappa Statics	0.78		0.81		0.80	

assess the reliability of land classification. Analysis of LULC accuracy assessment shows the classified map was 84.05% in 2000, 85.91% in 2010 and 86.22% in 2020. Kappa statistics of 0.78, 0.81 and 0.81, respectively.

16.4.4 SDG Alignment

The study's findings provide significant insight into land use changes and their implications for achieving SDGs. In the case of SGD 11, it is evident that urban areas have expanded over three decades, indicating rapid urbanization. Green areas like forest cover and open spaces are declining, raising concerns of inclusivity and resilience in the city. Loss of green space and rapid urbanization will reduce the quality of air and increase the average city temperature, leading to discomfort and lowering community well-being. There is a need to promote suitable mixed-use development that integrates green space in urban areas and prefers green infrastructure in urban planning.

SDG 13 is related to climate action. Degradation in forest cover, which impacts carbon sequestration capability, and urbanization have increased greenhouse gas emissions. The study showcases the urgent need for reforestation and afforestation practices. Developers must include land restoration and conservation strategies. SDG 15's life on land findings suggests that a decline in forest cover significantly reduces the natural habitat and local biodiversity. Conservation strategies should be included for restoration of degraded land.

Results of the analysis show that the water bodies are stressed due to overconsumption. Regarding an aspect of SDG 6, clean water, increased urbanization has led to water pollution and high water demand. There is a need for ensuring the availability of water and sanitation, so it becomes essential for the city to protect and restore water bodies. Policymakers and urban planners must incorporate water retention design and technologies in city planning. With urban expansion, there is a need to support population demands. Lack of mixed-use development can lead to inefficient transportation and resource distribution, affecting SDG 9. To address this, resilient infrastructure planning should be incorporated. Innovating urban design should be promoted to balance industrial development and environmental preservation. Unregulated land utilization has led to the exploitation of natural resources, so unchecked urban development has resulted in wasteful land consumption patterns. Urban planners must encourage more efficient and responsible land use in alignment with SDG 12 for responsible consumption and production of land resources. Policies should be targeted to sustainable agriculture, efficient land management and a circular economy.

16.5 Conclusion

This study involved utilizing the USGS Earth Explorer to access a variety of satellite images, each comprising different spectral bands. These bands were skilfully combined to create composite images that served as essential inputs for our analysis. Employing an unsupervised classification methodology, we systematically categorized the land use patterns within these cities for the years 2000, 2010, and 2020. These land uses were further classified into five distinct classes: urban land, agricultural area, water bodies, forest area and open space. This helps to gain valuable insight into the evolving landscape and change dynamics over three decades.

From analysis, it is evident that water bodies are consistently decreasing. This decline in available water resources indicates the need for water conservation. In Indore city, there is rapid expansion of urban areas, as noted, with time. This urban growth is due to rapid urbanization on a large scale; it poses challenges in infrastructure development and sustainable urban planning. Open, bare and exposed areas decrease with time over the edges, and the central region is open rocks that cannot be changed. Agricultural land increases with time.

The utilisation of the geospatial approach offers a clear insight into land use patterns; its changes and trend of dynamics over time. This technology supports sustainable development and decision-making over the long term. It aids the government, policymakers, authorities and researchers to achieve a balance between land use sprawl and land use conservation.

Incorporating artificial intelligence and machine learning methods can be investigated in future research to improve LULC classification. Furthermore, data from other sources like environmental and social factors can be incorporated into sustainable land use development. To understand insight into land use patterns multi-scenario simulation can be used and associated factors must be analysed.

References

1. Delazeri, L.M.M., Da Cunha, D.A., Oliveira, L.R., Climate change and rural–urban migration in the Brazilian Northeast region. *GeoJournal*, 87, 3, 2159–2179, 2021, https://doi.org/10.1007/s10708-020-10349-3.
2. Selod, H. and Shilpi, F., Rural-urban migration in developing countries: Lessons from the literature. *Reg. Sci. Urban Econ.*, 91, 103713, 2021, https://doi.org/10.1016/j.regsciurbeco.2021.103713.
3. Li, M., Cao, X., Liu, D., Fu, Q., Li, T., Shang, R., Sustainable management of agricultural water and land resources under changing climate and socio-economic conditions: A multi-dimensional optimization approach. *Agric. Water Manage.*, 259, 107235, 2021, https://doi.org/10.1016/j.agwat.2021.107235.
4. Mallucci, E., Natural disasters, climate change, and sovereign risk. *J. Int. Econ.*, 139, 103672, 2022, https://doi.org/10.1016/j.jinteco.2022.103672.
5. Cappelli, F., Costantini, V., Consoli, D., The trap of climate change-induced 'natural' disasters and inequality. *Global Environ. Change*, 70, 102329, 2021, https://doi.org/10.1016/j.gloenvcha.2021.102329.
6. Sloggy, M.R., Suter, J.F., Rad, M.R., Manning, D.T., Goemans, C., Changing opinions on a changing climate: The effects of natural disasters on public perceptions of climate change. *Clim. Change*, 168, 3–4, 2021, https://doi.org/10.1007/s10584-021-03242-6.
7. Bhattacharya, R.K., Das Chatterjee, N., Das, K., Land use and land cover change and its resultant erosion susceptible level: An appraisal using RUSLE and Logistic Regression in a tropical plateau basin of West Bengal, India. *Environ. Dev. Sustainability*, 23, 2, 1411–1446, 2021, https://doi.org/10.1007/s10668-020-00628-x.
8. Niquisse, S., Cabral, P., Rodrigues, A., Augusto, G., Ecosystem services and biodiversity trends in Mozambique as a consequence of land cover change. *Int. J. Biodivers. Sci. Ecosyst. Serv. Manag.*, 13, 297–311, 2017, https://doi.org/10.1080/21513732.2017.1349836.

9. Makumbura, R.K., Samarasinghe, J., Rathnayake, U., Multidecadal land use patterns and land surface temperature variation in Sri Lanka. *Appl. Environ. Soil Sci.*, 2022, 1–11, 2022, https://doi.org/10.1155/2022/2796637.
10. Alshari, E.A. and Gawali, B.W., Development of a classification system for LULC using remote sensing and GIS. *Global Trans. Proc.*, 2, 1, 8–17, 2021, https://doi.org/10.1016/j.gltp.2021.01.002.
11. Duong, P.C., Nauditt, A., Nam, H., Phong, N.T., Assessment of climate change impact on river flow regimes in The Red River Delta, Vietnam—A case study of the Nhue-Day River Basin. *J. Nat. Resour. Dev.*, 6, 81–91, 2016, https://doi.org/10.1016/s0034-4257(03)00079-8.
12. Jin, S., Yang, L., Zhu, Z., Homer, C., A land cover change detection and classification protocol for updating Alaska NLCD 2001 to 2011. *Remote Sens. Environ.*, 195, 44–55, 2017, http://dx.doi.org/10.1016/j.rse.2017.04.021.
13. Lv, Z., Liu, T., Wan, Y., Benediktsson, J.A., Zhang, X., Post-processing approach for refining raw land cover change detection of very high-resolution remote sensing images. *Remote Sens.*, 10, 472, 2018, https://doi.org/10.3390/rs10030472.
14. Wu, T., Luo, J., Fang, J., Ma, J., Song, X., Unsupervised object-based change detection *via* a Weibull mixture model-based binarization for high-resolution remote sensing images. *IEEE Geosci. Remote Sens. Lett.*, 15, 63–67, 2018, https://doi.org/10.3390/rs15071770.
15. Gashaw, T., Tulu, T., Argaw, M., Worqlul, A.W., Modeling the hydrological impacts of land use/land cover changes in the Andassa watershed, Blue Nile basin, Ethiopia. *Sci. Total Environ.*, 619–620, 1394–1408, 2018, https://doi.org/10.1016/j.scitotenv.2017.11.191.
16. Guzha, A.C., Rufino, M.C., Okoth, S., Jacobs, S., Nóbrega, R.L.B., Impacts of land use and land cover change on surface runoff, discharge and low flows: Evidence from East Africa. *Hydrol. Reg. Stud.*, 15, 49–67, 2018, https://doi.org/10.1016/j.ecolind.2023.110969.
17. Alawamy, J.S., Balasundram, S.K., Mohd Hanif, A.H., Boon Sung, C.T., Detecting and analyzing land use and land cover changes in the region of Al-Jabal Al-Akhdar, Libya using time-series Landsat data from 1985 to 2017. *Sustainability*, 12, 11, 4490, 2020, https://doi.org/10.3390/su12114490.
18. Ghayour, L., *et al.*, Performance evaluation of Sentinel-2 and Landsat 8 OLI data for land cover/use classification using a comparison between machine learning algorithms. *Remote Sens. (Basel)*, 13, 7, 1349, 2021, https://doi.org/10.3390/rs13071349.
19. Sang, X., *et al.*, The effect of DEM on the land use/cover classification accuracy of Landsat OLI images. *J. Indian Soc. Remote Sens.*, 49, 7, 1507–1518, 2021, https://doi.org/10.1007/s12524-021-01318-5.
20. Jia, K., Wei, X., Gu, X., Yao, Y., Xie, X., Li, B., Land cover classification using Landsat 8 Operational Land Imager data in Beijing, China. *Geocarto Int.*, 29, 8, 941–951, 2014, http://dx.doi.org/10.1080/10106049.2014.894586.

21. Alqadhi, S., *et al.*, Spatial and decadal prediction of land use/land cover using multi-layer perceptron–neural network (MLP-NN) algorithm for a semi-arid region of Asir, Saudi Arabia. *Earth Sci. Inf.*, 14, 3, 1547–1562, 2021, https://doi.org/10.1007/s12145-021-00633-2.
22. Kowalczyk, A.M., Bajerowski, T., Ogrodniczak, M., A method for hazard mapping illustrated with an example of fire service interventions. *Geomatics Environ. Eng.*, 14, 3, 21, 2020, https://doi.org/10.7494/geom.2020.14.3.21.
23. Martens, K. and Golub, A., Using principles of justice to assess the modal equity of regional transportation plans. *J. Transp. Geogr.*, 41, 10–20, 2014, https://doi.org/10.1016/j.jtrangeo.2014.07.014.
24. Farrington, J. and Farrington, C., Rural accessibility, social inclusion and social justice: Towards conceptualization. *J. Transp. Geogr.*, 13, 1–12, 2005, https://doi.org/10.1016/j.jtrangeo.2004.10.002.
25. Sharma, R., Xu, J., Sharma, G., Traditional agroforestry in the eastern Himalayan region: Land management system supporting ecosystem services. *Trop. Ecol.*, 48, 2, 189–200, 2007.
26. Gao, Y., Wang, Z., Xu, F., Geospatial characteristics and the application of land use functions in the Yangtze River Economic Belt, China: Perspectives on provinces and urban agglomerations. *Ecol. Indic.*, 155, 110969, 2023, https://doi.org/10.1016/j.ecolind.2023.110969.
27. Read, J.M. and Lam, N.S.N., Spatial methods for characterising land cover and detecting land-cover changes for the tropics. *Int. J. Remote Sens.*, 23, 12, 2457–2474, 2002, https://doi.org/10.1080/01431160110106140.
28. Petit, C., Scudder, T., Lambin, E., Quantifying processes of land-cover change by remote sensing: Resettlement and rapid land-cover changes in south-eastern Zambia. *Int. J. Remote Sens.*, 22, 17, 3435–3456, 2001, https://doi.org/10.1080/01431160010006881.
29. Voogt, J.A. and Oke, T.R., Thermal remote sensing of urban climates. *Remote Sens.*, 86, 3, 370–384, 2003, https://doi.org/10.1016/S0034-4257(03)00079-8.
30. Darem, A.A., *et al.*, Development of a map for land use and land cover classification of the Northern Border Region using remote sensing and GIS. *Egypt. J. Remote Sens. Space. Sci.*, 26, 341–350, 2023, https://doi.org/10.1016/j.ejrs.2023.04.005.
31. Alqurashi, A.F. and Kumar, L., Land use and land cover change detection in the Saudi Arabian Desert cities of Makkah and Al-Taif using satellite data. *Adv. Remote Sens.*, 3, 3, 106–119, 2014, https://doi.org/10.1016/j.ecolind.2023.110969.
32. Linda Theres, B. and Selvakumar, R., Comparison of land use/landcover classifier for monitoring urban dynamics using spatially enhanced Landsat dataset. *Environ. Earth Sci.*, 81, 5, 1–8, 2022, https://doi.org/10.1007/s12665-022-10242-x.
33. Mishra, P.K., Rai, A., Rai, S.C., Land use and land cover change detection using geospatial techniques in the Sikkim Himalaya, India. *Egypt. J. Remote Sens. Space. Sci.*, 23, 133–143, 2020, https://doi.org/10.1016/j.ejrs.2019.02.001.

17

Assessment of Urban Cities by Analysing Land Use Patterns *via* Satellite Images for Sustainable Development Goals

Abstract

The United Nations' Sustainable Development Goals (SDGs) include 17 objectives, of which Goal 11 focuses on Sustainable Cities and Communities. There is a need to measure the sustainability of the city based on UN and land use patterns to ensure balanced urban growth, resource distribution and resilience to environmental and socioeconomic factors. The study formulates a Composite Sustainability Index (CSI) incorporating economic, social, land use, and environmental parameters to assess urban sustainability. Score was calculated for cities like Bhopal, Indore, and Raipur for the years 2000, 2010, and 2020. To determine the land use pattern, Satellite images from Landsat 7 and 8 were used. The CSI scored from 1 to 100; the higher the value, the higher the sustainability. Results from the analysis show that the natural areas are declining due to human activities of urbanization and agricultural activities. Forest areas, with time, are covered in agriculture and later on in urban areas. This trend, along with social, environmental and financial parameters, has affected the sustainability of a city. To analyse the future trend, a CSI was projected for the future years of 2030 and 2040 for all the cities. This demonstrated a rise in urban areas and a decline in forest and open areas, which will aid policymakers, planners and government to assess the current development and take informed decisions and policy development to align the city growth with sustainable development.

Keywords: Cities, composite sustainability index, decision making, environmental monitoring, land use pattern, satellite images, sustainability

17.1 Introduction

United Nations 17 Goals focus on sustainability in multi-dimensional forms like economic, social and environmental aspects. It includes financial equality, conservation of environment, upliftment of living conditions,

Kundan Meshram and Chatrabhuj. A Sustainable Approach to Land Use, (409–450) © 2026 Scrivener Publishing LLC

reducing poverty in all forms, etc. Goal 11 focuses on sustainable and smart cities. It addresses the challenges and opportunities presented by rapid urbanization [1]. This goal underscores the need to create inclusive, safe, resilient, and sustainable cities that enhance the quality of life for residents. By promoting environmentally conscious urban planning, inclusive development, resilience to disasters, cultural heritage preservation, and sustainable transportation. SDG 11 aims to create cities that are not only vibrant but also equitable and environmentally responsible. The goal's measurable targets provide a framework for assessing progress and fostering global collaboration, recognizing the interconnected nature of urban development with broader sustainable development objectives [2]. Ultimately, the successful implementation of SDG 11 is instrumental in shaping a more sustainable and resilient future for urban areas worldwide.

Developing one SD Goal will also contribute to progress in other SDGs, as represented in Figure 17.1, ultimately advancing the collective well-being of humanity.

Recent landscape changes are predominantly attributed to man-made factors, with increased industrialization and urbanization playing a significant role. While urbanization is often considered an indicator of economic

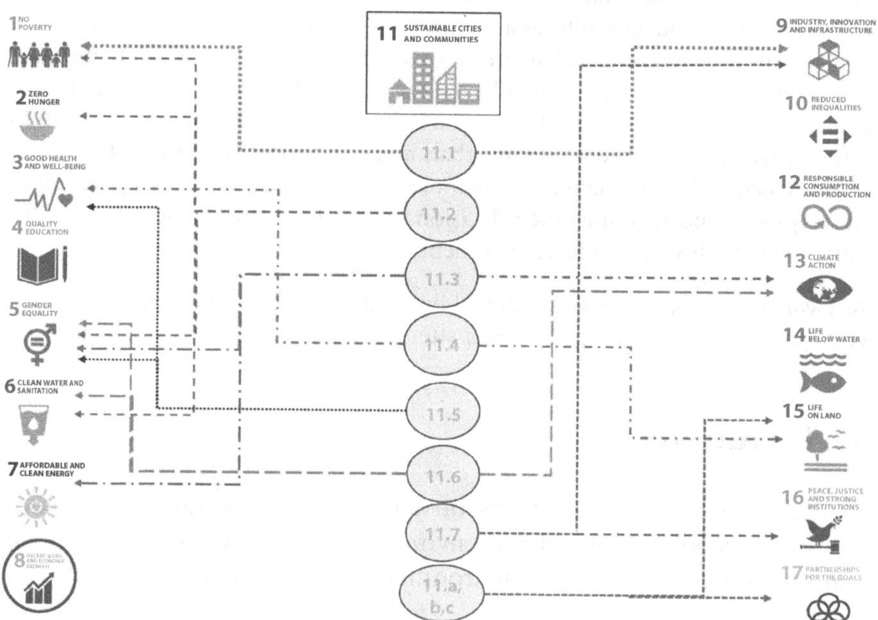

success and growth, it also brings about long- and short-term negative impacts on city development. Rapid urbanization profoundly influences land use patterns, encompassing the conversion of natural areas into urban infrastructure, expansion of built-up areas encroaching upon agricultural land, changes in land use mix, infrastructure development requiring land conversion, increased impervious surfaces altering local hydrology, intensification of land use with high-rise structures, and the potential for social and economic segregation. This transformation, driven by the demand for housing, amenities, and economic activities, not only alters the physical landscape but also strains infrastructure and poses environmental challenges. Effective urban planning and sustainable development strategies are crucial for mitigating the negative impacts of rapid urbanization, preserving natural resources, and fostering resilient and inclusive urban environments.

The UN predicts that world urbanization will rise from 55% in 2019 to 68% by 2050, in which 80% of the total population with reside in cities. This contributes to congestion, resource inequality, unregulated settlement, economic disparity, and the emission of greenhouse gases [3]. Rapid urbanization is due to better opportunities, better facilities, higher standard of living, industrialization, education and health care. Urbanization trends are more dominant in developing countries with high populations like India [4].

To map the rate of urbanization and its impact, land use patterns are an efficient tool. It is crucial to identify land use and land use change dynamics for multiple applications, such as changes in land patterns, floodplain detection, agriculture and deforestation, water bodies, disaster management, urbanization, coastal area erosion, and the analysis of wildfires. To analyze and investigate all these problems on a large scale, one requires satellite data. With the help of remote sensing technology, which is a technology to acquire information about an object without any physical contact, it is possible to extract data from satellite images that are obtained from satellite series like the Landsat series. Extracting features from satellite images, such as built-up areas, forest cover, cultivated land, water areas, open land, river plains, and vegetation cover, becomes a challenging task. To overcome these issues, Artificial Intelligence (AI) is used in image classification, analysis, and processing [5]. With classified features, one can easily visualize the change in the trend of different land use cover.

Considering socioeconomic data based on UN SD Goal Indicators, and incorporating land use patterns, an index can be made to measure the sustainability of a city. The measurement of sustainability is a crucial practice for evaluating the sustainability practices of a city. The primary parameters of sustainability encompass the realms of economics, environmental and social aspects [6]. There are several notable benefits associated with measuring sustainability:

- Policy Guidance for Climate Action: CSI will aid governments, policymakers, and organizations in setting better sustainable targets to tackle real-time challenges of pollution, waste management, and carbon emissions.
- Knowledge Sharing for Smart Cities: By evaluating sustainable data, cities will learn from efficient and successful models, like waste-to-energy model. It also helps in sharing successful strategies and possible solutions for sustainable development.
- Aligning Local Actions with Global Goals: Measuring the sustainability of a city helps to evaluate local standards with international standards. It also enables local government to evaluate local individual parameters.
- Competitive Comparison: It helps to compare the development of a city with another city with respect to the individual parameters of environment, social and financial aspects and overall sustainable development. It will foster better, healthy competition, and it will encourage improvements.

17.2 Literature Review

Land use/land cover deterioration is an environmental issue that leads to loss of landscape, disintegration of habitat, and climate irregularities. The major cause of environmental damage is the unregulated and unplanned growth of cities. Landscape metrics have been linked as an effective and valuable technique for monitoring, identifying, and analyzing the trend of change in the landscape. It gives a quantitative assessment of various landscape features and patterns [7, 8]. The combination of urban development research, land use change detection modelling, and geographic indicators benefits us in gaining insight into landscape knowledge and its driving variables. The combination of remote sensing technology with GIS aids in the investigation of changing landscape patterns, as depicted in Figure 17.2.

There has been a massive movement from rural to urban regions in recent decades. By 2050, it is expected that 67.5% of the world's population will be residing in urban land. With an expanding number, urban

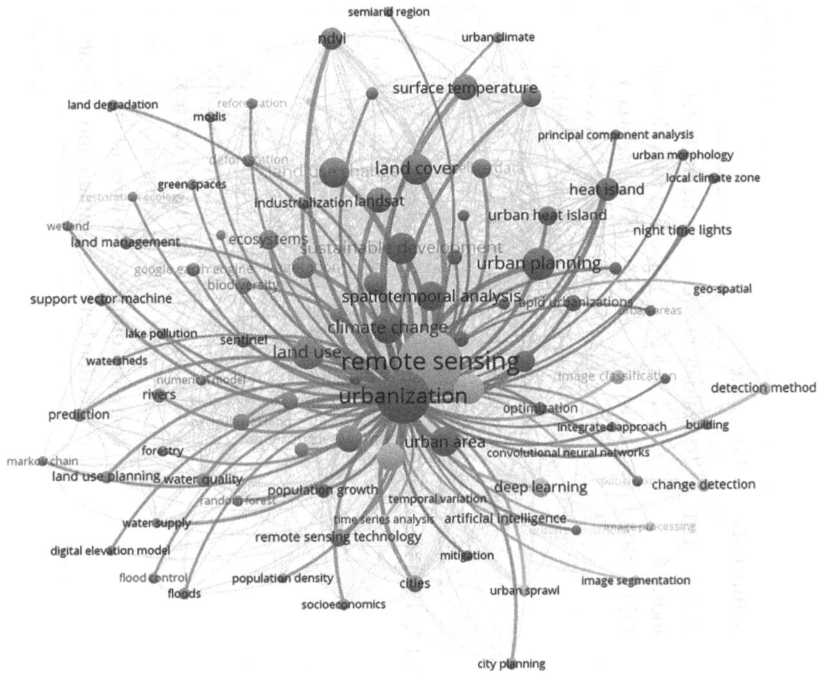

Figure 17.2 Interlinks between keywords.

expansion is inextricably linked to the need for sustainable development [9]. To address these global issues SDG 11, "Sustainable Cities and Communities," is an initiative to make cities more inclusive, safe, resilient, and sustainable within the larger framework of the 2030 Agenda [10–12]. Nonetheless, it is acknowledged that all other SDGs will also need to be met. As a result, there is increased interest in quantifying urban sustainability. Sustainable urban development evaluation approaches, such as projects, indexes, methodology, and tools, have become an active study subject [13]. Moreover, the bulk of these solutions have been concentrated on many environmental issues such as energy efficiency, waste recycling, clean environment, renewable resources, and carbon emission reduction [14, 15].

Satellite imagery with high resolution is now being utilized internationally by various governmental and many research organizations to monitor LULC changes as shown in Table 17.1. RapidEye imagery can offer more precise estimates of LULC changes than other satellite pictures (Landsat, MODIS, Sentinel, and so on), which can aid in identifying effective environmental protection methods. To construct effective land use management plans,

Table 17.1 Analysis of land use pattern.

Author	Area	Land type	Year	Satellite data	Outcome
[16]	Rib watershed from Ethiopia	Cultivated, waterbodies, shrubland, grassland, settlement	2000, 2010, 2020	Landsat 5 TM, Landsat 8 OLI_TIRS	To evaluate how changes in land use/ land cover (LULC) affect the values of ecosystem services (ESVs). Between 2000 and 2020, the value of ESVs lost was about US$ 9.3 million (13.5%). The specific ecosystem services have also been impacted by the noted LULC changes throughout this time.
[17]	Kaziranga National Park, Assam, India,	Water, River Bed, Woodland, Open Land, Agriculture, Degraded, Plantation, Settlement, Grassland	2002, 2013	IRS-P6, LISS- IV	A study on land use, land cover change, and forest fragmentation was conducted to identify the types of human disturbances to the forest cover inside the corridor complexes. Detailed field survey techniques along with high-definition imagery from satellites from 2002 and 2013 were used to record the current land use and cover patterns of the corridor complexes, to forecast an area of protection between humans and animals.

(Continued)

Table 17.1 Analysis of land use pattern (*Continued*)

Author	Area	Land type	Year	Satellite data	Outcome
[18]	Pishin, Balochistan	Built-up area, vegetation, water bodies, bare surfaces	1992, 2003 2013	Landsat- 5 TM Landsat-7 ETM Landsat-8 OLI	Multi-spectral photography may be used to identify changes in the geographical and temporal extents of land use and cover with great effectiveness. The rise in vegetation and built-up area that replaced the arid terrain in the research region was blamed for the decrease in bare surfaces.
[19]	Ashanti Region, Ghana	Bare, closed forest, cropland, open forest, savanna, settlement, water	1986, 2003, 2007, 2019	Landsat-5 TM Landsat-7 ETM+ Landsat 8 OLI_TIRS	For the period 1986 to 2019, LULC changes within the Atwima were classified using Landsat imagery. It was discovered that forest cover was steadily declining. Due to the area's rapid urbanization and high immigration rates, LULC has also had a detrimental effect on livelihoods by reducing the amount of forest cover and raising housing costs.

(*Continued*)

Table 17.1 Analysis of land use pattern. (*Continued*)

Author	Area	Land type	Year	Satellite data	Outcome
[20]	Yellow River Delta, China	Natural wetland, Constructed wetland, Non-wetland	2000, 2005 2010	HJ/CCD Landsat-5 TM Landsat-7 ETM+	Dynamics of Wetland Landscape Patterns in Space and Time. The research area's southern wetlands were severely fragmented, according to remote sensing data from the Yellow River Delta. Wetlands' dynamic was mostly brought about by causes produced by humans, particularly Urban built-up areas and GDP.
[21]	Watershed in Ethiopia	Farmland, barren area, forest cover, grazing area, settlement, water bodies	1984, 1999, 2015	Landsat-5 TM Landsat-8 OLI	In the previous three decades, the Beressa watershed has seen a dramatic change in land use and land cover due to rapid population increase that necessitated more land for farming, more trees for household fuelwood consumption, and more acreage for habitation.
[22]	Gilgel Abbay	Forest, wood and bushland, grassland, agriculture, settlement, water bodies, wetlands	1973, 1986, 1995, 2008	Landsat-5 TM,	Gilgel Abbay watershed situated in Lake Tana was used to examine the effects of land cover/use dynamics on climate variability for different years, 1973, 1986, 1995 and 2008.

(*Continued*)

Table 17.1 Analysis of land use pattern. (*Continued*)

Author	Area	Land type	Year	Satellite data	Outcome
[23]	Varanasi district, Uttar Pradesh, India	Agriculture, dense vegetation, sparse vegetation, fallow land, built up, water, sand area	2013	LISS IV Landsat 8-OLI Resourcesat-1 AWiFS	Accuracy of land use and land cover categorization was assessed using multi-resolution remote sensing imagery for the management of natural resources and sustainable planning of city-classified LULC with 83.28%, 77.93%, and 74.61% accuracy.
[24]	Ramsar Wetland, Assam, India	Water, forest, vegetation, cropland, fallow, barren land, built-up land, vegetation area	1989, 2001, 2012	Landsat 5 TM, Landsat 7 ETM+	From the period 1989–2012 an assessment of the Ramsar wetland ecosystem using indices generated from satellite data. The Landscape Development Index (LDI) showed a sharp rise in the nearby land use's influence on the wetland.
[25]	Hawalbagh block, district Almora, Uttarakhand, India	Vegetation, barren, forest, built-up area, water bodies	1990, 2010	Landsat 7 ETM+	The current work demonstrates the value of remote sensing and geographic information systems (GIS) for the temporal analysis and quantification of spatial phenomena, which cannot be accomplished using traditional mapping methods.

(*Continued*)

Table 17.1 Analysis of land use pattern. (*Continued*)

Author	Area	Land type	Year	Satellite data	Outcome
[26]	Murredu watershed in Telangana state, India	Barren land, water bodies, Forest, built-up, Agriculture	1996, 2019	Landsat 5 TM, Landsat 8 OLI	Using remote sensing and geographic information systems to categorize five primary LULC groups by identifying and classifying changes in the Murredu watershed in Telangana state, India, between 1996 and 2019. After a change analysis, the results demonstrate a very high degree of categorization accuracy was achieved.
[27]	Wami River Basin, Tanzania	Bushland, woodland, swamp, cultivated land, Settlements, grassland, forest, water, open, airfield	2000, 2006, 2011, 2016	Landsat 5 TM, Landsat 7 ETM+	LULC Change Detection was done in the Tanzania River Basin to evaluate the change in area with time. The degree of LULC changes that occurred in various LULC classes between 2000 and 2016 is displayed by the change-detection analysis and the change matrix table.

prospective future LULC changes may be easily recognized utilizing a simulation technique [28]. The LULC modeling methodologies are used to predict future transformations by taking into account previous LULC change patterns, population growth, geographical features of that area, and slope patterns. The simulation models detect future change in a probabilistic manner.

The most often utilized approach for modeling and simulating LULC changes is the Markov Chain (MC) model. This simulation has a random probability distribution or pattern that may be evaluated statistically but may not be predicted. Several researchers selected it as the best-suited model for short-time projections with only one direction transition [29]. The stochastic MC approach in conjunction with the probabilistic Cellular Automata (CA) model may readily replicate multi-directional change analysis.

A study [30] used maximum likelihood classification (MLC) and ISODATA algorithms to classify satellite images obtained from Landsat 5, 7 and 8, data to detect urban land development in the city of Al-Khobar, Saudi Arabia, between 1990 and 2013. Similar to this, [31] used the map to determine the built-up area and variations in LULC in Al-Baha, southwest Saudi Arabia, from 1984 to 2014. They did this by utilizing Landsat TM and ETM+. It is crucial to comprehend and measure the transition patterns in order. To figure out the expected future direction and rate of urban expansion [32] used Landsat photos and the OBIA technology to track the growth of 28 megacities worldwide. In Beijing, China [33], LULC changes were classified using the OBIA approach and Landsat TM data. In contrast to conventional pixel-based methods, OBIA classification produced excellent average class accuracies [34]. Table 17.2 describes the methods used for future prediction.

Accurately extracting LULC data from historical and present sources is necessary for modeling changes in land cover because it enables models to assess previous situations and forecast changes in the future. From satellite data, LULC change information has been extracted using a wide range of remote sensing techniques. The most popular techniques for creating LULC maps are pixel-based image analysis techniques. Object-based image analysis (OBIA) has garnered significant interest in image classification and has recently achieved recognition across several RS applications.

The objective of this study is to analyze the sustainability of a city. Our study is intricately designed to evaluate how these cities align with the principles and aspirations of United Nations Goal 11, which emphasizes sustainable cities and communities. To achieve this goal, we have thoughtfully selected a range of cities like Bhopal, Indore, and Raipur. These cities encompass a wide spectrum of environmental, economic, social and land

Table 17.2 Land use prediction.

Author	Area	Base year	Predicted year	Model	Accuracy	Land use type	Outcome
[35]	Bursa City, Turkey	1979, 1989, 2000, 2013	2040	SLEUTH model	_	Urban, Agriculture, Water bodies, Green, forest, Open, Bare, Road	The amount and quality of urban structure are reflected by the combination of high-quality remote sensing data, urban modeling, and landscape measurements with greater realism. Urban sprawl modeling will help to improve our comprehension of regional planning of the dynamics behind changes in land cover/use for sustainable evolution.
[36]	Manipur River basin	2007, 2014, 2017	2030	CA-Markov model	88%, 92% and 93%	Built-up, Forest, Water bodies, Wetland, Agriculture, Herbaceous wetlands	Research indicates that remote sensing data might yield useful geographical information that can be utilized to formulate appropriate planning and sustainable development strategies.

(Continued)

Table 17.2 Land use prediction. (*Continued*)

Author	Area	Base year	Predicted year	Model	Accuracy	Land use type	Outcome
[37]	Coka Watershed, Ethiopia	1988, 1998, 2008, 2018	2040, 2060	CA-Markov model	87, 85, 88 and 91%	Cultivated area, Rural area, Bare land, Built-up area, Forestland, Grassland, Bushland. and waterbodies	Examining trends in LULC from 1988 to 2018 and projected changes for the years 2040 to 2060. The Coka Watershed's LULC, transformation amply demonstrates the likelihood of significant environmental deterioration and associated issues that the region will experience shortly.
[38]	Ulaanbaatar, Mongolia	1974, 2018	2030, 2040	CA-ANN	90%	Urban, Non-Urban area	Urban expansion's spatiotemporal dynamics and its modelling in the Mongolian region using a CA-ANN model. This simulation has demonstrated a higher degree of accuracy and significance for the research findings.

(*Continued*)

Table 17.2 Land use prediction. (*Continued*)

Author	Area	Base year	Predicted year	Model	Accuracy	Land use type	Outcome
[39]	Bankura, West Bengal	1995, 2000, 2005, 2010, 2015, 2020	2030	SVM classifier, Cellular automata model, MOLUSCE	90%	Water bodies, Barren land, Forest cover, Built-up area, Agricultural land	Application of cellular automata and support vector machine techniques, which offer increased prediction efficiency, to assess the effect of irrigation projects on LULC.
[40]	Samsun, Turkey	1989, 2000 2013	2025	CA-MC, MLP-MC	80%	Water bodies, Open areas, Forest areas, Agriculture area, Artificial surface	Urban development models may forecast future changes based on historical and contemporary dynamics, giving rise to a deeper knowledge of existing patterns and temporal dynamics. The findings also demonstrate that the optimal model to use for simulating changes in a particular research area is the one that best conforms to those changes.

(*Continued*)

Table 17.2 Land use prediction. (*Continued*)

Author	Area	Base year	Predicted year	Model	Accuracy	Land use type	Outcome
[41]	Tripoli, Libya	1984, 1996, 2002, 2010	2020, 2025	CA–Markov model	85.02%	Agriculture Area, Built up Area, Restricted	A study of land use change conducted between 1984 and 2010 revealed a consistent decline in arable land. The study area's accelerating loss of rich land is alarmingly evident in the projected land use scenarios for 2020 and 2025.
[42]	Tons River Basin	1985, 1995, 2005 and 2015	2025	CA-Markov model	83%	Deciduous forest, Mixed Forest, Cropland, Barren land, Shrubs, Built-up, Water bodies	Used earth observation satellite data sets from the Tons River Basin to model changes in land use and cover.

(*Continued*)

Table 17.2 Land use prediction. (*Continued*)

Author	Area	Base year	Predicted year	Model	Accuracy	Land use type	Outcome
[43]	Saddle Creek Drainage Basin, Florida	1995, 1999 and 2006	2015	CA- Markov	84.33%	Urban land, Agricultural, Range area, Upland, Forest area, Water bodies, Wetlands, Barren land, and Transportation	The purpose of the study was to show how the CA-Markov model may be used to forecast changes in land use; nevertheless, care should be taken when applying the study's findings directly to urban planning or policy making.

use parameters [44–46]. The knowledge of the current land use and its implications will establish the level of sustainability. Based on our investigation of the years 2000, 2010, and 2020 for various cities, a Composite Sustainability Index (CSI) was developed [47, 48]. Based on the current scenario, future CSI was also developed by statistically predicting socio-economic parameters and using the Cellular Automata (CA) model to predict future land use patterns, resulting in projecting future CSI scores and aiding government bodies in taking informed decisions.

17.3 Study Area

Each of these cities has been chosen with a purpose. By examining a mix of metropolitan hubs, emerging urban centres, and regions with varying levels of development, the aim is to provide a comprehensive understanding of sustainable development practices. This selection of cities is carefully selected, representing a capital of a state, a major city and another a commercial capital city, which would ideally represent the characteristics of India. Our analysis will involve a holistic assessment of environmental sustainability, social inclusivity, economic prosperity, and urban planning strategies within these urban landscapes. Through rigorous data collection,

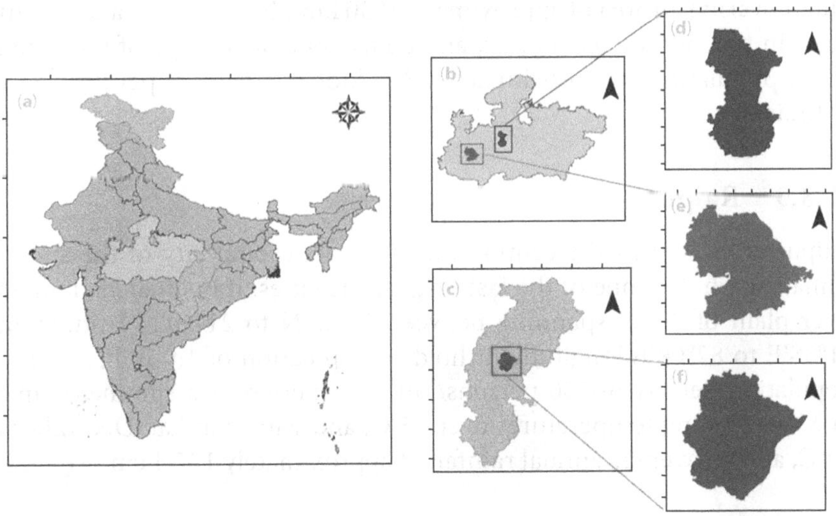

statistical analysis, and predictive modelling, we intend to uncover valuable insights and trends that can inform policymakers, city planners, and stakeholders. For this study, cities selected are Bhopal, Indore, and Raipur, which are shown in Figure 17.3.

17.3.1 Bhopal

Bhopal, a major city in Madhya Pradesh, India, is often referred to as the "City of Lakes." It extends between approximately 23°12′N latitude to 23°24′N latitude and from 77°24′E longitude to 77°50′E longitude, encompassing an area of around 3,055 km^2. Bhopal has a population of approximately 1.8 million residents and a population density of about 855 persons/km^2, as per the Census of India. The city's distinctive topography features a series of captivating lakes, including the Upper Lake (Bada Talab) and the Lower Lake. Bhopal experiences a subtropical climate with an annual temperature range from 24.2°C to 34.6°C. Relative humidity levels exhibit seasonal variations, fluctuating between 36.9% and 71.6%.

17.3.2 Indore

Indore is the Commercial Capital of Madhya Pradesh, India. This urban hub sprawls across the geographical coordinates of approximately 22°42′N latitude to 22°48′N latitude and from 75°50′E longitude to 76°12′E longitude, covering an area of approximately 530 km^2. Indore holds a distinctive place in the cultural, economic, and educational landscape of the region, with a population of approximately 3.3 million residents, as per the Census of India.

17.3.3 Raipur

Raipur is located in the central area; it is the capital city of the state of Chhattisgarh. It is one of the fastest-growing cities. It is situated along the river plain of Arpa, spanning between 21°23′N to 21°50′N latitude and 81°38′E to 82°00′E longitude. It holds a population of 1.9 million with a population density of 750 persons/km^2. It experiences a summer climate in May reaching temperatures over 45°C, and winter in late December at 10°C, and an average annual rainfall of approximately 127.4 cm.

Table 17.3 Sources of satellite images for all cities.

Sr. no.	City	Year	Satellite	Resolution	Path/row
1	Bhopal	2000	Landsat 7	30m	145/044
		2010	Landsat 7	30m	145/044
		2020	Landsat 8	30m	145/044, 146/043
2	Indore	2000	Landsat 7	30m	146/045, 146/045
		2010	Landsat 7	30m	146/045, 146/045
		2020	Landsat 8	30m	146/045, 146/045
3	Raipur	2000	Landsat 7	30m	145/046
		2010	Landsat 7	30m	145/046
		2020	Landsat 8	30m	145/046

17.4 Source of Data

To evaluate the sustainability of a city, many sources of information were procured, including satellite images and ground-related data.

17.4.1 Source of Satellite Images

A series of satellite images was obtained from USGS Earth Explorer for all the different cities, which is shown in Table 17.3.

The above table shows the resolution of all satellite images, which is 30 meters. It also shows the path and row of selected images. It shows the different satellite series, like Landsat 7 and Landsat 8 satellite series, for all the selected cities.

17.4.2 Source of Ground Data

To get all ground-related data for various selected indicators as per SD Goal 11, which includes socio-economic data [49–58]. The following government websites and reports were considered, which are given below.

1. Census of India - Report 2001
2. Census of India - Report 2011

3. District Domestic Product of Maharashtra - Report 2011-12 to 2021-22
4. Estimates of State Domestic Product Madhya Pradesh
5. Global indicator framework for the Sustainable Development Goals and targets of 2030
6. Central Pollution Control Board - Report 2010
7. Central Pollution Control Board - Report 2020
8. Ministry of Environment Forest and Climate Change
9. Ministry of Housing and Urban Affairs
10. Ministry of Statistics and Programme Implementation - Sustainable Development Goals -National Indicator Framework
11. National Crime Record Bureau - Report, Incidence and Rate of Violent Crimes During 2000
12. Road accidents in India - Report 2000-01
13. Road accidents in India - Report 2010-11
14. Road accidents in India - 2020-21
15. Slum data of ULBs in Chhattisgarh

17.5 Methodology

This comprehensive study embarked on a data-driven journey to construct a Sustainability Index for a diverse array of cities, including Bhopal, Indore and Raipur. At its core, this CSI was meticulously crafted in alignment with the United Nations Sustainable Development Goals, with a particular emphasis on Goal 11, which is depicted in Figure 17.4.

17.5.1 Methodology for Land Use Pattern Analysis

For this study, an extensive collection of satellite imagery was downloaded, spanning multiple time series, specifically the years 2000, 2010, and 2020. These satellite images were acquired for several cities. The source of these invaluable satellite datasets was the United States Geological Survey's (USGS) Earth Explorer platform. The base of this satellite imagery acquisition rested upon the utilization of the Landsat series satellite system such as Landsat 7 Enhanced Thermal Mapping and Landsat 8 Operational Land Imager.

This system provided high-quality images with a spatial resolution of 30 meters, thereby offering a detailed and comprehensive view of the selected urban landscapes. To gain insightful information from the acquired satellite

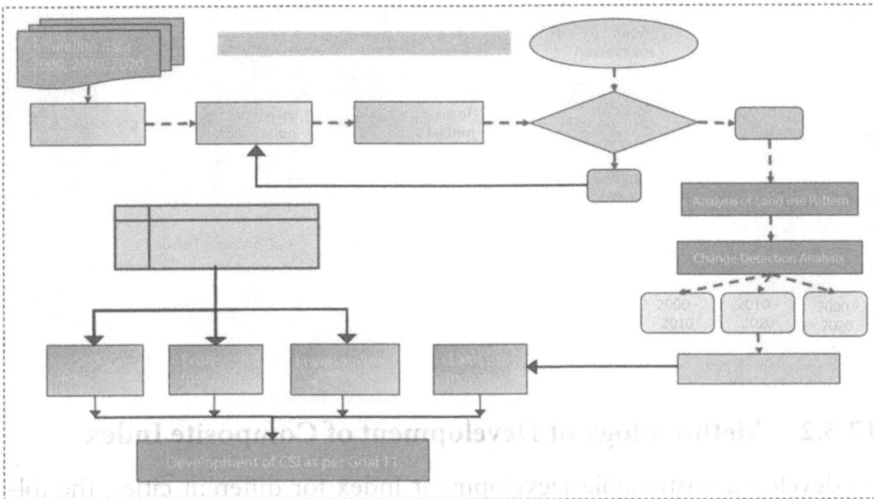

images, a sophisticated and integrated analytical approach was adopted. Advanced geospatial software tools, QGIS, were used for the comprehensive analysis and interpretation of the acquired imagery. All the images downloaded were refined by removing noise like cloud cover, and then the images were preprocessed by applying atmospheric correction and geometric correction. Suitable bands were selected and then merged to form a composite image, which is combining selected bands to get the desired output, like combining bands 1, 2 and 3 to get a True Colour Composite and combining bands 3, 4 and 5 to give a False Colour Composite (FCC).

Upon generation of composite images, the unsupervised classification of the images was done to classify land use into five types: Water bodies, Agricultural land, Urban areas, Open Spaces, and Forest cover. Post-classification, a rigorous accuracy assessment test was conducted utilizing Google Earth Pro images. This validation process involved the meticulous verification of 500 accuracy points, comparing them against the classified map types. This comprehensive accuracy assessment was systematically carried out for each consecutive year under investigation, as shown in Figure 17.5.

Specifically, change detection was performed for intervals spanning from 2000 to 2010, 2010 to 2020, and 2000 to 2020. The aim was to quantify and delineate alterations in land use patterns over the specified timeframes and to examine the observed trends and transformations in land use patterns over the studied years.

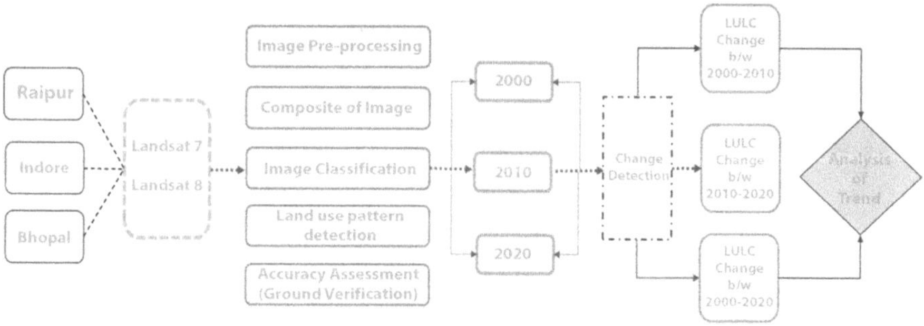

17.5.2 Methodology of Development of Composite Index

To develop a Sustainable Development Index for different cities, the following input parameters were used. All 18 Indicators are taken from SDG Goal no 11. All these indicators were further classified into four categories: social, economic, environmental, and land use patterns. Upon combining all individual parameters, a composite index reflecting the sustainability of a city is made.

Initially, we assessed sustainability across various dimensions, including the Economic Sustainability Index (ESI), Social Sustainability Index

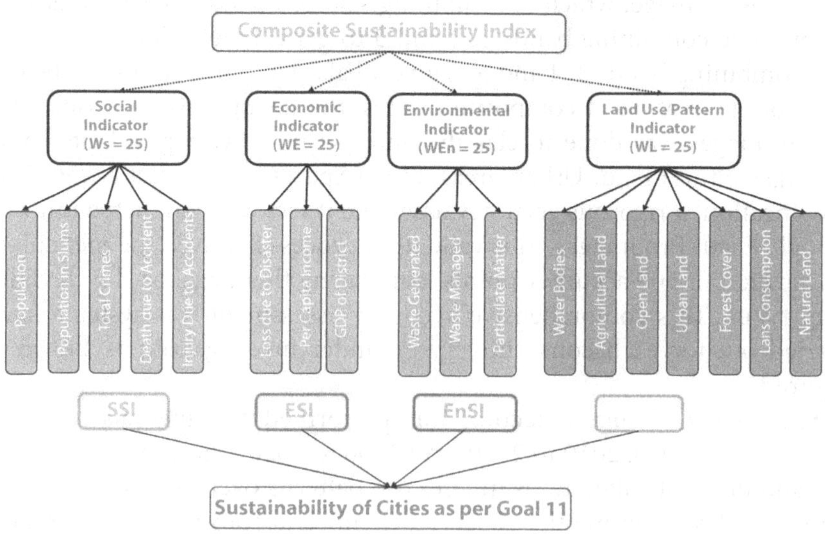

(SSI), Environmental Sustainability Index (EnSI) and Land Use Pattern Index (LSI). To construct the composite index, we adhered to the index construction guidelines outlined by the OECD (2008), involving a series of specific steps [37]. These steps encompass data normalization, the determination of weights for individual indicators, the calculation of intermediate indicators within each dimension, and the aggregation of data. Our approach assumed an equal significance for all sustainability dimensions, resulting in the computation of the Composite Sustainability Index (CSI) for Goal 11, by aggregating ESI, EnSI, LSI and SSI, each with evenly distributed weights.

Figure 17.6 represents the methodology of construction of the Composite Index, which is explained below.

Step 1: Normalization of Data

Given the multifaceted nature of the indicators, they encompass various dimensions. Consequently, before determining the weights for each indicator, a crucial step involves standardizing them into a consistent unit. Normalization serves as a methodology to derive unit-agnostic values for these indicators. In this research, the unitary method was used for data normalization, often referred to as the min-max approach, which confines normalized values within the range of zero to one. The equation for the unitary method in this context is as follows.

For positively associated indicators

$$\beta = \frac{(Xi - X_{min})}{(X_{max} - X_{min})} \tag{17.1}$$

For negatively associated indicators

$$\beta = \frac{(X_{max} - Xi)}{(X_{max} - X_{min})} \tag{17.2}$$

Where Xi is the value of the ith term. X_{min} is the minimum value of data and X_{max} is the maximum value.

Step 2: Assignment of Weights

To get better and effective results, there must be proper assignment of weight. Assignment of weight means the relevance of the factor to influence the overall results. Here, for this study, all the parameters were given equal weight, which is a common practice adopted by the UN Human Development Index.

Step 3: Calculation of Sustainability Index

After calculating the normalized values, calculating the individual Sustainability Index (SI) is as follows:

$$SI = \frac{((W_1\beta_1 + W_2\beta_2 + W_3\beta_3 + ... + W_j\beta_j)}{(W_1 + W_2 + W_3 + ... + W_j)} \tag{17.3}$$

Where $\beta_1 + \beta_2 + \beta_3 + ... + \beta_j$ is the normalized value

$W_1 + W_2 + W_3 + ... + W_j$ are the weights of the individual indicator.

Each of the indices is to be calculated separately to give its dimension, i.e., for environmental (EnSI), social (SSI), economic (ESI) and land use pattern (LSI).

Step 4: Equation for CSI (Composite Sustainability Index)

For this analysis, a Sustainable Development Index for all targeted cities is made. An indicator is developed for Sustainability, which is comprised of environmental indicators, social indicators, economic indicators, and land use indicators. Based on this, an indicator is given for all the analyzed cities. The equation for the composite sustainability index is given below.

$$CSI = \text{WE. (ESI)} + \text{WEn. (EnSI)} + \text{WS. (SSI)} + \text{WL. (LSI)} \tag{17.4}$$

Where,

CSI = Composite Sustainable Index

W_E, W_{En} W_S W_L are the weight assigned to each dimension showing their importance.

The equation calculates the CSI by multiplying each sustainability index by its respective weight and then summing these weighted values. In this way, the CSI provides a balanced assessment of sustainability by taking into account the unique importance of economic, environmental, social, and land use sustainability in the context being evaluated. This formula helps decision-makers and analysts evaluate sustainability comprehensively by considering multiple dimensions and assigning appropriate importance to each dimension based on their specific goals and priorities.

17.6 Results and Discussion

17.6.1 Analysis of Land Use Pattern

Various land use types involved in the analysis of land use patterns for the examined cities are given in Table 17.4.

All the selected land use types were depicted on the map; a colour code matrix was developed for visual narrative, enabling a deeper understanding of the dynamics of land use categories. The green colour is assigned for vegetation cover, Blue is for water bodies, Yellow is for agriculture colour, Red is for urban areas and purple shows open land.

Assigning individual colours for each classified map and a time series and change detection analysis gives a clear insight into the transformation and trend of urban development, environmental impacts and land use within the cities. Figure 17.7 depicts a classified map that provides a comprehensive visual representation of the diverse land use patterns for Bhopal for the years 2000, 2010, and 2020. This is shown in the colour code for the investigation of the change in land use patterns.

The below Figure 17.8 shows the trend of change in land use type for various decades. The area under agricultural and urban land is increasing due to urbanization. The area under open land, green vegetation and water bodies decreases due to the requirement of land. The area under water

Table 17.4 Descriptions of land use class.

Land use type	Description
Water Bodies	It includes pond water, streams, rivers, lakes, riverine, river banks and streams. It majorly includes surface water.
Urban Area	It includes houses, industries, roads, public buildings, structures, and human settlements.
Agricultural land	Land for crop cultivation of both annual and perennial crops.
Forest cover	It is the area covered with natural forest, trees, plantations, local forest area, etc.
Open area	It is a bare area, with no forest, vegetation, or constructed area. It also consists of exposed mountains and rocks.

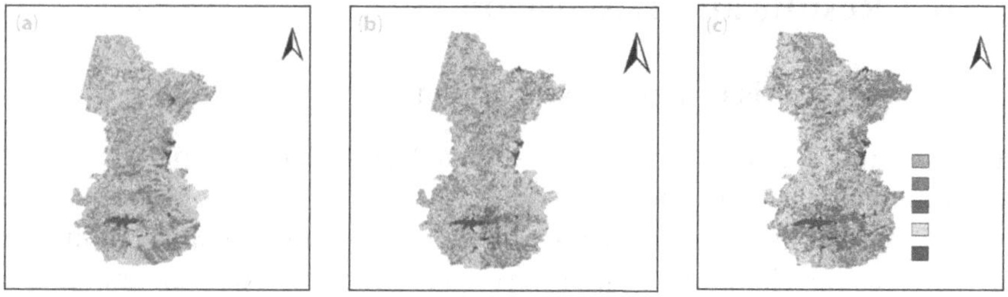

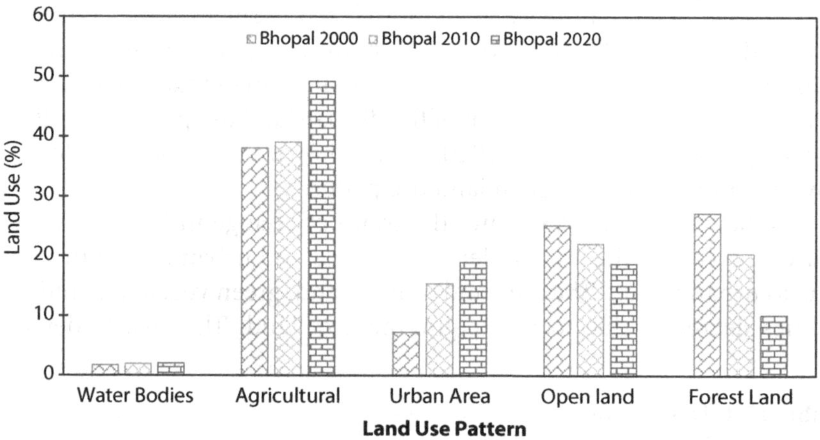

bodies is less than 3%, which is a matter of concern. The area under vegetation is decreasing rapidly at a rate of 9% per decade, which needs to be addressed.

Indore

The below Figure 17.9 shows a classified map of land use patterns for 2000, 2010 and 2020. Land use was further classified into five types: open area, vegetation, urban, agriculture and water bodies, which are indicated with a particular colour matrix. This classified map serves as a visual narrative of the ever-changing landscape of the study area, offering insights into how urbanization, environmental factors, and agricultural activities have evolved.

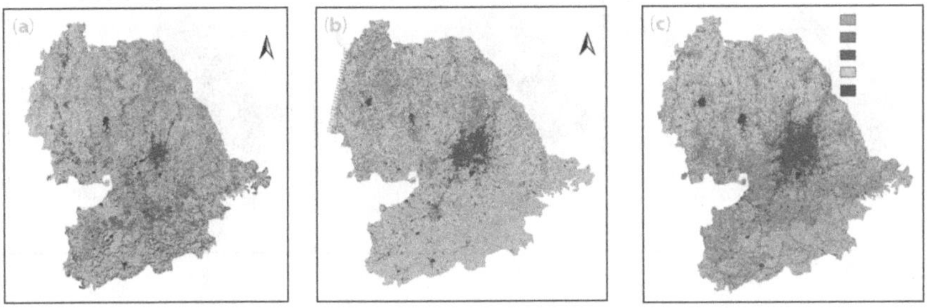

Analyzing the land use patterns of Indore, from Figure 17.10, it becomes apparent that the area dedicated to agriculture and urban development is expanding over time at 6% and urban land at 3%, while the available open land is diminishing to accommodate this group. Moreover, it is noticeable that the area covered by vegetation and forest land experienced a decrease in 2010 followed by a rise in 2020. Agricultural areas are expanding by over 7.2% per decade.

Raipur

In Figure 17.11, a constructed colour matrix gives changes in land use patterns observed across the years 2000, 2010, and 2020 within the study area. It shows the change in forest cover, the urbanization process, shifts in water landscapes, and areas under open spaces.

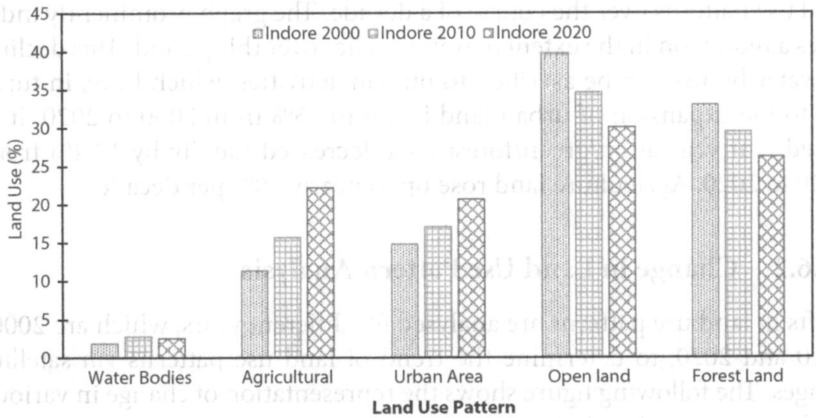

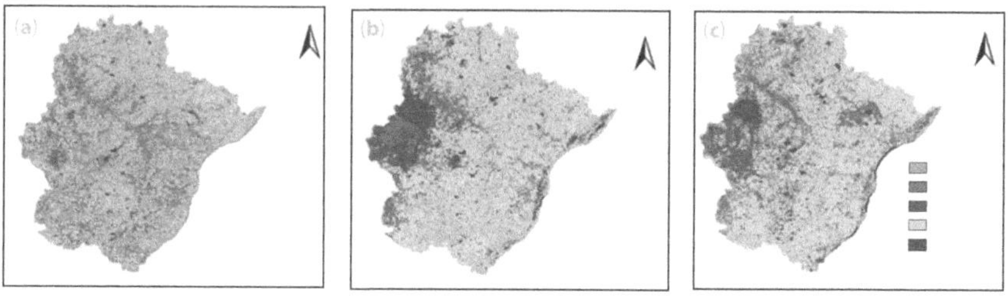

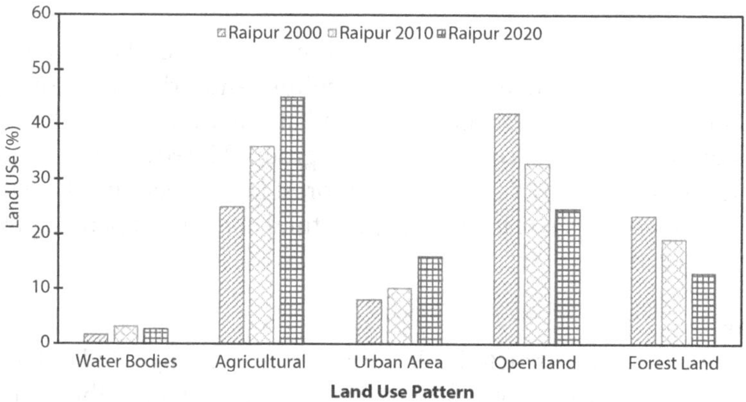

Figure 17.12 provides a visual portrayal of the dynamic shifts in Raipur land use patterns over the course of a decade. The graph prominently indicates a reduction in the extent of water bodies over this period. This decline in water bodies can be ascribed to human activities, which have, in turn, led to the expansion of urban land by up to 15% from 2000 to 2020. It is noted that vegetation/green/forest area decreased rapidly by 13.4% from 2000 to 2020. Agriculture land rose up to nearly 10% per decade.

17.6.2 Change in Land Use Pattern Analysis

Shifts in land use patterns are analysed for different years, which are 2000, 2010 and 2020, to determine the trend of land use patterns *via* satellite images. The following figure shows the representation of change in various land use patterns for all the cities.

17.6.2.1 Analysis of Water Bodies for the Cities

Figure 17.13 examines water body areas in Bhopal, Indore, and Raipur throughout time and reveals significant trends. The water bodies of Bhopal show a nearly straight curve with little improvement. Indore's water body areas were increasing till 2010, after that, a decline of 2.5% was observed. Raipur, on the other hand, shows a continuous improvement in the area of water bodies. The study highlights the dynamic character of these regional water ecosystems and stresses the significance of understanding the underlying processes affecting these changes. More research is needed to guide focused conservation efforts and guarantee the long-term management of water resources in these locations.

17.6.2.2 Analysis of Agricultural Land

Analysis of agricultural land from 2000 to 2020 indicates diverse land use trends. Figure 17.14 shows there was a significant rise in the year 2010 as well as in 2020, showing significant growth or changes in agricultural techniques. Indore saw expansion as well. Raipur had a similar upward trend, with agricultural land increasing by approximately 10%. These percentage increases reflect considerable agricultural expansion and extended land cultivation in these locations throughout the observed period. Natural vegetation cover is encouraged for cultivation purposes. It is noted that all cities had percentage gains, suggesting more land area is used for agricultural purposes.

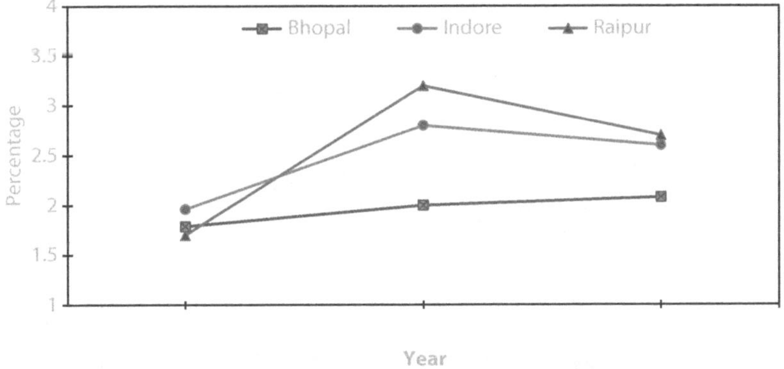

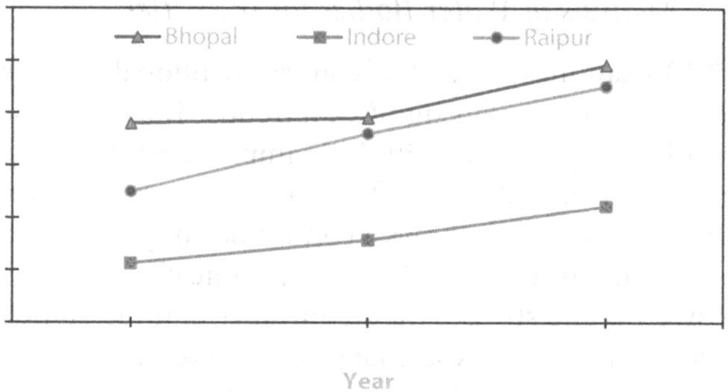

17.6.2.3 Analysis of Urban Area

In terms of percentage growth, Figure 17.15 concludes that Bhopal's urban area expanded by around 10% in 2020, whereas the urban area of Indore also increased at a high rate. Raipur grew at a continuous rate with a slight increase in 2020. These numbers highlight the varied degrees of urban expansion in these cities, underlining the importance of careful urban planning and sustainable development methods in dealing with the problems of fast urbanization. It is critical to address the socioeconomic and environmental consequences of these changes, such as infrastructural requirements, resource use, and the influence on nearby ecosystems. A thorough examination of these elements is required for informed urban development policies and the building of resilient, habitable urban environments in these locations.

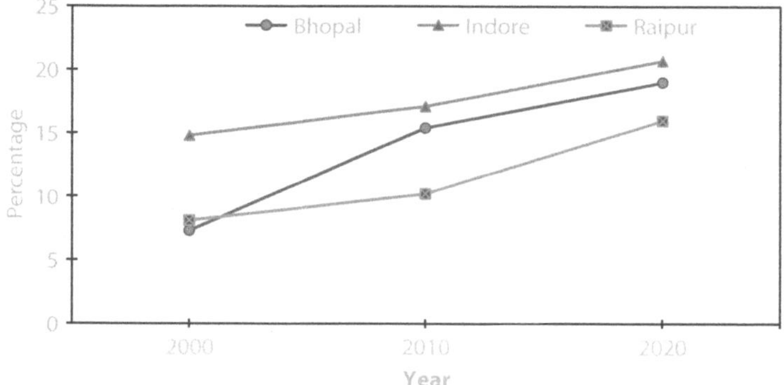

17.6.2.4 *Analysis of Open Area*

Over the studied time in Figure 17.16, it is observed that the open area is decreasing continuously at various locations. In terms of percentage loss, the loss rate is slow in Bhopal and Indore, while in Raipur, the open area decreases at a very rapid rate. These percentage reductions highlight the difficulties associated with urbanization and other land use changes, underlining the importance of sustainable land management techniques and conservation initiatives. It is critical to examine the consequences of shrinking open space, such as the effects on biodiversity, ecosystems, and general environmental quality. Comprehensive planning and regulations that balance urban expansion with open space preservation are critical for developing resilient and sustainable communities in these locations.

17.6.2.5 *Analysis of Forest Cover*

Examining statistics on forest land in Bhopal, Indore, and Raipur throughout time in Figure 17.17 gives insights into the changes in these regions of forest areas. There is a huge drop for Bhopal as well as Raipur, showing a large loss of wooded lands. Indore's total change is rather moderate. Concerns have been raised regarding the possible implications on biodiversity, ecosystem services, and general environmental health as a result of the loss of wooded regions. To preserve the long-term viability of forest ecosystems in these places, comprehensive plans that balance development and conservation are required.

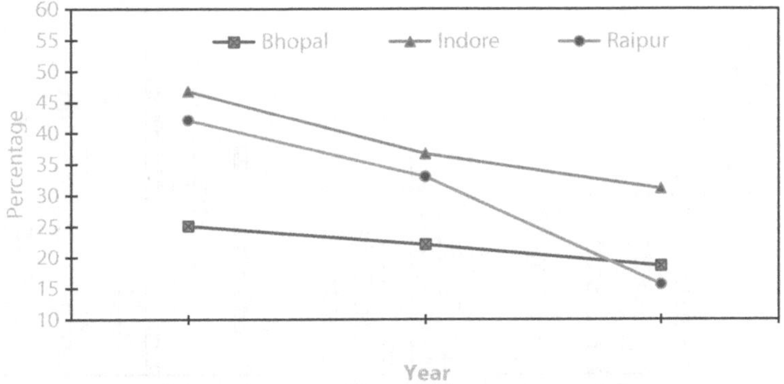

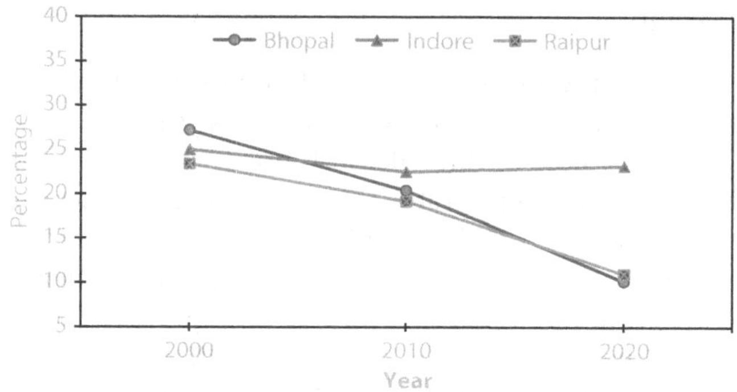

17.6.3 Composite Sustainability Index

All the Social, Economic, Environmental and land use parameters are used for the construction of the Composite Sustainable Index (CSI).

The below Figure 17.18 shows the individual Index of Economic Sustainability Index (ESI), Social Sustainability Index (SSI), Environmental Sustainability Index (EnSI) and Land Use Pattern Index (LSI), which is made from combining all the factors. Each index score is shown for every year of analysis. It is noted that social and environmental parameters are high for each city due to education and awareness among people for

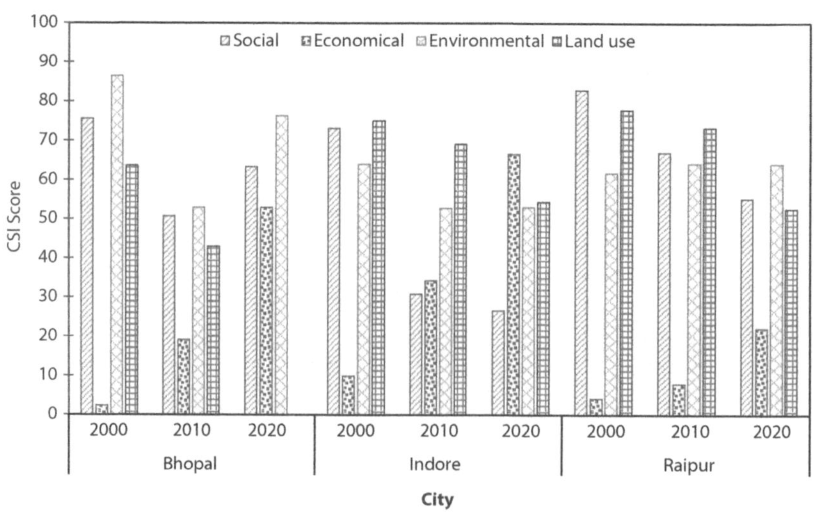

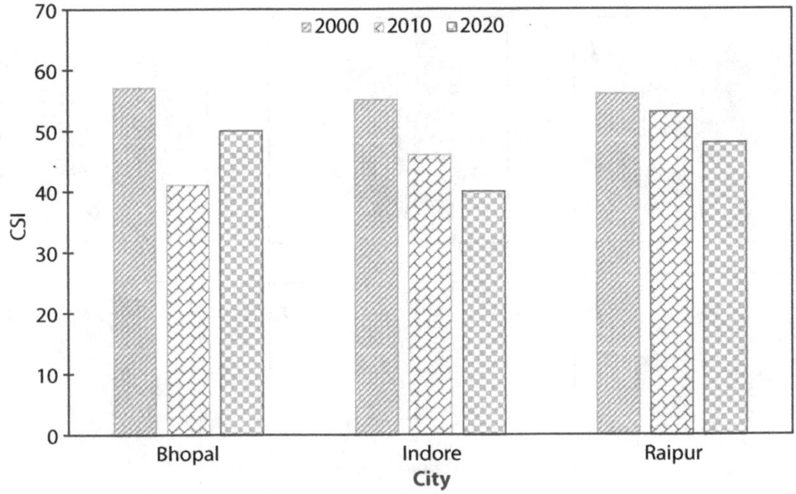

Figure 17.19 CSI score for all cities.

eco-friendly policies. The economic factor is very low due to low per capita income. The land use index is decreasing with time due to the high rate of urbanization and degradation of land in the city.

An analysis of sustainability trends in cities over the years 2000, 2010, and 2020 provides valuable insights. Figure 17.19 depicts CSI, which is made by combining ESI, EnSI, SSI and LSI. CSI decreases for all the cities due to less sustainable practices, low expenditure, rapid urbanization, decrease in forest cover, and unregulated land use planning. CSI for Bhopal decreased in 2010 and increased in 2020 due to economic and environmental factors. CSI for Raipur and Indore decreases with time due to no improvement in their conditions.

17.6.4 Future CSI Prediction

To get valuable insight and to study the case scenario, if all the parameters are as it is, like no change in government policies, future CSI are predicted for the examined cities.

Prediction of land use pattern

Predictions for future land use patterns in 2030 and 2040 are made using the Cellular Automata model, relying on land use patterns observed in the years 2000, 2010, and 2020, as illustrated in Figure 17.20.

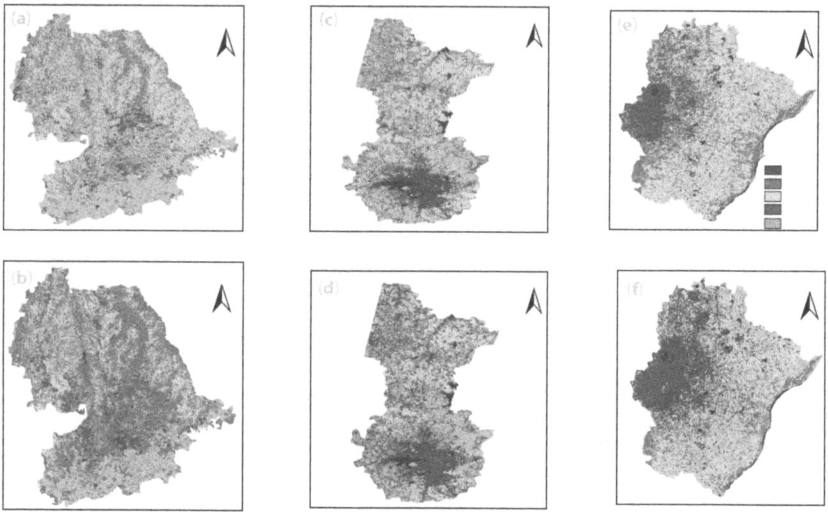

Understanding the future land use pattern is crucial as it directly impacts the Land Sustainability Index (LSI), a determinant for the CSI. Predicting future land use based on the previous year's data is a key aspect of this process, and various tools are available for this purpose. In this context, the Cellular Automata (CA) model is employed. Land use patterns of 2000 and 2010 were used to predict for 2020, and it was verified by satellite images for 2020, achieving an accuracy of more than 70%. Based on it, the future model was predicted for 2030 and 2040 to get LSI.

The above figure shows that over time, in 2030 and 2040, urban areas experience rapid expansion across all the examined cities. Concurrently, agricultural areas tend to remain relatively constant, while forest and open areas decrease progressively. The pattern that has been seen can be attributed to the growing demand for land resources as a result of the world's population growth.

Prediction of CSI

A basic statistical tool was used to forecast other indexes like ESI, EnSI and SSI from the existing data. LSI was obtained from a future predicted land use model. By combining all the indexes a CSI was predicted for the future for all studied cities.

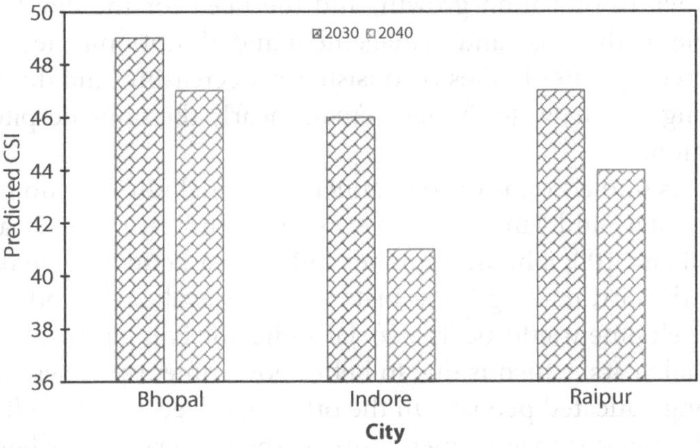

Figure 17.21 Predicted CSI values.

The above Figure 17.21 shows the trend of projected CSI for all cities. The decline in CSI is due to many factors like the rise in urbanization, high rate of waste generation, conversion of natural land to urban land, decline in forest cover, unplanned development of cities, depleting surface water, rise in crime rate, decline of open areas for human settlement and unhealthy land sustainable practices.

If these conditions persist in the coming decades, it is anticipated that chaos may ensue. Consequently, there is a pressing need for the implementation of sustainable land use practices, guided by government policies. Such measures are crucial not only for addressing the identified issues but also for aligning with the objectives of the UN SD Goal 11, emphasizing the importance of creating sustainable and resilient cities and communities.

17.7 Conclusion

Through a comprehensive analysis of LULC and the sustainable development indicators for SDG Goal 11, a sustainability index was developed for multiple study areas, including the cities of Bhopal, Indore, and Raipur for the years 2000, 2010 and 2020. While examining various cities at different time intervals, a noticeable trend emerges. Urban areas experience rapid expansion across all the examined cities. Concurrently, agricultural areas rise with time while forest and open areas decrease progressively. The pattern that has been seen can be attributed to the growing demand for land resources because of population growth. The strain on available land increases with population growth, and the need for the development of sustainable methods of land management and allocation is necessary. The area covered by water bodies is consistently decreasing, and due to awareness among people, water bodies remain nearly the same despite massive urbanization.

CSI takes into account various dimensions, including economic, social, environmental, and land use patterns. This indicator yields a number between 0 and 100; a higher number indicates a better degree of sustainability in the city, making it a valuable tool for evaluating and comparing urban development efforts. The social indicator is high for all examined cities for all years, which is due to better government policies and awareness among educated people. On the other hand, economic indicators are low for all cities due to less expenditure on the conservation of heritage and less per capita income due to the massive population in India. With time, environmental indicators are increasing due to awareness among people and opting for environment-friendly practices and policies. As far as the land use parameter is concerned, it decreases with time for all examined cities due to the rise in urbanization and encroachment of natural land for human activities. In this research work, Bhopal was found to have a high CSI score compared to Indore followed by Raipur, indicating Bhopal is more aligned to sustainability as per UN SDG Goal 11.

Projected CSI trends reveal a declining pattern across all cities, due to rapid urbanization, waste generation, and deforestation. These factors led to environmental degradation. If this trend continues, it will lead to severe socio-environmental disruptions. Therefore, immediate action is necessary, with a strong emphasis on sustainable land-use policies, conservation efforts, and responsible urban planning.

In future, this research can be improved by adding other parameters, projecting to other cities, for different years. Additionally, machine learning techniques can be used to analyse land use patterns with high accuracy. Better normalization techniques can be incorporated to improve its effectiveness in real-case scenarios. This will provide valuable insights into future geospatial trends, enabling planners to formulate strategies and make informed decisions aligned with anticipated changes in the landscape.

References

1. Nakhle, P., Stamos, I., Proietti, P., Siragusa, A., Environmental monitoring in European regions using the sustainable development goals (SDG)

framework. *Environ. Sustainability Indic.*, 21, 100332, 2024, https://doi.org/10.1016/j.indic.2023.100332.

2. Reyers, B. and Selig, E.R., Global targets that reveal the social–ecological interdependencies of sustainable development. *Nat. Ecol. Evol.*, 4, 8, 1011–1019, 2020, https://doi.org/10.1038/s41559-020-1230-6.
3. Martí-Ballester, C.-P., Analysing the financial performance of sustainable development goals-themed mutual funds in China. *Sustainable Prod. Consumption*, 27, 858–872, 2021, https://doi.org/10.1016/j.spc.2021.02.011.
4. Rashid, M.A.U.H., Manzoor, M.M., Mukhtar, S., Urbanization and its effects on water resources: An exploratory analysis. *Asian J. Water Environ. Pollut.*, 15, 1, 67–74, 2018, doi: 10.3233/ajw-180007.
5. Myint, S.W., Gober, P., Brazel, A., Grossman-Clarke, S., Weng, Q., Per-pixel vs. object-based classification of urban land cover extraction using high spatial resolution imagery. *Remote Sens. Environ.*, 115, 1145–1161, 2011, https://doi.org/10.1016/j.rse.2010.12.017.
6. Talukder, B., Hipel, K.W., vanLoon, G.W., Developing composite indicators for agricultural sustainability assessment: Effect of normalization and aggregation techniques. *Resources*, 6, 4, 66, 2017, https://doi.org/10.3390/resources6040066.
7. Alqurashi, A.F. and Kumar, L., Investigating the use of remote sensing and GIS techniques to detect land use and land cover change: A review. *Adv. Remote Sens.*, 2, 193–204, 2013, https://doi.org/10.4236/ars.2013.2202.
8. Hansen, M.C. and Loveland, T.R., A review of large area monitoring of land cover change using Landsat data. *Remote Sens. Environ.*, 122, 66–74, 2012, https://doi.org/10.1016/j.rse.2011.08.024.
9. Berrone, P., Rousseau, H.E., Ricart, J.E., Brito, E., Giuliodori, A., How can research contribute to the implementation of sustainable development goals? An interpretive review of SDG literature in management. *Int. J. Manage. Rev.*, 25, 2, 318–339, 2023, https://doi.org/10.1111/ijmr.12331.
10. Avramov, D., Cheng, S., Lioui, A., Tarelli, A., Sustainable investing with ESG rating uncertainty. *J. Financ. Econ.*, 145, 2, 642–664, 2022, https://doi.org/10.1016/j.jfineco.2021.09.009.
11. Bauckloh, T., Schaltegger, S., Utz, S., Zeile, S., Zwergel, B., Active first movers vs. late free-riders? An empirical analysis of UN PRI signatories' commitment. *J. Bus. Ethics*, 182, 3, 747–781, 2023, https://doi.org/10.1007/s10551-021-04992.
12. Berg, F., Kolbel, J.F., Rigobon, R., Aggregate confusion: The divergence of ESG ratings. *Rev. Finance*, 26, 6, 1315–1344, 2022, https://doi.org/10.1093/rof/rfac033.
13. Brandts, J., Huber, S.J., El Baroudi, S., Rott, C., Gender differences in private and public goal setting. *J. Econ. Behav. Organ.*, 192, 222–247, 2021, https://doi.org/10.1016/j.jebo.2021.09.012.

14. Karimi, A., *et al.*, New developments and future challenges in reducing and controlling heat island effect in urban areas. *Environ. Dev. Sustainability*, 25, 10485–10531, 2023, https://doi.org/10.1007/s10668-022-02530-0.
15. Blaschke, T., Object-based image analysis for remote sensing. *ISPRS J. Photogramm. Remote Sens.*, 65, 2–16, 2010, https://doi.org/10.1016/j.isprsjprs.2009.06.004.
16. Anley, M.A., Minale, A.S., Ayehu, N.H., Gashaw, T., Assessing the impacts of land use/cover changes on ecosystem service values in Rib watershed, Upper Blue Nile Basin, Ethiopia. *Trees For. People*, 7, 100212, 2022, https://doi.org/10.1016/j.tfp.2022.100212.
17. Areendran, G., *et al.*, Documenting the land use pattern in the corridor complexes of Kaziranga National Park using high-resolution satellite imagery. *Trees For. People*, 2, 100039, 2020, https://doi.org/10.1016/j.tfp.2020.100039.
18. Khan, S., Qasim, S., Ambreen, R., Syed, Z.U.H., Spatio-temporal analysis of land use/land cover change of District Pishin using satellite imagery and GIS. *J. Geogr. Inf. Syst.*, 8, 3, 361–368, 2016, https://doi.org/10.4236/jgis.2016.83031.
19. Forkuo, E.K., Biney, E., Harris, E., Quaye-Ballard, J.A., The impact of land use and land cover changes on socioeconomic factors and livelihood in the Atwima Nwabiagya district of the Ashanti region, Ghana. *Environ. Challenges*, 5, 100226, 2021, https://doi.org/10.1016/j.envc.2021.100226.
20. Liu, G., Zhang, L., Zhang, Q., Musyimi, Z., Jiang, Q., Spatio-temporal dynamics of wetland landscape patterns based on remote sensing in Yellow River Delta, China. *Wetlands*, 34, 4, 787–801, 2014, https://doi.org/10.1007/s13157-014-0542-1.
21. Meshesha, T.W., Tripathi, S.K., Khare, D., Analyses of land use and land cover change dynamics using GIS and remote sensing during 1984 and 2015 in the Beressa Watershed, Northern Central Highland of Ethiopia. *Model. Earth Syst. Environ.*, 2, 4, 1–12, 2016, https://doi.org/10.1007/s40808-016-0233-4.
22. Minale, A.S. and Rao, K.K., Impacts of land cover/use dynamics of Gilgel Abbay catchment of Lake Tana on climate variability, Northwestern Ethiopia. *Appl. Geomatics*, 4, 3, 155–162, 2012, https://doi.org/10.1007/s12518-012-0092-2.
23. Mishra, V.N., Rai, P.K., Kumar, P., Prasad, R., Evaluation of land use/land cover classification accuracy using multi-resolution remote sensing images. *Forum Geogr.*, 15, 1, 45–53, 2016, https://doi.org/10.5775/fg.2016.137.i.
24. Mozumder, C., Tripathi, N.K., Tipdecho, T., Ecosystem evaluation (1989–2012) of Ramsar wetland Deepor Beel using satellite-derived indices. *Environ. Monit. Assess.*, 186, 11, 7909–7927, 2014, https://doi.org/10.1007/s10661-014-3976-2.
25. Rawat, J.S. and Kumar, M., Monitoring land use/cover change using remote sensing and GIS techniques: A case study of Hawalbagh block, district Almora, Uttarakhand, India. *Egypt. J. Remote Sens. Space Sci.*, 18, 1, 77–84, 2015, https://doi.org/10.1016/j.ejrs.2015.02.002.

26. Shekar, P.R. and Mathew, A., Detection of land use/land cover changes in a watershed: A case study of the Murredu watershed in Telangana State, India. *Watershed Ecol. Environ.*, 5, 46–55, 2023, https://doi.org/10.1016/j.wsee.2022.12.003.
27. Twisa, S. and Buchroithner, M.F., Land-use and land-cover (LULC) change detection in Wami River Basin, Tanzania. *Land*, 8, 9, 136, 2019, https://doi.org/10.3390/land8090136.
28. Hussain, M., Chen, D., Cheng, A., Wei, H., Stanley, D., Change detection from remotely sensed images: From pixel-based to object-based approaches. *ISPRS J. Photogramm. Remote Sens.*, 80, 91–106, 2013, https://doi.org/10.1016/j.isprsjprs.2013.03.006.
29. Singh, A., Digital change detection techniques using remotely-sensed data. *Int. J. Remote Sens.*, 10, 6, 989–1003, 1989, https://doi.org/10.1080/01431168908903939.
30. Rahman, M., Detection of land use/land cover changes and urban sprawl in Al-Khobar, Saudi Arabia: An analysis of multi-temporal remote sensing data. *ISPRS Int. J. Geo-Inf.*, 5, 2, 15, 2016, https://doi.org/10.3390/ijgi5020015.
31. Pellikka, P. and Alshaikh, A.Y., Remote sensing of the decrease of juniper woodlands in the mountains of southwestern Saudi Arabia—Reasons and consequences. *Arabian J. Geosci.*, 9, 1–12, 2016, https://doi.org/10.1007/s12517-016-2481-z.
32. Taubenböck, H., *et al.*, Monitoring urbanization in mega cities from space. *Remote Sens. Environ.*, 117, 162–176, 2012, https://doi.org/10.1016/j.rse.2011.09.015.
33. Yu, W., Zhou, W., Qian, Y., Yan, J., A new approach for land cover classification and change analysis: Integrating backdating and an object-based method. *Remote Sens. Environ.*, 177, 37–47, 2016, https://doi.org/10.1016/j.rse.2016.02.030.
34. Platt, R.V., Ogra, M.V., Badola, R., Hussain, S.A., Conservation-induced resettlement as a driver of land cover change in India: An object-based trend analysis. *Appl. Geogr.*, 69, 75–86, 2016, https://doi.org/10.1016/j.apgeog.2016.02.006.
35. Akın, A. and Erdoğan, M.A., Analysing temporal and spatial urban sprawl change of Bursa City using landscape metrics and remote sensing. *Model. Earth Syst. Environ.*, 6, 1331–1343, 2020, https://doi.org/10.1007/s40808-020-00758-7.
36. Anand, V. and Oinam, B., Future land use land cover prediction with special emphasis on urbanization and wetlands. *Remote Sens. Lett.*, 11, 3, 225–234, 2020, https://doi.org/10.1080/2150704X.2019.1704304.
37. Buraka, T., Elias, E., Lelago, A., Analysis and prediction of land-use and land-cover changes and driving forces by using GIS and remote sensing in the Coka Watershed, Southern Ethiopia. *Indian J. Agric. Res.*, 56, 5, 581–587, 2022, https://doi.org/10.18805/IJARe.A-699.

38. Gantumur, B., *et al.*, Spatiotemporal dynamics of urban expansion and its simulation using CA-ANN model in Ulaanbaatar, Mongolia. *Geocarto Int.*, 37, 2, 494–509, 2020, https://doi.org/10.1080/10106049.2020.1723714.
39. Halder, H., Das, S., Basu, S., Use of support vector machine and cellular automata methods to evaluate the impact of irrigation project on LULC. *Environ. Monit. Assess.*, 195, 1, 6, 2023, https://doi.org/10.1007/s10661-022-10588-6.
40. Ozturk, D., Urban growth simulation of Atakum (Samsun, Turkey) using cellular automata–Markov chain and multi-layer perceptron–Markov chain models. *Remote Sens.*, 7, 5, 5918–5950, 2015, https://doi.org/10.3390/rs70505918.
41. Al-Sharif, A.A.A. and Pradhan, B., Monitoring and predicting land use change in Tripoli Metropolitan City using an integrated Markov chain and cellular automata models in GIS. *Arabian J. Geosci.*, 7, 10, 4291–4301, 2014, https://doi.org/10.1007/s12517-013-1119-7.
42. Singh, S.K., *et al.*, Modelling of land use land cover change using Earth observation data-sets of Tons River Basin, Madhya Pradesh, India. *Geocarto Int.*, 33, 11, 1202–1222, 2018, https://doi.org/10.1080/10106049.2017.1343390.
43. Subedi, P., Subedi, K., Thapa, B., Application of a hybrid cellular automaton–Markov (CA–Markov) model in land-use change prediction: A case study of Saddle Creek drainage basin, Florida. *Appl. Ecol. Environ. Sci.*, 1, 6, 126–132, 2013, https://doi.org/10.12691/aees-1-6-5.
44. Abdar, Z.K., Amirtaimoori, S., Mehrjerdi, M.R.Z., Boshrabadi, H.M., A composite index for assessment of agricultural sustainability: The case of Iran. *Environ. Sci. Pollut. Res.*, 29, 31, 47337–47349, 2022, https://doi.org/10.1007/s11356-022-19154-6.
45. Priya, S. and Singh, S.P., Assessing and comparing the sustainability of organic and conventional wheat farming in India: An indicator-based approach. *J. Cleaner Prod.*, 423, 138652, 2023, https://doi.org/10.1016/j.jclepro.2023.138652.
46. Fallah-Alipour, S., Boshrabadi, H.M., Mehrjerdi, M.R.Z., Hayati, D., A framework for empirical assessment of agricultural sustainability: The case of Iran. *Sustainability*, 10, 12, 4823, 2018, https://doi.org/10.3390/su10124823.
47. Lo-Iacono-Ferreira, V.G., Garcia-Bernabeu, A., Hilario-Caballero, A., Torregrosa-López, J., Measuring urban sustainability performance through composite indicators for Spanish cities. *J. Cleaner Prod.*, 359, 131982, 2022, https://doi.org/10.1016/j.jclepro.2022.131982.
48. Lu, D., Mausel, P., Brondizio, E., Moran, E., Change detection techniques. *Int. J. Remote Sens.*, 25, 12, 2365–2401, 2004, https://doi.org/10.1080/0143116031000139863.
49. Census of India 2001, Available online: https://censusindia.gov.in/.
50. Census of India 2011, Available online: https://censusindia.gov.in/.
51. Central Pollution Control Board (CPCB), Annual Report 2010, Available online: https://cpcb.nic.in/annual-report.php.

52. Central Pollution Control Board (CPCB), Annual Report 2020, Available online: https://cpcb.nic.in/openpdffile.php?id=UmVwb3J0RmlsZXMvMTQwM18xNjU1MzU0NzkxX21lZGlhcGhvdG8xNjQ3MS5wZGY=.
53. Ministry of Road Transport and Highways of India, Annual Report on Road Accidents in India, 2020, Available online: https://morth.nic.in/road-accident-in-india.
54. Ministry of Statistics and Programme Implementation, Sustainable Development Goals—National Indicator Framework, 2015, Available online: https://unstats.un.org/sdgs/hlg/.
55. National Crime Records Bureau (NCRB), Report: Incidence and Rate of Violent Crimes during 2000, Available online: https://ncrb.gov.in/en/crime-india.
56. Regional Centre for Urban and Environmental Studies (RCUES), Slum Data of Urban Local Bodies in Chhattisgarh, 2001, Available online: http://rcueslucknow.org/states/Compendium%20of%20Urban%20Data%20(Chhattisgarh)/41%20SLUM%20DATA%20OF%20ULBs%20IN%20CHHATTISGARH.pdf.
57. United Nations Environment Programme (UNEP), Measuring Progress: Environment and the SDGs, 2021, Available online: https://www.unep.org/resources/publication/measuring-progress-environment-andsdgs.
58. World Bank, World Development Indicators, 2023, Available online: https://datatopics.worldbank.org/world-development-indicators/wdi-and-the-sustainable-development-goals.html.

18

Ethics and Values in Land Use

Abstract
When land use is handled ethically, it greatly supports the sustainable growth of the community. This way, the economy will support fairness and moral values. To use land ethically, one must analyze principles of right and wrong and incorporate them into how decision-making takes place in organizations. This will ensure that economic goals can be achieved while maintaining a focus on ethical values, primarily centred on openness, equal treatment, and justice. The idea of environmental justice is central here, pointing out that it is necessary to ensure good community well-being by dealing with imbalances in land and resource access. One must see how our own choices and society's laws both influence the way land is used. Ethically driven urban planning supports the use of land in a way that helps the community, boosts clean energy, and ensures everyone gets fair chances for growth. Since dealing with environmental protection and competing interests is not easy, strong ethics in leadership are vital. Because of this style, leaders encourage people to discuss varied viewpoints, take a long-term approach, and act based on principles. When one considers land use in an ethical way, it helps to form reliable and eco-friendly communities that deal with honestly and respect.

Keywords: Ethical land use practices, ethics, values, sustainability

18.1 Introduction

Making decisions about land use in an ethical manner is very important for achieving sustainability, equity, and well-being for cities and towns. When societies experience change and development, how land is managed has a major impact on the environment, the economy, and the structure of society. Guiding principles for fair land use show how people should use, manage, and look after the land. These guidelines are important because they deal with issues such as growth, fairness towards all people, and care for the environment. They try to find a way to handle the challenges of moving forward with development and still protecting nature. It is often hard to reduce conflicts between protecting nature and its inhabitants

Kundan Meshram and Chatrabhuj. A Sustainable Approach to Land Use, (451–468) © 2026 Scrivener Publishing LLC

and fostering development in the economy [1–3]. For communities to overcome these difficulties, they must depend on open and value-based decision-making. They have to be prepared in a way that long-term social trust and support are achieved, so that economic, ethical, and social issues are all addressed during development decisions.

Focusing on environmental ethics in land use mainly involves looking after native resources and minimizing damage to nature. The goals are mainly to combat climate change, protect different kinds of life, and decrease pollution. Adding these ethical ideas to planning encourages people to make eco-friendly decisions and provide a healthy environment for coming generations. It is also important to think ethically about social justice, as this looks at the fairness of how land is shared and used by individuals or groups [4–7]. This way of thinking looks at how people from different groups get different amounts of land perks and challenges. Throughout history, it is often marginalized groups who must suffer more from city building or pollution, being evicted from places where their ancestors once lived under the cover of progress or claims for safety. Such planning encourages everyone in the community to be involved in decision-making so that their rights and interests are taken care of [7]. Land use policies help create just, strong, and sustainable communities by following both environmental care and social fairness.

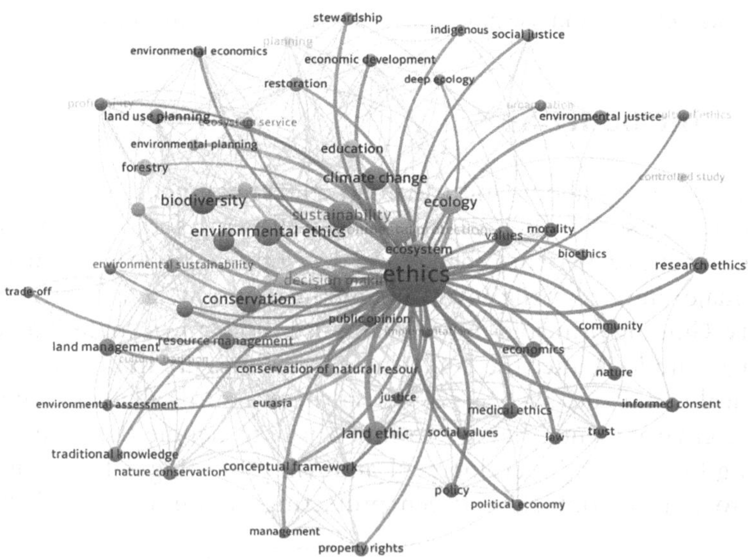

Figure 18.1 Relationship between ethics and its aspects.

Economic considerations often lead to important ethical questions in land use planning. Getting the economy to work well is important, yet doing so should not require sacrificing ethics like accountability, transparency and fairness. Land use practices are meant to provide jobs and business support that benefit different generations at the same time. Many land use decisions must be made while following ethical guidelines for them to be effective [8–10]. Achieving this requires teamwork by government agencies, private developers, environmental organizations and local communities to agree on a shared plan, as seen in the following Figure 18.1.

The formation of communities that are healthy for the environment and provide equal opportunities is built on ethical land use policies. Considering ethics during land use planning allows societies to avoid harming either humans or the environment while improving development. It looks at the different issues of ethical land use, providing tips and guidance for decisions that help the common good.

18.2 Significance of Values in Land Use Decisions

Influencing land use decisions toward outcomes that benefit the environment, society and are equitable and just depends on values related to environmental stewardship, equity and justice. As a result of these values, the environment is preserved for future use, and everyone involved shares the advantages and disadvantages of land handling fairly [11, 12]. How important these values are in land use decisions is discussed below.

18.2.1 Environmental Responsibility

Environmental responsibility means being a good environmental steward, which is shown by taking care of the natural environment with the help of environmentally friendly measures. The point is made that preserving ecosystems is essential for their environmental contribution as well as for protecting the well-being of people. When cities expand and resources are strained, making land use planning environmentally responsible is key to protecting the environment for the long run.

Sustainable development:
A key part of being environmentally responsible is practising sustainable development, which means decisions about land use today do not limit possibilities for future generations. Making environmental protection goals in planning supports careful use of resources, reduces climate change

harm and helps protect plants and animals. Such strategies for sustainable use of land might introduce solar, wind, green building and low-impact development methods. All of these practices together help decrease carbon footprints, improve the environment and further global sustainability goals.

Resilience:
Taking care of the environment improves community resilience, which makes it possible for ecosystems and settlements to handle and recover from things such as climate change, severe weather, and disasters. So, saving important wetlands, forests, and coastal areas helps block floods, reduce sea-level increase, and protect us from windstorms. Green roofs, bioswales, permeable asphalt, and urban trees are used in urban areas to keep the climate comfortable, control the rise in temperature, and help with handling scattered rainwater. Such interventions save the environment and strengthen both the community as a whole and its readiness for disasters.

Quality life:
Green spaces and thriving ecosystems enhance the quality of life by promoting mental and physical health, enhancing air and water quality, and providing recreational opportunities. Being close to parks, community gardens, nature reserves, and recreational trails has a positive effect on a person's mental health, physical health, and the community's unity. Public health is supported by fresh air and water, plus experiences in nature, since they lower stress, lift the mood, and increase physical activity. In addition, green spaces in cities help build places where all residents can socialize, work out, and rest, making their communities more attractive and unified. Having biodiverse and green cities is helpful for the environment, as it provides a place for local wildlife and plants as well as people.

Economic advantages:
Environmental stewardship can also generate economic advantages. The preservation of natural areas and the promotion of eco-tourism have the potential to generate revenue and create employment. Conserving nature can lead to growth in eco-tourism, bring jobs in the conservation industry, and make a property more valuable when making it green and energy-efficient. Skyrocketing healthcare costs from polluted air and the results of climate-related emergencies are lessened by using sustainable land management. Even though creating green infrastructure may present costs at first, in most cases, it leads to energy and water savings over the long term and makes the infrastructure more resistant to climate-related problems.

So, environmentally friendly development stabilizes the economy, increases the country's competitive advantage, and helps attract both people and investors who care about the environment.

18.2.2 Fairness

When land use planning promotes equity, development ensures that its rewards and risks are not biased according to who someone is or their income. Discussions about social equality and justice usually include the value of fairness in ethical land use planning. It pays attention to ensuring that the good and bad aspects of development are distributed evenly between all members of society and all important groups. If fairness is part of planning how land is used, it unites people, reduces differences between them, and guarantees that development is fair. The most important aspects that show fairness in land use planning are described below.

Inclusive growth:
Having a fair land use strategy helps include all members of the community in any economic growth, as everyone has easy access to better roads, jobs and public services. With inclusionary zoning, developers are asked to reserve certain affordable units to ensure that residents from multiple income groups are all part of one area. A community benefit agreement might enable a project to help the surrounding area by requiring developers to pay for new schools, build parks or improve local transportation.

Preventing displacement:
Expanding cities without proper controls and gentrification commonly make it difficult for low-income people to keep their houses and apartments, as both property and rental prices go up. A good land use plan includes ways to keep housing accessible and secure. Among these are policies for rental caps, subsidies meant for housing, increased protections for renters, and systems that keep residents from being evicted forcibly. Also, including affordable housing in redevelopment areas stops low-income residents from leaving and prevents them from missing out on new city plans. They ensure that everyone has fair access to living space and helps residents stay long term.

Health and safety:
For a long time, areas where low-income or minority people live have faced a greater risk of getting polluted air, tainted water, and a shorter lifespan due to housing polluting industries, waste sites, and roads with high traffic.

This task is handled by requiring strong environmental standards for dangerous activities on land, carrying out in-depth studies of their impact on the environment, and cleaning up polluted areas. When everybody enjoys clean air, safe water, parks, and a clean environment, health justice and good public health are possible.

Access to services and amenities:
It is important in land use to ensure that everyone can use necessities and everyday services. Underserved communities often encounter problems in getting quality education, proper healthcare, fresh food, affordable transportation, and places to have fun. To deal with these issues, officials should use comprehensive land use strategies and invest in community infrastructure in spots that have long been left out. Making sure that areas get the services they need increases living conditions, aids in human development, and includes all communities in growth. In addition, making sure that digital infrastructure is easily available to all contributes to fair use of land in today's world.

18.2.3 Justice

To achieve justice in using land, it is important to be both fair with resources and fair in the way decisions are made. It calls for rules that help the weakest, respect all people's rights, and promote the future of the planet. Managing land justly gains public support, strengthens democracy in the community, and results in cities being both well-run and having a sense of social and moral foundation. How justice is included in land development is given through the following dimensions.

Fair processes:
Fairness expects that making decisions about land use involves everyone through participation. To do this, all stakeholders, especially those who often do not have a voice, need to have their opinions represented in planning and choices. Public hearings, participatory budgeting, community forums and consultation workshops make sure the views of many people are taken into account. Taking different perspectives into account while making decisions increases their quality and raises people's trust and respect for urban governance.

Accountability:
Proper accountability helps to ensure that land use planning cannot be manipulated by big political or economic groups. Land use choices need to

work for everyone and follow all appropriate legal and ethical rules. For this to work, there should be clear government structures, unbiased regulation and systems to assess the results of planning. Making conflict-of-interest policies easy to understand, giving independence to officials and providing opportunities for feedback by citizens helps to improve both ethics and accountability in urban development.

Moral responsibility:
A society ensures justice in land use by putting fairness, rights and attention to future generations ahead of its own interests. When making decisions about land, planners keep in mind how these decisions will affect the environment, public health and the sense of community. It means putting sustainable actions first, upholding indigenous and local rights and seeing that development help with social justice. Policies should be directed by economic success and by what is fair to all members of the society.

Conflict resolution:
Because urban land use is complicated, confrontations between stakeholders are often seen. Dealing with these issues appropriately and politely is a prerequisite for justice. Using types of conflict resolution like mediation, arbitration and community advisory panels can support a better way to manage those conflicts. As a result, these platforms give each party a voice, encourage everyone to understand one another and contribute to peaceful and democratic decisions.

18.3 Incorporating these Values into Land Use Planning

Including elements of environmental stewardship, equity, and justice in the planning of land use leads to more environmentally friendly and lasting urban changes. Through a value-centered strategy, plans become about the current needs, as well as the protection of future generations and the dignity of all participants. Using the following methods, one can help to include these values.

Comprehensive planning:
Creating wide-ranging land use plans that include economic, social and environmental factors forms the most important initial step. Any such plans need to result from proper reviews of existing needs and future challenges, along with discussions on what the economy indicates, other social

matters and any environmental concerns. Both Environmental Impact Assessments (EIAs) and Geographic Information Systems (GIS) give planners the means to understand and evaluate the spatial results of their land use decisions. Smart use of data helps planners develop plans that allow urban, peri-urban and rural areas to grow steadily and fairly for years to come.

Stakeholder engagement:
For planning to be fair and successful, everyone must take part. When residents, environmental activists, businesspeople, developers, marginalized groups and policymakers are involved, the planning process includes many voices. People in the community can influence the outcomes of decisions by joining charrettes, participating in hearings, joining focus groups or using digital tools. When people are involved, they can follow the land management process, gain trust and share the responsibility for its results.

Policy frameworks:
Having strong and versatile policies is necessary to establish environmental stewardship, fairness and justice. Their scope should concern zoning, building in an environmentally friendly manner, creating low-cost housing, guarding nature and making neighbourhoods accessible. Environmental justice requires policy changes so that nobody is exposed to more pollution or deprived of needed resources. When cities regularly review their land use planning policies, they are able to adjust to ongoing trends, technical progress and population changes.

Monitoring and evaluation:
Continuously keeping an eye on land use planning efforts supports their compliance with the set goals. To do this, one should watch the developments in social, environmental and economic conditions, based on indicators, benchmarking and impact studies. Planners use data to guide their decisions, increase transparency and assure people that the organization is accountable. Evaluations occur at regular intervals to find where to succeed and what requires changes, which allows the plan to respond quickly and positively to new pressures.

Education and awareness:
Building a culture where one takes care of the environment starts with informing people. Information about the importance of ethical land use helps both people and government officials take action. Widespread knowledge of ethics in city development can be created by holding workshops in

the community, schools, advertising through the media and training public planners and officials. Educated communities can take part in planning properly and advocate successfully for themselves.

18.4 Balancing Economic Interests with Ethical Values

Practical and ethical questions often arise when economic and moral ideas meet in planning land use. These issues happen since there may be clashes between driving economic progress and protecting the environment, ensuring all are treated fairly and making life better for everyone. A number of important ethical problems come out of this combination, as presented below.

18.4.1 Economic Growth vs. Environmental Degradation

A key problem in land use planning is that supporting economic growth might threaten the environment. Improving living standards, creating employment and advancing infrastructure are all important results of economic development, but it often damages the environment, for example, by removing forests, polluting the air and water, affecting soil and decreasing the diversity of life on Earth. The problem arises because economic goals such as developing industry, building cities and increasing farming are not given enough consideration for the Earth's future. In rural or semi-urban places, establishing an industrial area can result in a much stronger local economy, due to job creation, better transportation and new investment. While extreme development can deliver significant benefits, if environmental rules and sustainable planning are not followed, it may permanently spoil the environment, pollute lakes and rivers, increase air pollution and push wildlife out of their habitats.

Besides, when pollutants from industries reach the community, people can experience more health risks. To achieve balance between different interests, ethical land use planning should include green infrastructure, develop new environmentally friendly industries, add renewable energy and focus on restoring the environment. Officials are required to fully examine costs and environmental consequences to make certain that decisions today do not harm people living in the future. Dealing with the economic growth versus environmental damage dilemma requires considering both the economic and ecological results together.

18.4.2 Profit Maximization vs. Equity and Fairness

It is often challenging in land use planning because developers and investors look for lucrative opportunities while everyone needs to be treated fairly. Private sector development projects are commonly motivated by earning money and usually involve remaking city districts, starting up new commercial spaces, and building upscale homes for wealthy people. At the same time, these developments can be really risky for poor communities, who usually do not get to enjoy these new changes.

Gentrification is one of the main effects of redevelopment aimed at profits, since it leads to an increase in property and rental prices, which means more profit for landlords, property owners, and real estate investors. Improving infrastructure, strong economy, and more livable cities are results of gentrification, while for the long term, low-income individuals are pushed out of their familiar communities. The presence of displaced people disrupts the community, reduces its sense of identity, and increases economic differences among people. It creates difficulties in determining the roles played by both winners and losers as developments take place in cities.

18.4.3 Long-Term Sustainability vs. Short-Term Gains

A major ethical concern in land use planning comes from trying to balance future security with the temptation of past financial benefit. While a lot of people feel that the ongoing success of ecosystems, smart use of renewable resources, and care for those to come should be the key goals, others are mainly focused on making financial gains. Such a divide becomes clear in situations that involve expanding infrastructure, cities growing outward, removing forests and large-scale use of natural resources.

For instance, picking out hydrocarbons, minerals, or timber can lead to growth for the economy, due to the formation of local and foreign jobs, an increase in taxes and new investments from abroad. They can attract interest from governments and industries wanting to boost the economy by fulfilling their political or financial needs over a short period. Yet, getting these benefits often results in damage that cannot be fixed, environmental harm, less diversity of species, contaminated soil and water and climate change. Damage that is caused when these things are exhausted or the ecosystems are hurt can sometimes be permanent and, as a result, reduce the ability of people in the future to support themselves. When land development is based only on fast profit, the actual environmental damage goes unnoticed and poses a threat to sustainability goals. As a result, responsible land use must incorporate studies on nature, set up renewable energy programs, plan green infrastructure and establish rules that prevent pollution and straining of resources. Making sure land use planning benefits sustainability over the long run requires switching from dealing with development problems as they arise to planning ahead to prevent them. Such policies must even out the demands of increasing profits and responsible care for the environment, because the real success of development depends greatly on the lasting strength and health of our natural and man-made surroundings.

18.4.4 Corporate Interests vs. Community Interests

Disputes between the goals of corporations and the rights, values and needs of the local community are a major ethical issue in land use planning. Developers and international enterprises tend to shape the use of land, especially in ways that are profitable for their business interests. Although corporate development often leads to economic growth, more jobs and improved infrastructure, it can also overlook or reduce the social, cultural and environmental goals of communities. This could involve a big business suggesting a large project, such as a shopping mall or an industrial center; the creation of jobs and new business opportunities can damage local agriculture, push farmers away, eradicate their land and reshape the area's lifestyle and heritage.

Much of the profits from real estate in cities are taken out of the community and given to external investors, but the downsides like pollution, more traffic and less green space, affect city residents. Planning for how land will be used should reflect the needs and opinions of all those who are influenced by the decision. Local communities need to be included so that development benefits them over the long term, more than just satisfying the needs of rich investors. Also, frameworks for ethics should give priority to fairness, cultural protection, caring for the environment and open government, so that any decision about land use does not unfairly favour big companies instead of people who are especially vulnerable. When city and regional development plans include community values, knowledge of the area and a common goal, it promotes equality, inclusion and sustainability that equally values economic options and the integrity of the place.

18.4.5 Expediency vs. Transparency and Accountability

One significant ethical issue for land use planning arises when trying to make quick profits jeopardizes the main principles of openness, accountability, and public involvement. Often, the requirements of investors, strict deadlines, or money-related incentives push leaders to decide without consulting the public or checking their decisions. While this focus on quick successes can satisfy current desires, it can reduce the public's trust in the government, damage the environment, and reduce equality among all those affected. By doing this, planning processes may lose trust, and social acceptance of development projects can be undermined. One example is that governments or planning departments can speed up building a highway, airport or industrial area by arguing it will bring in investments and new jobs.

Even so, if the project goes ahead without properly reviewing its effects on nature, talking to the public or listening to community grievances, it may create serious problems that last. Ecological damage, conflicts between residents, moving communities or lawsuits may be examples of social problems. In addition, not including marginalized individuals in decision-making grows existing inequalities and reduces support for democracy. Good ethical land use planning should put transparency and accountability above all, helping ensure that every decision is also both fair and evidence-driven. In order to do this, companies must obey the law, openly disclose important information, be involved with affected communities and set up methods for public comments and complaints.

18.5 Methods for Harmonizing Economic Interests and Moral Principles

It is important for land use planning to be ethical as well as focused on economic development for communities to become inclusive, sustainable and resilient. The strategies below allow the combination of economic needs with moral beliefs.

Inclusive decision-making:
The process of land use planning should always begin with open and involving decision-making that brings all stakeholders, such as underprivileged and marginalized communities, into the discussion. Having a variety of people involved in policymaking and enactment adds many valuable perspectives to the process. The use of discussions in public meetings, community boards, citizen-led mapping and online platforms gives people chances to express themselves and become part of the planning process.

Sustainable practices:
Protecting the environment while boosting economic development can be reached by using sustainable methods. This can be done by pushing for more renewable energy, supporting agriculture that protects our soil and water and making sure all new buildings are built with environmentally friendly practices. Such practices help the environment and provide new opportunities for work, growth and a stable economy in these areas.

Equitable policies:
Decisions about using land should always consider ways to help those who are vulnerable, prevent exclusions and limit damage to the environment. For example, laws requiring affordable housing, measures to stop evictions, groups to set rental fees, and adjusting zoning rules to ensure that economic development does not displace present communities. In addition, it is necessary to ensure that transportation, schooling and medical resources are shared fairly to everyone, rather than mainly helping the wealthy.

Long-term planning:
Ethical land use planning requires us to consider the long-term benefits for society and the planet, more than how quickly one can make a profit. Taking into account resilience, climate solutions, and equity between generations allows us to secure the future ability of others to manage their needs. Smart growth and the circular economy are examples of new approaches that help plan development for the next decades while considering both ecological factors and people.

Clear honesty and responsibility:
Successful land use planning depends on how open and answerable the government is and how straightforward the rules are. Every decision should be open so that people can access information, data and explanations for them. Mechanisms that hold people in charge accountable, conduct ethics reviews, report details in public, and reveal conflicts of interest can help stop corruption and secure and maintain the fairness of land use policies.

Placing ethical values into the planning of land use is needed for creating fair, safe and successful communities. Justice, sustainable practices and inclusivity can help planners and policymakers make correctly chosen decisions on complicated situations that do not endanger the interests of the public or the world.

18.6 Morality and Land Use Ethics

Land use planning relies on strong morals to help determine what is fair or unfair and what is right or wrong. It is a tool for seeing that no group, particularly disadvantaged groups, is unfairly disadvantaged by land use decisions. Policymakers must use justice, equality and inclusivity as their main guiding beliefs. Land use morality rests on giving higher priority to the community's interests than to individual personal goals.

Public officials must focus on the environment, support sustainability efforts, and work hard to ensure people's health so that communities survive and prosper in the long term. Still, land use planning with ethics involves not only the practical side, but also the personal duty to take care of the natural resources and the Earth, both now and for the benefit of people in the future. Here, it is stressed that long-term environmental outcomes should be more important than fast economic results, since our choices now affect the future.

Besides, focusing on human rights and dignity means land use planning must ensure all people have enough housing, necessary services, and their heritage is not damaged. It is important in this field to deal with individuals fairly, show respect, include them, and avoid decisions that result in people being pushed out or discriminated against. Having a set of ethics means that decisions avoid any harm to those around us or the world we share. A good way to use land ethically is to assign duties for certain actions, help resolve conflicts when interests do not match, and encourage everyone to join. Ethics should guide the process, all stakeholders should be involved, decisions should be clear and accountable, and everyone in society should learn about green strategies. All these concepts increase trust, encourage openness, and make decision-making better. Having ethical considerations in land use planning makes communities more prepared and helps ensure that all generations will have equal, lasting, and inclusive development.

18.7 Interplay between Personal and Collective Morality

Personal values and collective morality are very important in affecting decisions about land use. They impact the development of laws, determination of land divisions and evaluation of how changes take effect on everyone and nature. The link between ethics and resilience plays an important role when designing land use plans.

Personal moral values are the ethical standards that help a stakeholder decide on what to do. They are influenced by a range of things like a person's faith, opinions, culture, education, wealth or poverty and their life experiences. Such personal views impact the order in which planners, developers, policymakers, activists and citizens regard goals, including good economic growth, caring for the environment, including all people and preserving culture. If a developer is guided by entrepreneurial ethics, they might want to use all the land available and earn maximum profit, but an environmentalist influenced by ecological thinking would like to see houses built gently on the land and the environment preserved.

Collective morality includes the general ethical rules, deeply seated culture and written laws society holds in place. The presence of community land use appears in formal documents, such as constitutions, environmental laws, zoning codes, planning requirements and community. Collective morality strengthens society by trying to adjust actions so that land use supports the benefit of everyone.

The main values in ethical land use planning are justice, equity, sustainability, and responsible stewardship. It offers guidelines for dealing with conflicts, protecting those who need help most, and making people in communities feel responsible for everyone around them. There are often problems when it comes to mixing personal morals and the morals of a group. Even though these steps can lead to great achievements for everyone, they may result in disputes among people. As an illustration, when economic policies focus only on the economy and not on culture, some members of the community may feel that they are not ethically right.

It is crucial to create strategies that honour people's rights and the values that everyone in the community shares when dealing with land use planning. Land management can be judged according to ethics using utilitarian analysis, virtue ethics, and rights-based ideas. Because of these frameworks, managers make sure their decisions are right by law and morals. Moral dialogue in planning becomes possible by involving many different stakeholders. Through participation in planning, hearings, discussions, and cooperation workshops, a range of people get to express and settle their ethical issues. If marginalized and indigenous communities are represented in these matters, democracy becomes stronger, and many ethical opinions are included.

At the same time, programs that educate and inform the public help create an ethical society. Any planning program that covers environmental ethics, social justice, intergenerational equity, and sustainable development has the chance to create people who feel they can engage in discussions about land use. When the public is ethically informed, it is more likely to spot the consequences of planning actions and back initiatives for fairness. Taking care of ethical matters in business helps maintain public trust and increases responsibility, and also leads to better decision-making. This way of thinking sees land as an asset that every person shares and must protect, introducing different values and benefiting future generations as well.

18.8 Environmental Justice

Environmental justice means that everyone is treated fairly in how environmental laws, regulations and policies are made, applied and carried out. Everyone is guaranteed the same defence against environmental and health dangers and the opportunity to be part of decisions relevant to their living area. Essentially, environmental justice works to address the reasons why some communities face more environmental harm than others, such as industrial pollution, nearby sites for waste, worsening air and water quality and difficulties using green spaces and getting essential services. Often, those who face the worst of environmental injustice are made up of low-income, indigenous or racially minoritized groups who have missed out on land use and governing options for many years. With regard to land use planning, environmental justice urges us to act on the ethical and fairness concerns of these imbalances. It makes planners, policymakers, and developers think about land use in terms of both how it functions and how it is distributed and decided.

It is concerned about not just avoiding harm, but also creating fairness and good results for everyone involved. Ensuring that all populations gain from sustainable infrastructures, renewable energies, easy access to public transport, safe places to live, access to green areas in cities, and improved climate resistance counts as well. It encourages putting justice considerations into land use plans that eliminate growing social differences and improve community health, environment and social life over time. Environmental justice can be added to land use planning through inclusive and cooperative management. Making sure historically unrepresented groups are active in planning processes is very important.

18.9 Conclusion

Proper use of land is very important, so it is necessary to balance economic interests against important ethical values. Communities often design urban spaces to suit all members equally and explore sustainable options due to how people's personal values interact with those of the whole community. If urban planning follows openness, fairness, and durability, it becomes possible to use strategies that support the environment and improve living conditions. It is also important for leaders to be ethical since this helps build trust and guarantees meaningful involvement of the stakeholders in planning activities. Following honesty, accountability, and sustainability guides, decisions about land will better reach the goals important to all members of the community. Ensuring that property is used ethically is needed now for social growth and also for keeping natural resources safe for people in the future.

References

1. Satterthwaite, D. and McGranahan, G., Urbanization and its implications for food and farming. *Philos. Trans. R. Soc. London, Ser. B*, 375, 1554, 2809–2820, 2010, doi: 10.1098/rstb.2010.0137.
2. Beatley, T., *Ethical Land Use: Principles of Policy and Planning*, Johns Hopkins University Press, Baltimore, MD, 1994.
3. Lipton, M. and Saghai, Y., Food security, farmland access ethics, and land reform. *Global Food Secur.*, 12, 59–77, 2017, https://doi.org/10.1017/j.gfs.2017.03.004.
4. Allam, Z., The city of the living or the dead: On the ethics and morality of land use for graveyards in a rapidly urbanised world. *Land Use Policy*, 87, 104037, 2019, https://doi.org/10.1017/j.landusepol.2019.104037.
5. Golany, G.S., *Ethics and urban design: Culture, form, and environment*, John Wiley & Sons, New York, 1995.
6. Kaufman, J.L., Land planning in an ethical perspective, in: *Ethics in Planning*, 1st ed., pp. 291–298, Routledge, 2017.
7. Pauleit, S., Ennos, R., Golding, Y., Modeling the environmental impacts of urban land use and land cover change—a study in Merseyside, UK. *Landscape Urban Plann.*, 71, 2-4, 295–310, 2005, https://doi.org/10.1017/j.landurbplan.2004.03.009.
8. Dempsey, N., Bramley, G., Power, S., Brown, C., The social dimension of sustainable development: Defining urban social sustainability. *Sustainable Dev.*, 19, 5, 289–300, 2011, https://doi.org/10.1002/sd.417.
9. Box, T., Community, Science, Ethics, and Professional Land Stewards. *Rangelands Arch.*, 34, 4, 71–72, 2012, https://doi.org/10.2111/RANGELANDS-D-12-00041.1.
10. Abumoghli, I., The Role of Religions, Values, Ethics, and Spiritual Responsibility in Environmental Governance and Achieving the Sustainable Development Agenda. *Religion Dev.*, 1, 1–11, 2023, https://doi.org/10.30975/27507955-20230008.

11. Keong, C.Y., The nexus of environmental ethics and environmental sustainability: An empirical assessment, in: *Global environmental sustainability—Case studies and analysis of the United Nations' Journey toward sustainable development*, pp. 253–288, 2021, https://doi.org/10.1017/B978-0-12-822419-9.00005-9.
12. Melvani, K., Stacey, N., Bristow, M., Crase, B., Moles, J., Farmers' values for land, trees and biodiversity underlie agricultural sustainability. *Land Use Policy*, 117, 105788, 2022, https://doi.org/10.1017/j.landusepol.2021.105788.

Index

Also of Interest

Check out these other related titles from Scrivener Publishing

Agricultural Supply Chain Optimization Using Federated Learning, Edited by Abhishek Kumar, Pooja Dixit, J.P. Ananth, S. Oswalt Manoj, and S. Panneerselvam, ISBN: 9781394461264. Master the next evolution of agricultural intelligence with this definitive guide to federated learning, providing the decentralized, privacy-preserving strategies needed to optimize global supply chains without compromising data sovereignty.

Climate Smart Agriculture, Edited by Anitha Velu, Prasanth Aruchamy, Raghu Ramamoorthy, Rajesh Kumar Dhanaraj, and Seifdine Kadry, ISBN: 9781394336333. Transform the future of sustainable farming with this guide to mastering the deep reinforcement learning architectures and algorithms that turn complex environmental data into precise, high-yield decisions for climate-smart agriculture.

Sustainable Agriculture Production Using Blockchain Technology, Edited by Rekh Ram Janghel, Rajesh Doriya, Jaykumar Lachure, and Yogesh Kumar Rathore, ISBN: 9781394248674. Revolutionize the agricultural supply chain with this essential guide, which provides the practical knowledge to leverage blockchain technology for transparency, traceability, and trust, alongside AI for overcoming modern farming challenges.

Mathematical Modeling in Agriculture, Edited by Sabyasachi Pramanik, Niranjanamurthy M., Ankur Gupta, and Ahmed J. Obaid, ISBN: 9781394233717. The main goal of the book is to explore the idea behind data modeling in smart agriculture using information and communication technologies and tools to make agricultural practices more functional, fruitful and profitable.

Agroforestry, Edited by Abhishek Raj, Manoj Kumar Jhariya, Arnab Banerjee, Ramesh Kumar Jha, and Krishan Pal Singh, ISBN: 9781394231133. This book offers comprehensive insights into the management of agroforestry for livelihood security and sustainable development in the tropics, addressing ecological interactions, productivity, and the monetization of carbon credits, while also outlining a future roadmap and policy challenges.

Smart Agritech: Robotics, AI, and Internet of Things (IoT) in Agriculture, Edited by Santosh Kumar Srivastava, Durgesh Srivastava, Korhan Cengiz, and Pramod Gaur, ISBN: 9781394302959. The main goal of *Smart Agritech: Robotics, AI, and the Internet of Things (IoT) in Agriculture* is to explore how emerging technologies such as robotics, artificial intelligence (AI), and IoT can be leveraged to improve efficiency, sustainability, and productivity in agriculture.

DIGITAL AGRICULTURAL ECOSYSTEM: Revolutionary Advancements in Agriculture, Edited by Kuldeep Singh and Prasanna Kolar, ISBN: 9781394242931. The book comprehensively explores the dynamic synergy between modern technology and agriculture, showcasing how advancements such as artificial intelligence, data analytics, and smart farming practices are reshaping the landscape to ensure food security in the era of climate change, as well as bridging the gap between cutting-edge research and practical implementation.

THE DIGITAL AGRICULTURAL REVOLUTION: Innovations and Challenges in Agriculture through Technology Disruptions, Edited by Roheet Bhatnagar, Nitin Kumar Tripathi, Nitu Bhatnagar and Chandan Kumar Panda, ISBN: 9781119823339. The book integrates computational intelligence, applied artificial intelligence, and modern agricultural practices and will appeal to scientists, agriculturists, and those in plant and crop science and management.

HANDBOOK OF AGRICULTURAL BIOTECHNOLOGY: Volume I Nanopesticides, Edited by Charles Oluwaseun Adetunji and Julius Kola Oloke, ISBN: 9781119836148. The book provides detailed information on the application of nanopesticides for the management of numerous agricultural pests and pathogens, as well as to foster an innovative pathway toward future sustainable agriculture and food systems.

HANDBOOK OF AGRICULTURAL BIOTECHNOLOGY: Volume II Nanobioherbicides, Edited by Charles Oluwaseun Adetunji, Julius Kola Oloke, ISBN: 9781119836155. The book provides a detailed examination of the application of nanobioherbicides that come from plants including information on the different metabolites derived from numerous plants that could become bioherbicides.

HANDBOOK OF AGRICULTURAL BIOTECHNOLOGY: Volume III Nanofungicides, Edited by Charles Oluwaseun Adetunji and Julius Kola Oloke, ISBN: 9781119836162. This book provides state-of-the-art information about recent advancements in the application of nanobiofungicides for effective management of post-harvest diseases and agricultural fungal diseases, including blights, mildews, molds, and rusts.

HANDBOOK OF AGRICULTURAL BIOTECHNOLOGY: Volume IV Nanoinsecticides, Edited by Charles Oluwaseun Adetunji and Julius Kola Oloke, ISBN: 9781119836179. The book provides detailed information about the application of repellent products that contain plant-based ingredients known as nanobioinsecticides, including the pesticide evaluation scheme guidelines for repellent testing, relevant information about the procedures to evaluate several repellent compounds, the development of new products that offer high repellency, and guidelines for consumer safety.

ECORESTORATION FOR SUSTAINABILITY, Edited by Arnab Banerjee, Manoj Kumar Jhariya, Surendra Singh Bargali, and Debnath Palit, ISBN: 9781119879718. With climate change and land degradation advancing at an alarming rate, the state of the global ecosystem and how it can be restored in our march toward sustainability is incredibly important, and this new, groundbreaking volume tackles these important issues head on.

FUNCTIONAL POLYMERS IN FOOD SCIENCE, Part 1: Food Packaging, Edited by Giuseppe Cirillo, Umile Gianfranco Spizzirri, and Francesca Iemma, ISBN: 9781118594896. Provides an extensive overview of the link between polymer properties and food quality. *NOW AVAILABLE!*

FUNCTIONAL POLYMERS IN FOOD SCIENCE, Part 2: Food Processing, Edited by Giuseppe Cirillo, Umile Gianfranco Spizzirri, and Francesca Iemma, ISBN: 9781118595183. Provides an extensive overview of the link between polymer properties and food quality. *NOW AVAILABLE!*

ADVANCES IN FOOD SCIENCE AND TECHNOLOGY Volume 1, Edited by Visakh, P.M, Sabu Thomas, Laura B. Iturriaga, and Pablo Daniel Ribotta, ISBN: 9781118121023. Written in a systematic and comprehensive manner the book reports recent advances in the development of food science and food technology areas. *NOW AVAILABLE!*

ADVANCES IN FOOD SCIENCE AND NUTRITION Volume 2, Edited by Visakh P.M, Laura B. Iturriaga, and Pablo Daniel Ribotta, ISBN: 9781118137093. This important book comprehensively reviews research on new developments in all areas of food chemistry/science and nutrition. *NOW AVAILABLE!*

Applied Water Science Volume 1: Fundamentals and Applications, Edited by Inamuddin, Mohd Imran Ahamed, Rajender Boddula and Tauseef Ahmad Rangreez, ISBN: 9781119724766. Edited by one of the most well-respected and prolific engineers in the world and his team, this is the first volume in a two-volume set that is the most thorough, up-to-date, and comprehensive volume on applied water science available today. *COMING IN SUMMER 2021*

Sustainable Water Purification, By Safiur Rahman and M. R. Islam, ISBN: 9781119650997. This is the only book that takes a zero-waste approach to propose 100% sustainable water purification techniques. *NOW AVAILABLE!*

Books in the series, "Bioprocessing in Food Science"

NUTRACEUTICALS FROM FRUIT AND VEGETABLE WASTE, Edited by Vidisha Tomer, Navnidhi Chhikara, Ashwani Kumar, and Anil Panghal, ISBN: 9781119803508. Nutraceuticals from Fruit and Vegetable Waste, the latest volume in the series, "Bioprocessing in Food Science," is an essential reference on the valorization of fruit and vegetable waste and the state of the art for fruit and vegetable processing.

NUTRITIONAL SCIENCE AND TECHNOLOGY: Concept to Application, Edited by Tejpal Dhewa, Anil Kumar Puniya, and Anil Panghal, ISBN: 97811198089. This new volume in the groundbreaking series, "Bioprocessing in Food Science," Nutritional Science and Technology, is a comprehensive resource covering topics such as human nutrition, food safety concerns, and emerging technologies like nanotechnology, and it also addresses the need for innovative food products, contamination, and regulatory challenges in the global food market.

HARVESTING FOOD FROM WEEDS, Edited by Prerna Gupta, Navnidhi Chhikara, and Anil Panghal, ISBN: 9781119791973. An exciting new volume in the groundbreaking new series, "Bioprocessing in Food Science," Harvesting Food from Weeds focuses on the latest processes, industrial applications, and leading research on this important, evolving new area, for engineers, scientists, students, and other industry professionals.

MICROBES IN THE FOOD INDUSTRY, Edited by Navnidhi Chhikara, Anil Panghal, and Gaurav Chaudhary, ISBN: 9781119775584. This newest volume in the groundbreaking new series, "Bioprocessing in Food Science," focuses on the latest processes, industrial applications, and leading research on microbes in the food industry, for engineers, scientists, students, and other industry professionals.

NOVEL TECHNOLOGIES IN FOOD SCIENCE, Edited by Navnidhi Chhikara, Anil Panghal, and Gaurav Chaudhary, ISBN: 9781119775584. The third book in the series, "Bioprocessing in Food Science," covers the latest and greatest new technologies in food science.

FUNCTIONAL FOODS, Edited by Navnidhi Chhikara, Anil Panghal, and Gaurav Chaudhary, ISBN: 9781119775560. Presenting cutting-edge information on new and emerging food engineering processes, *Functional Foods*, the second volume in the groundbreaking new series, "Bioprocessing in Food Science," is an essential reference on the modeling, quality, safety, and technologies associated with food processing operations today.

THERMAL FOOD ENGINEERING OPERATIONS, Edited by Nitin Kumar, Anil Panghal, and M. K. Garg, ISBN: 9781119775591. Presenting cutting-edge information on new and emerging food engineering processes, Thermal Food Engineering Operations, the first volume in the new series, "Bioprocessing in Food Science," is an essential reference on the modeling, quality, safety, and technologies associated with food processing operations today.

Printed in the USA/Agawam, MA
August 26, 2026

909668.005